Cardiovascular Outcomes
Collaborative, Path-Based Approaches

Dominick L. Flarey, PhD, MBA, RN, CS, CNAA, FACHE
President
Dominick L. Flarey & Associates
and
The Center for Medical-Legal Consulting
Niles, Ohio

Suzanne Smith Blancett, EdD, RN, FAAN
Editor-in-Chief
Journal of Nursing Administration
Bradenton, Florida

AN ASPEN PUBLICATION®
Aspen Publishers, Inc.
Gaithersburg, Maryland
1998

The authors have made every effort to ensure the accuracy of the information herein. However, appropriate information sources should be consulted, especially for new or unfamiliar procedures. It is the responsibility of every practitioner to evaluate the appropriateness of a particular opinion in the context of actual clinical situations and with due considerations to new developments. Authors, editors, and the publisher cannot be held responsible for any typographical or other errors found in this book.

Library of Congress Cataloging-in-Publication Data

Cardiovascular outcomes: collaborative, path-based approaches/
[edited by] Dominick L. Flarey, Suzanne Smith Blancett.
p. cm.
Companion v. to: Health care outcomes: collaborative, path-based
approaches. 1998.
Includes bibliographical references and index.
ISBN 0-8342-1138-6 (alk. paper)
1. Cardiology.   2. Cardiovascular system—Diseases—Treatment.
3. Outcome assessment (Medical care).   4. Medical protocols.
5. Critical path analysis.   6. Medical cooperation.   I. Flarey,
Dominick L.   II. Blancett, Suzanne Smith.   III. Health care
outcomes.
[DNLM: 1. Cardiovascular Diseases—therapy.   2. Outcome Assessment
(Health Care)   3. Critical Pathways.   WG 166 C26663 1998]
RC669.C28756   1998
616.1'06—dc21
DNLM/DLC
for Library of Congress
98-13313
CIP

Orders: (800) 638-8437
Customer Service: (800) 234-1660

About Aspen Publishers • For more than 35 years, Aspen has been a leading professional publisher in a variety of disciplines. Aspen's vast information resources are available in both print and electronic formats. We are committed to providing the highest quality information available in the most appropriate format for our customers. Visit Aspen's Internet site for more information resources, directories, articles, and a searchable version of Aspen's full catalog, including the most recent publications: **http://www.aspenpub.com**
Aspen Publishers, Inc. • The hallmark of quality in publishing
Member of the worldwide Wolters Kluwer group.

Editorial Services: Lenda P. Hill
Library of Congress Catalog Card Number: 98-13313
ISBN: 0-8342-1138-6

Printed in the United States of America

1   2   3   4   5

Table of Contents

4—Data Driven Critical Pathway Development and Outcomes: The Case of Heparin in Thromboembolic Disease 39

Carlos A. Estrada, Joan Wynn, and Taffy Klaassen

5—Heart Failure: Managing Care and Outcomes Across the Continuum 48

Linda D. Urden, Susan Casamento, Mary Mitus, Marsha Terry, and E. Thomas Arne Jr.

6—Improving Outcomes Related to the Cooperative Cardiovascular Project 62

Bette Keeling, Joycelyn Weaver, and Emily Murph

7—Pathways and Outcomes Across the Continuum: A Community Medical Center's Experience with Congestive Heart Failure 81

Judy Conarty, Rita Zenna, Harriet V. Werkman, and Kathleen A. Shafer

Contributors

Rella Adams, RN, PhD, CNAA
Senior Vice President
Valley Baptist Medical Center
Harlingen, Texas

E. Thomas Arne, Jr., DO, FACC
Gene E. Myers Cardiac and Vascular
    Consultants, Inc.
Sarasota, Florida
Former Medical Director
CHF Service
Butterworth Hospital
Grand Rapids, Michigan

Dawn A. Bailey, RN, MAOM
Director
Patient Care Services
Internal Medicine/Hypertension/
    Nephrology Nursing
Cleveland Clinic Foundation
Division of Nursing
Cleveland, Ohio

Jean Barry-Walker, MS, RN, CNA
Clinical Director
Department of Nursing
University of Iowa Hospital & Clinics
Iowa City, Iowa

**Suzanne Smith Blancett, EdD, RN,
    FAAN**
Editor-in-Chief
Journal of Nursing Administration
Bradenton, Florida

**Annabelle R. Borromeo, RN,
    CCRN, CNS, CPAN**
Staff Nurse
PACU
St. Luke's Episcopal Hospital
Houston, Texas

Suzanne M. Burton, BSN, MEd
Vice President
Outpatient Services Development
Sunrise Children's Hospital
Sunrise Hospital & Medical Center
Las Vegas, Nevada

Patty Calver, RN, BSN
Nurse Manager
CICU/MICU
Harborview Medical Center
University of Washington
Seattle, Washington

Susan Casamento, MSN, RN, ACNP
Cardiology Nurse Practitioner
CHF Service
Coordinator
Transplant Program
Butterworth Hospital
Grand Rapids, Michigan

Lawrence H. Cohn, MD
Chief, Cardiac Surgery
Professor of Surgery
Brigham and Women's Hospital
Boston, Massachusetts

Judy Conarty, BSN, MA
Disease Management Coordinator
Clinical Resources
Community Medical Center
Toms River, New Jersey

Dana Danielson, RN
Case Manager
Valley Baptist Medical Center
Harlingen, Texas

Carlos A. Estrada, MD, MS
Assistant Professor of Medicine
Department of Internal Medicine
East Carolina University
Greenville, North Carolina

Dorothy Goulart Fisher, MS, RN, CS
Program Manager
Systems Improvement
Brigham and Women's Hospital
Boston, Massachusetts

**Dominick L. Flarey, PhD, MBA, RN,
    CS, CNAA, FACHE**
President
Dominick L. Flarey & Associates and
    The Center for Medical-Legal
    Consulting
Niles, Ohio

Donna J. Gilski, BSN, MSN
Cardiology Care Manager
Chicago, Illinois

Becky Hassebrock, BSN, RN
Assistant Nurse Manager
Case Manager
Department of Nursing
University of Iowa Hospital & Clinics
Iowa City, Iowa

Cynthia Hinojosa, RN, MSN, CNAA
Administrative Director of Nursing
Valley Baptist Medical Center
Harlingen, Texas

Diane L. Huber, PhD, RN, FAAN, CNAA
Associate Professor
College of Nursing
The University of Iowa
Iowa City, Iowa

Bette Keeling, MSN, RN, CNAA
Nursing Coordinator
Nursing Administration
Hillcrest Baptist Medical Center
Waco, Texas

Taffy Klaassen, PharmD
Clinical Pharmacist
University Health Systems of Eastern
  Carolina
Greenville, North Carolina

David Litaker, MS, MD, FACP
Assistant Professor
The Cleveland Clinic Foundation
  Campus of the Ohio State
  University School of Medicine
Cleveland, Ohio

Jean M. Mau, BSN, MSN
Care Coordinator
Telemetry
Chicago, Illinois

Kelley McLaughlin, RN
Case Manager
Department of Nursing
University of Iowa Hospital and Clinics
Iowa City, Iowa

Cynthia H. McMahon, RN, MN
Director
Cardiovascular Services
Mission Hospital Regional Medical
  Center
Mission Viejo, California

Delores S. Meyer, BSN
Care Coordinator
CICU
Chicago, Illinois

Lorraine Mion, RN, PhD
Director
Department of Nursing Research
Division of Nursing
The Cleveland Clinic Foundation
Cleveland, Ohio

Mary Mitus, MSN, RN
Vice President
Clinical Services
Visiting Nurse Association of Western
  Michigan
Grand Rapids, Michigan

Emily Murph, MS, PT
Deceased

Lynne Nemeth, RN, MS
Project Manager
Clinical Pathways and Coordinator
  Care
Harborview Medical Center
Patient Care Services
University of Washington
Seattle, Washington

Marilyn Oermann, PhD, RN, FAAN
Professor
College of Nursing
Wayne State University
Detroit, Michigan

Barbara E. Parlotz, MSW
Supervisor
Social Work Department
Harborview Medical Center
University of Washington
Seattle, Washington

Celine Peters, RN, MN
Director
Outcomes Management
Mission Hospital Regional Medical
  Center
Mission Viejo, California

Hermie F. Robles, RN, BSN, CPAN
Staff Nurse
PACU
St. Luke's Episcopal Hospital
Houston, Texas

Donna M. Rosborough, MS, RN, CCRN
Care Coordination Team Manager
Cardiology/Cardiac Surgery
Brigham and Women's Hospital
Boston, Massachusetts

Cynthia L. Russell, BSN, MS
Coordinator of Evaluation Services
University of Missouri-Columbia
  Hospitals & Clinics
Columbia, Missouri

Constance Ryjewski, MSN
Unit Resource Coordinator
CICU
Lutheran General Hospital
Advocate Health Care
Chicago, Illinois

Marilyn Rymer, MD
Neurologist
Saint Luke's Hospital of
  Kansas City
Kansas City, Missouri

Kelli King Sagehorn, BSN, MS
Cardiothoracic Clinical Nurse
  Specialist
University of Missouri-Columbia
  Hospitals & Clinics
Columbia, Missouri

Laura Savage, RN, MSN
Cardiovascular Clinical Nurse
  Specialist
Department of Cardiovascular
  Transplant Nursing
Medical College of Virginia
  Hospitals
Richmond, Virginia

Mary Porter Schooler, RN, MSN
Cardiovascular Case Manager
Department of Continuum of Care
University of Kentucky Hospital
Lexington, Kentucky

Kathleen A. Shafer, BSN, MA
Director of Case Management
Community & Kimball Medical
  Centers
Toms River, New Jersey

Eric Six, MD, FACS, AB, AURP
Neurosurgeon
Medical Director
Valley Baptist Medical Center
Harlingen, Texas

Krista Smeins, RN
Case Manager
Department of Nursing
University of Iowa Hospital &
    Clinics
Iowa City, Iowa

Carol Smith, MOT, OT
Occupational Therapist
Department of Rehabilitation
    Medicine
Harborview Medical Center
University of Washington
Seattle, Washington

Patricia A. Soper, RN, MBA
Senior Associate Director
Chief Nursing Officer
Executive Staff
Saint Luke's Hospital of
    Kansas City
Kansas City, Missouri

Patrice L. Spath, BA, ART
Health Care Quality and Resource
    Management Consultant
Brown-Spath & Associates
Forest Grove, Oregon

Johnese Spisso, RN, MPA
Associate Administrator
Patient Care Services
Director of Nursing
Harborview Medical Center
University of Washington
Seattle, Washington

Linda C.H. Stennett, RN, CCRN
Manager
Cardiovascular Outcomes
Outcomes Management
Mission Hospital Regional Medical
    Center
Mission Viejo, California

Sherri Stevens, RN, MSN, CS
Clinical Nurse Specialist
Case Manager
Vascular Surgery
Center for Clinical Evaluation
Saint Thomas Health Services
Nashville, Tennessee

Deborah R. Summers, RN, MSN
Clinical Nurse Specialist
Saint Luke's Hospital of
    Kansas City
Kansas City, Missouri

Marsha Terry, BSN, RN
Cardiac Program Manager
Visiting Nurse Association of Western
    Michigan
Grand Rapids, Michigan

**Denise Tucker, BS, RN, CMC,
    ABQAUR**
Director
Utilization Management/Social
    Services
Valley Baptist Medical Center
Harlingen, Texas

Linda Urden, DNSc, RN, CNA
Administrative Director
Patient Care Services
Butterworth Hospital
Grand Rapids, Michigan

Pamela M. Warner, RN, CPHQ
Administrative Director
Performance Improvement
Valley Baptist Medical Center
Harlingen, Texas

Joycelyn Weaver, MSN, RN
Director
Patient Focused Care
RHD Medical Center
Dallas, Texas

Harriet V. Werkman, BSN, MSN
Administrative Director
Home Health/Hospice Program
Community Medical Center
Toms River, New Jersey

Gayle R. Whitman, RN, PhDc, FAAN
Visiting Associate Professor
University of Pittsburgh, School of
    Nursing
Pittsburgh, Pennsylvania

**Pamela Windle, RN, MS, CNA,
    CPAN, CAPA**
Nurse Manager
St. Luke's Episcopal Hospital
PACU & Day Surgery Center
Houston, Texas

Joan Wynn, MScN, RNCS
Project Manager
Patient Focused Care
University Health Systems of Eastern
    Carolina
Greenville, North Carolina

Rita Zenna, RN
Manager
Pathways & Outcomes
Clinical Resources
Community Medical Center
Toms River, New Jersey

Foreword

"Give us the tools and we will finish the job," was a sentiment expressed by Gilbert Chesterton, an English essayist over a century ago. And this is in many ways the same lament currently emanating from the voices of today's health care leaders who struggle daily to restructure and rebuild America's health care system. Give us the tools! Give us the tools to define, measure, and achieve outcomes. Give us tools that are relevant to the patient's reality. Give us the tools that move us past multidisciplinary actions to interdisciplinary practices. Give us the tools so that we can continue to provide increasingly higher quality health care to our patients in a cost-effective manner. This book, *Cardiovascular Outcomes: Collaborative, Path-Based Approaches,* is a perfect answer to this lament. For this text provides rich and detailed examples and samples of effective tools—pathways, care maps, and processes of interdisciplinary care—which have been shown to be effective in producing outstanding outcomes in patients experiencing America's leading cause of death and disability—cardiovascular disease.

As the American health care delivery system undergoes its current revolution and the management of outcomes continues to be paramount, the tools used to help achieve quality outcomes have also captured center stage and the interest of providers and payers. Over the past decade, with continued use these tools have evolved from their first generation versions as blunt and crude tools to more refined and precise second-generation instruments. The content of the chapters in this book reflects and highlights this honing and refinement. For example, the traditional cardiac medical and surgical categories have been further divided into diagnostic categories of acute myocardial infarction, stroke, carotid

endarterectomy, coronary interventions, coronary artery bypass surgery, and abdominal aortic aneurysm. But even more importantly, some of these diagnostic categories have been further stratified based on risk. There is a chapter on uncomplicated cardiac surgery patients and another on uncomplicated myocardial infarction patients. First-generation pathways did not have the insight to dichotomize patients as uncomplicated or complicated. And as such, all patients were managed with a broad stroke. With experience it became clear that the applicability of a single pathway did not extend across all patients with a given disease entity. Rather, underlying risk greatly impacted the appropriateness of a given pathway to a given patient. Having learned that lesson, these second-generation pathways provide a better-matched prescription of care for these targeted patients.

Another difficulty with early pathway tools was that care was prescribed to patients predominantly from the provider's perspective. They were written as tools to be used within the provider's environment rather than the patient's experience or reality. The difference is that the provider's environment is most often unidimensional, occurring in one type of unit or practice setting and in a circumscribed period of time; while the patient experiences and lives a reality crossing a number of settings and occurring over a longer period of time. Pathways that reflect this continuum of illness, rather than a continuum of individual provider practice, provide a sharper focus on the patient's needs. The chapters on managing the patient with congestive heart failure across the continuum and care of the coronary intervention patient from the catheterization laboratory to the home are two excellent examples of providing truly patient-focused prescriptions of care. And finally, the path-based practices discussed in this text provide excellent examples of interdisciplinary approaches to these patients. By focusing on the processes of care that lead to optimal patient outcomes, rather than the roles of each individual provider, they provide a "how-to" for teams to move from a multidisciplinary approach to a more holistic provider collaboration approach. With this latter approach the team focuses more on collaborating and combining their expertise and actions to achieve patient outcomes rather than focusing singularly on what their discipline can bring to the patient in an isolated fashion. At the interdisciplinary level the goals truly become patient goals rather than practitioner goals.

The chapters throughout this text are replete with examples of how this level of interdisciplinary collaboration can be attained. It is these attributes of specificity and targeting to circumscribed categories of patients, the comprehensive view of patients' needs across practice settings, and the articulation of an interdisciplinary practice that truly move this book into a second-generation cutting edge view of cardiovascular outcomes. It is a useful and essential tool for any health care leader concerned with managing the outcomes for cardiovascular patients. Whether in executive or operational roles, or implementing case management and quality outcomes programs or evaluating them, or whether working with a single health care facility or a complex integrated system this text will provide valuable insights and roadmaps to health care providers. And so for the health care leader who is lamenting "Oh if only I had the tools!" I give them this book.

Gayle R. Whitman RN, PhDc, FAAN
Visiting Associate Professor
University of Pittsburgh, School of Nursing
Pittsburgh, Pennsylvania

Preface

In the companion book to this work, *Health Care Outcomes: Collaborative, Path-Based Approaches,* we presented a compilation of approaches to care delivery from a collaborative, path-based perspective—from the new beginnings of case management and critical pathways, introduced in the 1980s to the new focus on measuring and managing outcomes in the late 1990s. Its chapters tell the story of how we have taken innovations in health care delivery and created newer, more effective and efficient models and systems of care based on the management of care and the definition, analysis, and management of outcomes. The companion book presented a wide array of approaches to managing high risk, high cost health problems, with the exception of those related to cardiovascular.

In this book, using the same framework, we focus on cardiovascular care and practice. Why a separate book for this specialty? The reason is clear as we look at the current health care practice environment and one of society's most challenging health care problems. Cardiovascular disease is a leading cause of death in the United States; cardiovascular care is a large and growing specialty. The cardiovascular patient population challenges health care professionals both clinically and financially. The scope of care is very broad—fragmented, uncoordinated care management of these patients can quickly lead organizations into financial distress. Cardiovascular processes of care need to be well managed with a focus on continually improving outcomes.

As such, we believe that this book, dedicated to outcomes resulting from the use of path-based care for cardiovascular care, makes a significant contribution to more fully understanding and addressing one of the most compelling health care issues today. Cardiac-related diagnoses often result in long lengths of acute care stay, high readmission rates, and marked stress on patients, families, and society. The assessment of outcomes and the development of new and innovative care strategies for this patient population are critical to the viability of acute care hospitals. Addressing this patient population's complex needs also extends well beyond the acute care hospital—into settings such as subacute care, long-term care, home health care, ambulatory, and preventative care. Care needs of patients with cardiovascular problems are great as evidenced by our long time focus on developing plans of care that will benefit them fully. Measuring outcomes in this population, however, has been traditionally lacking from our approaches until recently. This book provides specific approaches, and their outcomes, to cardiovascular care—both general and surgical—to assist health practitioners meet the challenges and demands of this specialty practice.

We have compiled some of the best case scenarios from organizations that are successfully practicing in a collaborative model with a focus on path-based care. Your challenges in caring for cardiovascular patients are their challenges; the broad care issues faced by providers and their patients are relatively similar across the health care spectrum. We believe that by studying the approaches of others to cardiovascular care, others can gain greater insight into some of the best strategies and innovations being applied today in this practice specialty.

You will learn how these organizations worked collaboratively to build models of care for cardiovascular patients, and most importantly you will be able to see the outcomes of such models of care. Your knowledge base will grow tremendously as you use this book as a study guide and reference manual. You too can achieve maximum outcomes in cardiovascular care by learning from your colleagues who have generously shared their experiences to make this book a reality.

The world of outcomes is the new world in which we live. The importance of identifying, measuring, and enhancing outcomes in patient care is destined to grow in importance as we enter the next century. We are truly on a journey to higher levels of professional, collaborative practice and the road to this end is outcomes. The outcomes addressed in this book illustrate (1) how patients are responding to our plans of care; (2) how clinically sound and realistic goals can be identified and achieved; (3) how reliable data can be generated to help consumers, payers, and other third parties to more fully evaluate the effectiveness of our care plans; (4) how collaborative pathways can be designed to reach predetermined goals for individual patients; (5) how data can be generated for further analysis and contribution to benchmarking and continuous quality improvement efforts; and (6) how outcomes can enhance our knowledge and understanding of patient and family needs.

This book opens with a chapter on the history of outcomes. It also appears in the companion book. We have reprinted it here for several reasons; it is an excellent introduction for approaching the new world of outcomes, and it provides those readers who have not yet studied book 1 to obtain a fundamental perspective for appreciating the approaches used in this book. The authors of this first chapter are highly regarded experts in outcomes management and we are pleased to present their work for your continued learning.

We acknowledge and thank this book's contributors, whose efforts to bring you the best in cutting edge information have been meritorious. We thank Aspen Publishers for supporting such an important and innovative project. Our goal is to have you apply, refine, and advance the valuable and useful information presented in this book for the purpose of improving care process in general, and specifically, care given to patients with cardiovascular problems. Our hope is that this book will make a positive contribution to a health care system in transition and stimulate actions that will reinforce the benefits of collaborative, coordinated care processes.

As in the companion book, we dedicate this book to the memory and legacy of Mother Teresa of Calcutta and Princess Diana of Wales, two great humanitarian women of our time whose life's journeys ended just as we completed this project. Finally, we dedicate this companion book in the loving memory of J. Worthington Smith, whose life was taken unexpectedly at the beginning of this project.

Dominick L. Flarey, PhD, MBA, RN, CS, CNAA, FACHE
Suzanne Smith Blancett, EdD, RN, FAAN

■ Part I ■
Overview

1

The Evolution of Outcomes Management

Diane Huber and Marilyn Oermann

Of the many images of nursing, perhaps the most enduring is that of the "Lady with the Lamp." Reflecting the vision of Florence Nightingale caring for the troops during the Crimean War, this image has come to symbolize the caring presence of the nurse to the sick or injured and has been instrumental in building the trust and esteem that nurses hold in the public's view.[1] Yet nurses know that nursing is and can be much more than holding the lamp. In fact, one of the best kept health care secrets is that Florence Nightingale was the original nurse researcher and administrator who pioneered the systematic use of patient outcomes, in the form of mortality data, to improve health care.[2,3]

Nurses trace the history of outcomes initiatives back to Nightingale's work in the 1850s. However, nurses are not generally credited with either having an outcomes orientation or being the professionals with the best sets of skills to manage quality-of-care and cost-effectiveness initiatives within health care systems. Perhaps it is time to reclaim our roots and reemphasize the strategies used by Nightingale to decrease the mortality rate in a military hospital from 60% to almost 1%, using data collection, analysis, and care process improvement techniques specific to nursing care.[1] An argument can be made that the original nurse's role included outcomes management and that this has reemerged as a crucial aspect of contemporary health care.

By tradition, nurses are seen and often view themselves as caregivers. However, McClure identified the nurse role as having two dimensions: that of caregiver and that of care integrator.[4] Caregiver role functions are related to the augmentation of patients' dependency needs, comfort, education, therapeutics, and monitoring. The role of care integrator involves the linking, synthesizing, and integrating of the work output of multidisciplinary differentiated providers so that patient outcomes occur as needed and desired despite the complexity of the health care environment. Nurses possess the knowledge to perform as integrators and to effect change. This is the ultimate in outcomes: to coordinate and integrate care processes into a seamless continuum of cost-effective, high-quality health care that has a positive effect on the health of the public. Nurses do this by diagnosing patient and system problems, taking action to intervene, and evaluating by measuring and monitoring outcomes.

With such a crucial role for nurses, it is important to explore and build a knowledge base about outcomes: the meaning of related terms, an understanding of their historical development, and the relationships among outcomes measurement, outcomes management, quality initiatives, and performance issues. The purpose of this chapter is to build an understanding of outcomes and to chronicle the evolution of outcomes measurement and outcomes management.

DEFINITIONS

Fundamental to an understanding of outcomes and their measurement and management is a clear definition of terms. Definitions were culled from the literature for the following terms: *outcomes, outcomes measurement, outcomes management, pathways,* and *variance* (see Exhibit 1–1).

At the most basic level, outcomes are the results obtained from the efforts to accomplish a goal. *Outcomes* has been defined simply as "end results, or that which results from something"[2(p.158)] and as conditions to be achieved.[5] Lang and Marek noted that the complex construct of outcomes is usually defined as an end result of a

Exhibit 1–1 Definitions

Outcomes	End results, or that which results from something
Outcomes Measurement	Observing, describing, and quantifying indicators of outcomes
Outcomes Management	The multidisciplinary process designed to provide quality health care
Pathways	Using predetermined care activities and mapping timelines to form a guide to usual treatment patterns for a group of individuals with similar needs
Variance	The deviation or departure from the expected clinical trajectory

treatment or intervention.[6] They identified four components of outcomes: the outcome measure that represents the end result; the determination of when the end point occurs; the treatment or intervention used; and the identification of the problem, diagnosis, or population that gives rise to the treatment and is linked to the outcomes. Donabedian considered outcomes to be changes in the actual or potential health status of individuals, groups, or communities that could be attributed to either prior or concurrent health care.[7] Of the three measures of health care quality, structure, process, and outcomes, he considered outcomes to be the ultimate measure of care.[8]

Important to the definition of outcomes is an understanding of the multiple perspectives from which definitions are approached. While some definitions view outcomes globally as end results or changes in status, other definitions are more specific to a segment of health care. Outcomes, like the measures of the quality of health care, can be defined from medical, nursing, organizational, patient, or other perspectives.[9] For example, Lohr defined outcomes as the end result of medical care in terms of palliation, control of illness, cure, or rehabilitation of the patient.[10] She noted that the classic list of outcomes was death, disease, disability, discomfort, and dissatisfaction. Another medically focused framework is the Medical Outcomes Study (MOS), which identified clinical end points, functional status, general well-being, and satisfaction with care as the outcomes of physician care.[11] The MOS has been critiqued for lacking a multidisciplinary collaborative perspective.[12] Whether global or specific, outcomes really answer the basic question of whether the individuals, groups, or communities of concern were benefited by the care provided.[13]

Outcomes measurement has been defined simply as measuring the results of health care.[14] The measurement of outcomes encompasses activities of observing, de-

scribing, and quantifying indicators of outcomes.[15] This can be distinguished from outcomes monitoring, which is the repeated observation, description, and quantification of outcome indicators for the purpose of improving care.[15] Outcomes measurement draws upon the problem-solving and scientific inquiry processes. Outcomes measurement involves the following steps:

1. Determine the measures of interest (key elements or indicators).
2. Gather the necessary data.
3. Aggregate and analyze the data.
4. Interpret the results and subsequent actions.
5. Make changes.
6. Measure again to evaluate the effectiveness.[15]

Outcomes management relates to a variety of activities designed to use the data gathered from outcomes measurement and monitoring to continuously improve care toward an ideal achievement of outcomes.[15] *Outcomes management* has been defined as a multidisciplinary process designed to provide quality health care, decrease fragmentation, enhance outcomes, and constrain costs.[14,16] The core idea of outcomes management is the use of process activities to improve outcomes.

Pathways refers to critical paths or clinical pathways that form a structured multidisciplinary action plan outlining the critical or key events and activities and the expected outcomes of care for each discipline during each day of a care episode.[17] As introduced in the late 1980s by Boston's New England Medical Center, critical paths were designed for standardization of key events and time frames for a patient's hospitalization.[18] They are a method of planning and documenting care.[17] Pathways use predetermined care activities and map timelines to form a guide to usual treatment patterns for a group of individuals with similar needs. These paths of care to be delivered reflect the most appropriate path to take, with room for variance and corrections. Elements include client problems, expected outcomes, and intermediate outcomes or milestones per visit or day for an entire episode of care.[5]

Variance refers to a deviation or departure from the expected clinical trajectory. In the case of clinical pathways, the common sequencing of events is delineated for a similar group of individuals. Variance analysis is used to examine when and how an individual patient differs from the expected norm.[19] The goal is to reduce provider practice variations, minimize delays in treatment, and decrease resource use while maintaining or improving the quality of care.[20]

The above definitions form a context for understanding outcomes. Clearly, measurement of the quality of care revolves around the determination of outcomes and efforts to use this knowledge to improve the processes of care through outcomes measurement, monitoring, and management. Pathways and variance analysis are key strategies in shaping financial and clinical outcomes. An overview of the history of outcomes measurement and management gives a context for understanding the movement from traditional indicators to a contemporary focus on pathways and a major focus on outcomes.

HISTORY OF OUTCOMES MEASUREMENT

Dramatic changes in health care have become the norm in recent years. These changes have affected access to care, who provides care and in which settings, the structuring of health care organizations, and the costs of care. Many of these changes have arisen from the continuing need to control the costs of health care and from consumers' desire for information about the quality of care they are purchasing. From individual consumers through major corporations, the pressure of huge health expenditures has resulted in a call for greater accountability by health care organizations in providing quality and cost-effective care. Provider organizations need to demonstrate that their practices are sound and that their services are superior.[21]

Outcomes measurement provides a means of verifying the success of a provider's care in terms of predetermined outcomes. The focus on measuring outcomes is a response to the need to determine internally the quality of health care provided by an organization, to compare it with that of similar organizations, and to assess the cost-effectiveness of care. Ellenberg suggested that outcomes measurement is the direct result of the need to provide comparative databases on the effectiveness of treatment protocols, evaluations of health-related quality of life, and cost containment measures.[22] While the focus of early outcomes measurement was on the costs of health care, current efforts examine clinical, functional, patient satisfaction, and other types of outcomes. Outcomes measurement, therefore, has shifted from a more narrow focus on costs to encompass broader outcomes that are more clinically based and provide a measure of both the quality and effectiveness of care.

Factors Influencing Measurement of Patient Outcomes

A number of factors are creating the demand for outcomes measurement (see Exhibit 1–2). One is the tremendous variability in medical treatments in different geographic areas of the United States. Wennberg found in his landmark study comparing the cost and effectiveness of different invasive procedures for prostate surgery

Exhibit 1–2 Factors That Create Demand for Outcomes Measurement

- The tremendous variability in medical treatments in different geographic areas of the United States
- The need to measure the effectiveness of treatments with heterogeneous populations
- The priority given health care reform by the state and federal government, insurers, employers, and consumers
- The demand by purchasers for quantifiable information about the value of the health care dollars they are spending

that there was great variability in the procedures used for treating patients.[23] Less invasive techniques were found to be as successful as more invasive ones and were most cost-effective. The less invasive and less costly procedures, however, were not necessarily used by physicians. Other studies have documented variability in medical treatments for the same diagnosis. The related question is what effect these differences in treatments have on patient outcomes. The variability in medical practice has been one major impetus for outcomes measurement, fueled by questions among the public as to why such differences exist in certain geographic areas and sometimes across institutions in the same area.

Current efforts focus on determining the most appropriate and cost-effective medical treatments and translating these findings into practice. "The interest of payers and policymakers in evidence-based medicine clearly derives from a desire to identify sources of cost without benefit to free resources for other uses."[24(p.329)] Titler and Reiter suggested that the need to measure the effectiveness of treatments with heterogeneous populations is another factor underlying the movement toward outcomes measurement.[25] Patient outcomes data can be used to determine the effectiveness of medical treatments, nursing interventions, and care of other providers.

A third factor that has increased the demand for outcomes measurement is the priority given health care reform by the state and federal government, insurers, employers, and consumers. Yet demands for controlling costs of health care need to be balanced with a careful consideration of the quality of that care. Measurement of patient outcomes provides an opportunity to examine quality of care in relation to costs.

A fourth and related factor is that purchasers want quantifiable information about the value of the health care dollars they are spending. In particular, employers offering health care coverage as a benefit seek information about costs of care, services provided, and patient outcomes. Outcomes measurement provides quantifiable information for use by purchasers. One related issue, however, is the variety of definitions of quality and methodologies for measuring it, limiting the availability of comparable and reliable data.[26]

Significant Events in the Development of Outcomes Measurement

Nurses have always been concerned about patient outcomes and evaluating outcomes as a means of determining the effectiveness of care. Much of the philosophical basis for outcomes measurement is derived from the early work of Florence Nightingale. Nightingale used mortality statistics to portray the low quality of care provided to British soldiers during the Crimean War.[3] Through data collected by herself and her nurses on preventable deaths related to changes in sanitation, Nightingale developed compelling arguments for needed reforms. The reforms were effective in reducing the mortality rate at the Barrack Hospital in Scutari from 60% to approximately 1%.[27]

From Nightingale's time through the 1960s, patient outcomes data consisted mainly of mortality rates. Later, morbidity rates also were used for measuring the quality of care.[28] In nursing in the early 1960s, Aydelotte studied patient welfare as an outcome of nursing care.[29] Multiple outcomes measures were used, such as number of postoperative days, doses of medications, instruments to measure the patient's behavioral characteristics and physical state of health, and time the patient spent in certain activities. This was an important study in that the impact of nursing care on outcomes was examined.

The 1970s marked a significant period in outcomes measurement. The Joint Commission on Accreditation of Healthcare Organizations (Joint Commission) recommended the use of outcome criteria in nursing audits, and many health care organizations developed their own criteria for measuring the quality of their services.[30] A number of notable projects at this time influenced the development of outcomes measurement in nursing. Hover and Zimmer proposed five outcomes for measuring the quality of nursing care: the patient's knowledge of illness and treatments to be performed, knowledge of medications, skills, adaptive behaviors, and health status.[31] Horn and Swain developed an extensive set of outcomes measures to evaluate the quality of nursing care; a total of 348 outcome measures were then categorized based on Orem's self-care theory.[32] Daubert developed a patient classification system that included five levels of outcomes from recovery to terminal care.[33] These beginning projects provided a framework for outcomes measurement in nursing that extended beyond the traditional measures of mortality and morbidity.

In the area of community health, the Visiting Nurse Association of Omaha in the 1970s began development of the Omaha System. The intent of the Omaha System was to improve the agency's patient record system. As part of this system, the Problem Rating Scale for Outcomes was developed to measure three outcomes: (1) *knowledge,* the ability of the patient to remember and interpret information; (2) *behavior,* observable responses, actions, or activities of the patient; and (3) *status,* the condition of the patient in terms of signs and symptoms.[34] The Problem Rating Scale for Outcomes is an instrument for use by nurses and other health professionals to measure change in the client in terms of these outcomes. The testing of the Omaha System has continued from the beginning phase of development in the early 1970s through the present. The System has evolved into a model for practice, documentation, and information management for use not only by home care and public health nurses but also by other health personnel.[35]

The Past 20 Years

The 1980s set the stage for continued emphasis on patient outcomes as a measure of quality of care. The use of outcome criteria to evaluate care paralleled the introduction of various mechanisms for controlling health expenditures, such as prospective payment, and related concern about the quality of care.[10] The Health Care Financing Administration (HCFA) introduced an initiative to measure the effectiveness and appropriateness of health care services for both Medicare and Medicaid programs.[36] This initiative included the development of a large database to monitor outcomes of certain treatments.

In 1989, the Agency for Health Care Policy and Research (AHCPR) was created with a commitment to examine the effectiveness of medical treatments for various conditions. One goal was to control Medicare spending by developing practice guidelines to enhance the quality and effectiveness of care. Clinical practice guidelines, developed from research and expert opinions, provide a blueprint for managing the care of patients. These research-based and standardized guidelines assist practitioners and patients in arriving at decisions about health care for specific clinical conditions.

Another significant impetus for outcomes measurement during the 1980s was the Joint Commission's Agenda for Change, approved in 1986, aimed at improving quality of health care through the accreditation process.[37] One important component of the Agenda for Change was the development of outcome indicators. An indicator is a quantitative measure of patient care that can be used to monitor and evaluate quality of care delivery.[37] Nurses can use such indicators to assess performance and evaluate patient outcomes. The indicators were developed by expert groups with nurse members and tested extensively, resulting in the development of the *Indicator Measurement System* (IMSystem) in 1994. Participating hospitals submit patient data and receive comparative reports with risk-adjusted information. On the basis of these reports, an organization can compare its actual rates with its predicted rates and with rates of other hospitals enrolled in the IMSystem.[34]

Outcomes Measures Sensitive to Nursing

From the 1980s to the present, nurses have been increasingly interested in identifying patient outcomes to measure the effectiveness of nursing care. The focus of these studies has been on describing outcomes sensitive to nursing care.

A number of these studies have been in home care. Lalonde identified seven outcome measures for patients receiving home care: taking prescribed medications, symptom distress, discharge status, caregiver strain, functional status, physiological indicators, and knowledge of health problems.[38] Similarly, Rinke attempted to categorize outcomes for home health care into five areas: physical, behavioral, psychosocial, knowledge, and functional.[39] During this same period, an outcome-based home care quality assurance program was developed in Alberta.[40] The program included both client and family outcomes important for monitoring in home care. The outcomes included pain management, symptom control, physiological status, activities of daily living, instrumental activities of daily living, well-being, goal attainment, knowledge and ability to apply that knowledge, patient and family satisfaction, family strain, and home maintenance.

Marek described types of outcomes appropriate for measuring the effectiveness of nursing care: physiological, psychosocial, functional, behavioral, knowledge, home functioning, family strain, safety, symptom control, quality of life, goal attainment, patient satisfaction, cost and resource utilization, and resolution of nursing diagnoses.[2,30,41] While there is still no consensus as to outcomes to measure, these early studies provided a beginning point for many of the current outcomes measurement projects in nursing.

The measurement of outcomes in long-term care also began during the 1980s. The Omnibus Budget Reconciliation Act of 1987, with the goal of improving the quality of care in nursing homes, mandated the development of outcomes measures for long-term care.

An extensive research program to develop and test outcome measures for home care was initiated by Shaughnessy in the 1980s, resulting in the standardized Outcomes and Assessment Information Set (OASIS) for home health care. In an early study, Shaughnessy examined the outcomes of Medicare nursing home and home health patients. From 1989 through 1994, a series of studies was conducted to develop and test outcome-based measures of quality in home care.[42] The goal of the research was to develop outcome measures that home health care agencies could incorporate into their quality improvement programs. In this way, a partnership would be established between home care agencies and the Medicare program for collecting and processing information to improve patient outcomes, improve agency performance, and enhance the efficiency of the Medicare approach to quality assurance.[43] The new Medicare Conditions of Participation for certified home care agencies include a requirement that they collect outcomes data using OASIS.[44]

Outcomes Measures of the Quality of Health Care

The 1990s marked a period of intense scrutiny of the quality of services provided by health care agencies. Many proposals were developed for measuring the quality of health care and reporting data about service, quality, and cost. Some proposals emphasized practice guidelines, such as the AHCPR guidelines; more rigorous and outcome-based reporting requirements came from bodies such as the Joint Commission and HCFA; new standards emerged for health plan reporting, such as the Health Plan Employer Data and Information Set (HEDIS); and individual health care organizations developed their own systems for documenting health care quality in report cards and instrument panels.

The AHCPR clinical practice guidelines were established to assist practitioners and patients to arrive at decisions about appropriate health care for specific clinical conditions.[45] They also were intended to serve as quality review criteria and to set quality improvement goals.[46] To date, AHCPR has published 19 clinical practice guidelines. The focus now is on implementation of the guidelines by practitioners; nurses have assumed an important leadership role in implementing the guidelines in practice.[47] Brown et al suggested that tailoring the guidelines to local conditions may be critical for implementation of them.[46] This creates a paradox for practitioners attempting to use the guidelines in that their validity may be compromised by modifying them, yet without this modification the guidelines may not be used by practitioners.

While many nurses are using the AHCPR guidelines in their own practice, others are developing practice guidelines at the local level. The intent of these locally developed guidelines is similar to that of the national ones; to assist nurses and patients in making informed decisions about health care. Dean-Baar suggested that the guidelines provide approaches to managing client conditions.[48] In addition, the guidelines may be used to evaluate patient care, identify future care needs, identify potential variations in nursing practice, reduce costs, and educate staff.[49,50]

The 1990s were a time in which groups such as the Joint Commission and HCFA shifted their attention to developing quality indicators that emphasized outcomes rather than the structures and processes of care. The Joint Commission's IMSystem is a performance measurement system for voluntary participation by health care organizations. An indicator is a valid and reliable quantitative process or outcome measure related to one or more dimensions of performance.[51] Indicators indicate an element of the process being measured or an outcome of that process. Traditionally, accreditation agencies such as the Joint Commission surveyed organizations' structure and process capabilities rather than their outcomes. Today, accreditation bodies focus on measures of actual performance.[52]

In the initial development of the IMSystem, the goal was to design an outcome-focused performance assessment to assist organizations to improve the quality of their care. Following extensive testing, indicators were gradually added to the IMSystem for a total of 33 indicators.[52] In 1995, the focus of the IMSystem shifted to create a broader group of performance indicators including those developed by health care organizations that would meet predetermined criteria.

Patient outcomes data are increasingly important in evaluating the quality of care provided by home health agencies. HCFA has focused its attention on measuring patient outcomes rather than organization and provider activities.[53] As indicated earlier, Medicare's OASIS data set is outcomes based. OASIS is designed to measure outcomes for adult home care patients; home care agencies will receive three annual outcome profile reports comparing their patient outcomes to a national sample of patients as well as to monitor their own improvement over time. Three types of outcomes are included in OASIS: (1) *end result,* a change in the patient's health status between two or more points in time; (2) *intermediate,* a change in patient behavior, affect, or knowledge that might affect end-result outcomes; and (3) *types of health care utilization,* such as hospital admission.[42,53]

HEDIS was developed by the National Committee for Quality Assurance to provide managed-care organizations with a standardized system for measuring quality performance indicators and for reporting this information. In addition, it allows organizations to monitor improvement activities over time, provides information to employers and purchasers for comparing managed-care organizations, and enables organizations to determine priorities for prevention.[54] There are a number of versions of HEDIS; the most recent version, 3.0, is intended for collecting data on commercial and Medicare and Medicaid risk populations. HEDIS 3.0 focuses more on outcomes than earlier versions. The performance indicators are categorized into eight domains: effectiveness of care, access and availability of care, satisfaction, health plan stability, use of services, costs of care, informed health care choices, and health plan descriptive information.[55] HEDIS provides a systematic measurement process, thereby enabling health plans to compare their outcomes nationally.

Report cards provide information about the quality and costs of health care to meet the needs of consumers and purchasers. While report cards are a positive response to the need for comparative information, there may be problems with the validity of the data gathered.[56] Uses of report cards include providing guidance as to which providers of care achieve the best clinical, functional, and satisfaction outcomes at lowest cost; holding providers accountable for achieving outcomes and maintaining costs; identifying opportunities for improvement; and identifying benchmarking sources.[56] The information in report cards, however, must be clear as to the outcomes that are measured and must be valid and reliable. Some health care organizations issue report cards on themselves to demonstrate their quality, costs, and services and how well they are scoring in these areas.

Nelson et al recommended designing instrument panel data collection systems to feed directly into report cards.[56] Instrument panel data have dual aims of learning about variation in performance in a system and meeting external information needs for purchasers. They provide for a balanced review of outcomes, such as clinical outcomes, functional health status, patient satisfaction, and costs.[56-58] One goal of instrument panels is to provide knowledge for improvement within the organization.

Nursing Outcomes Classifications

Efforts to describe and classify nursing-sensitive outcomes continue; these efforts are critical to measuring the impact of nursing care on patient outcomes. The Omaha System, which began development in the 1970s, includes standardized classifications for various conditions, interventions, and ratings of patient problems.[34] The Problem Rating Scale for Outcomes is designed specifically for measuring outcomes of care, knowledge, behavior, and status, thereby providing data for clinicians and administrators. Recent studies suggest the usefulness of the Omaha System in describing and quantifying nursing practice and in providing a systematic way of collecting client outcome data from diverse home health agencies.[59]

In 1989, the National League for Nursing and its Community Health Accreditation Program began its project to develop outcome measures of care quality for elderly patients receiving home care and to design a report card for collecting and reporting the data.[60] The outcomes are consumer based, such as knowledge and family support; clinically based, such as functional ability; and organization based, such as financial viability.

At the University of Iowa, work began first in the 1980s on developing a nursing intervention classification. In the 1990s, research was initiated to develop a comprehensive classification of nursing-sensitive patient outcomes. The Nursing-Sensitive Outcomes Classification (NOC) completes the third of the four patient-level nursing process elements of the Nursing Minimum Data Set. The NOC includes 190 patient outcomes sensitive to nursing interventions. Each outcome is labeled and defined and includes indicators, a measurement scale, and references.[61] The NOC also is significant because of its ability to be included in nursing clinical data sets in varied settings and in large regional and national databases used to assess the effectiveness of nursing interventions and inform policy makers.[62]

In 1994, the American Nurses Association began its project to identify quality indicators and measurement tools to measure the quality of nursing care in acute care settings. The indicators included outcome, process, and structure measures. Eight of the indicators originally identified were subjected to more intensive study because of their specificity to nursing quality and ability to be tracked.[63] The outcome indicators include nosocomial infection rate; patient injury rate; and patient satisfaction with nursing care, pain management, educational information, and care during the hospital stay.[63]

The 1990s will be remembered as a period that emphasized outcomes as indicators of quality. Continued efforts to determine the effectiveness of nursing interventions in achieving patient outcomes are critical. Without such efforts, the focus of outcomes measurement will remain decidedly on medical practice, with limited opportunity for nursing's contributions to be recognized.

HISTORY OF OUTCOMES MANAGEMENT

Outcomes management is the use of information collected through the measurement of outcomes to continually improve processes of care. Outcomes management is a multidisciplinary approach to promote quality health care, decrease fragmentation, improve patient outcomes, and control costs.[16] In outcomes management, data are collected and analyzed over a period of time, trends are identified, and decisions are made as to changes necessary to enhance patient outcomes and improve the effectiveness of care delivery.

In 1988, Ellwood proposed outcomes management as a "technology of patient experiences designed to help patients, payers, and providers make rational medical care-related choices based on better insight into the effect of these choices on the patient's life."[64(p.1551)] He proposed that outcomes management would enable health providers to analyze clinical, financial, and health outcomes, drawn from a national database, to identify relationships among medical interventions, outcomes, and cost. Ellwood envisioned outcomes management as a system that could be modified continuously and improved upon through advances in science, changes in patients' expectations regarding their care, and availability of resources.[64]

With this as a framework, one of the goals of an outcomes management program is to collect a standardized set of data on patient outcomes and then use this information to improve care processes, develop new research on effectiveness, and determine clinical policies.[65] Outcomes management has come to be recognized as a method for measuring and then improving performance on the basis of the results obtained.

Outcomes management provides a systematic approach to linking outcomes data with continuous quality improvement (CQI). The process begins with the measurement of outcomes, followed by continuous improvement of work processes to achieve better outcomes.[14]

Early quality assurance efforts focused mainly on resolving problems to improve system effectiveness, but there was limited concern as to whether these efforts actually improved patient outcomes. Standards of care were developed, and audits were initiated to ensure that there was compliance with these standards. The Joint Commission's Agenda for Change, however, required hospitals to develop systems for measuring quality of processes, structures, and patient outcomes. The perspective has now shifted away from quality *assurance* to continuous quality *improvement*. Houston and Miller described outcomes management as an essential part of CQI and other quality enhancement programs.[66]

Wojner developed an outcomes management quality model that clearly shows this relationship between outcomes management and quality improvement.[67] In phase 1, an interdisciplinary team of providers is organized to initiate the outcomes management process, outcomes are identified for measurement at specified points in time, and instruments for outcomes measurement are selected. In phase 2, interventions are created and tested, or structured care methodologies, such as critical pathways, protocols, and algorithms, are designed to standardize practices. Phase 3 involves the implementation of these methods to standardize practice within each discipline; this phase also includes the collection of data. During phase 4, data are analyzed, leading to the identification of opportunities for practice enhancement and new research questions. The process then recycles through phases 2 through 4. The continuous nature of outcomes management provides for cyclical measurement and practice improvement targeted at outcomes enhancement.[67]

Other types of improvement efforts are ongoing in individual health care organizations. Outcomes management, while similar in process to these, begins with identifying outcomes for measurement, then seeks to examine the processes and practices for achieving them. Through outcomes management, data gained from measuring outcomes provide the basis for improving processes and in turn achieving optimal patient outcomes.[68]

TRACKING AND EVALUATING OUTCOMES

It is no longer sufficient to put a clinical program, process, or delivery system in place and assume that all is well unless a sentinel event occurs. This is a strategy sometimes referred to as "shoot first and call whatever you hit the target." Today, consumers and payers expect providers to demonstrate effective, efficient, minimally costly, and outcomes-based care services. This is a strategy of identifying a target and then shooting. Thus, health care providers need to track and evaluate their care outcomes. Critical pathways and variance analysis are two major care management tools designed to meet outcomes management needs by structuring an identified target and making any necessary midcourse corrections.

Critical Pathways

Providers search for ways to have some assurance of providing the best care. Providing the best care at the least cost in a complex, multidisciplinary environment includes orchestrating the proper care components in appropriate time frames. This involves multidisciplinary teamwork to plan and develop a standardized written map of care activities and the sequencing of interventions. A critical pathway, also called a *clinical pathway* or *critical path,* is one major tool for outcomes tracking and evaluating. It is often used in conjunction with case management systems.

A critical pathway is a document designed to organize interventions and activities for an episode of care. It incorporates process and outcomes components. A critical path can form a standard of care or a care plan or both. The idea is to use critical thinking by all members of a care team to identify critical and predictable clinical inci-

dents needed to achieve desired outcomes. Thus, the critical pathway is a practical form of practice accountability and documentation that can be used to track health outcomes, patient and provider activities, complications, and teaching/learning 'outcomes. Critical pathways may also be used as education tools, to prepare and orient patients before treatment, and to negotiate expectations and care roles with patients and families.[69]

Critical pathways can be thought of as protocols of interdisciplinary treatments that are based on professional standards of practice and placed in order on a decision tree.[69,70] Critical pathways should incorporate daily expected patient outcomes as subgoals. The daily subgoals form a system of incremental patient outcome targets for reaching the final goal that easily lend themselves to real-time planning and evaluating patient care. Critical pathways are now an accepted tool for clinical care planning.[71] Critical pathways also can be used to accomplish financial and systems goals. These can include the reduction of clinical practice variations, minimization of delays in treatment, and a decrease in resource use and costs.[20] Research has indicated that critical pathways are an important determinant of improved quality in a managed-care environment.[20] Further, the combination of case management, critical pathways, an outcomes database, and a report card approach to program evaluation was a synergistic method to improve care via secondary prevention of heart disease in one population studied.[71] Critical pathways, with their focused outcomes management design, are a key feature of improved quality-of-care efforts and deliberative outcomes management systems.

Variance Analysis

Variance analysis is a second major outcomes tracking and evaluation tool. A variance is anything that varies or differs. In relation to critical pathways, a variance occurs when there is a deviation from the standard, expected treatment path. When some aspect of care "falls off the path," the clinician monitoring and coordinating care delivery can note this, evaluate it, and take corrective action. Variance data can be aggregated over populations or groups to analyze trends having an impact on the larger scope of care delivery. Variance data analysis can be used to improve health care effectiveness, reduce risks, understand and strengthen provider interventions, modify care protocols, trigger research, and identify the impact of care processes on patient outcomes. Thus, provider decision making can be enhanced, and the real-time care of an individual patient can be improved or kept on target. Clearly, variance analysis is a key strategy for incorporation into CQI initiatives.

A variance, or departure, from a critical pathway can be either positive or negative. The direction is determined in relation to the desired outcome. For example, if reduced length of hospital stay is desirable, then any occurrence that contributes to lengthening stay time is considered negative. For individuals receiving health care, variances should be assessed concurrently with the care process. For aggregated populations, variances are combined and retrospectively analyzed. Variances can be derived from patients, practitioners, or systems.[72]

Variance data analysis begins with careful measurement of outcomes for population specificity, sensitivity, reliability, and validity. Outcomes may be clinical, functional status, financial, performance, or service quality related. For example, length of stay, patient satisfaction, patient knowledge, delays in treatment, errors, complications, preexisting conditions, social complications, or charge per case may be outcomes of interest. A time frame for data analysis needs to be chosen. For example, 100% of all records for a population can be monitored for a year, or data can be analyzed quarterly. Data for analysis will need to be selected from among the mass of data gathered. Variance data may only point to the need to gather additional data. Aggregated data, displayed via graphs, scatter plots, or other data display methods, can be examined for trends and patterns. Clinicians then can analyze the data by assessing the significance of variance patterns or sentinel events and the need to take corrective action. In this way, variance analysis becomes a key link in the CQI chain.

CONCLUSION

The use of outcomes measurement and management to magnify quality improvement and cost reduction activities remains the best strategy for genuine care improvement. These efforts also promise to help assuage the lingering public skepticism about managed care initiatives. Nurses have a major role to play in all aspects of outcomes measurement and management. The knowledge and insights about care management that nurses have to offer are central to effecting positive health care outcomes. Critical pathways and variance analysis are two essential tools. In the role of care integrator, nurses can demonstrate effective and efficient care contributions to vital health care processes of collaborative CQI and outcomes management that make a difference in the health care of individuals, groups, and communities.

REFERENCES

1. Kalisch PA, Kalisch BJ. *The Advance of American Nursing.* Boston: Little, Brown & Co; 1978.
2. Lang NM, Marek KD. The classification of patient outcomes. *J Prof Nurs.* 1990;6:158–163.
3. Nightingale F. *Notes on Matters Affecting the Health, Efficiency, and Hospital Administration of the British Army.* London: Harrison & Sons; 1858.
4. McClure M. Introduction. In: Goertzen I, ed. *Differentiating Nursing Practice: Into the Twenty-First Century.* Kansas City, MO: American Academy of Nursing; 1991:1–11.
5. Peters DA. Outcomes: the mainstay of a framework for quality of care. *J Nurs Care Qual.* 1995;10(1):61–69.
6. Lang NM, Marek KD. Outcomes that reflect clinical practice. In: *Patient Outcomes Research: Examining the Effectiveness of Nursing Practice. Proceedings of the State of the Science Conference Sponsored by the National Center for Nursing Research.* Washington, DC: NIH; 1992:27–38. NIH Pub. No. 93-3411.
7. Donabedian A. *The Methods and Findings of Quality Assessment and Monitoring: An Illustrated Analysis.* Vol 3. Ann Arbor, MI: Health Administration Press; 1985.
8. Donabedian A. Evaluating the quality of medical care. *Milbank Q.* 1966;44(3, part 2):166–206.
9. Gardner DL. Measures of quality. *Series on Nursing Administration.* 1992;3:42–58.
10. Lohr KH. Outcome measurement: concepts and questions. *Inquiry.* 1988;25(1):37–50.
11. Tarlov A, Ware J, Greenfield S, Nelson E, Perrin E, Zubkoff M. The medical outcomes study: an application of methods for monitoring the results of medical care. *JAMA.* 1989;262(7):925–930.
12. Kelly KC, Huber DG, Johnson M, McCloskey JC, Maas M. The medical outcomes study: a nursing perspective. *J Prof Nurs.* 1994;10(4):209–216.
13. Shaughnessy PW, Crisler KS. *Outcome-Based Quality Improvement: A Manual for Home Care Agencies on How to use Outcomes.* Washington, DC: National Association for Home Care; 1995.
14. Nadzam DM. Nurses and the measurement of health care: an overview. In: *Nursing Practice and Outcomes Measurement.* Oakbrook Terrace, IL: Joint Commission on Accreditation of Healthcare Organizations; 1997:1–15.
15. Oermann M, Huber D. New horizons. *Outcomes Manage Nurs Prac.* 1997;1(1):1–2.
16. Moss MT, O'Connor S. Outcomes management in perioperative services. *Nurs Econ.* 1993;11:364–369.
17. Cohen EL, Cesta TG. *Nursing case management: from concept to evaluation.* St. Louis, MO: CV Mosby; 1993.
18. Zander K. Care maps™: the core of cost/quality care. *The New Definition.* 1991;6(3):1–3.
19. Gardner K, Allhusen J, Kamm J, Tobin J. Determining the cost of care through clinical pathways. *Nurs Econ.* 1997;15(4):213–217.
20. Ireson CL. Critical pathways: effectiveness in achieving patient outcomes. *J Nurs Adm.* 1997;27(6):16–23.
21. Lansky D. The new responsibility: measuring and reporting on quality. *J Qual Improve.* 1993;19:545–565.
22. Ellenberg DB. Outcomes research: the history, debate, and implications for the field of occupational therapy. *Am J Occup Ther.* 1996;50:435–441.
23. Wennberg J. Outcomes research, cost containment and the fear of health care rationing. *N Engl J Med.* 1990;323:1202–1204.
24. Clancy CM, Kamerow DB. Evidence-based medicine meets cost-effectiveness analysis. *JAMA.* 1996;276:329–330.
25. Titler MG, Reiter RC. Outcomes measurement in clinical practice. *MEDSURG Nurs.* 1994;3:395–398, 420.
26. Wilson AA. The quest for accountability: patient costs & outcomes. *CARING.* 1996;XV(6):24–28.

27. Strodtman LKT. The historical evolution of nursing as a profession. In: Oermann MH, ed. *Professional Nursing Practice*. Stamford, CT: Appleton & Lange; 1997:38.

28. Bergner M. Measurement of health status. *Med Care*. 1985;23: 696–704.

29. Aydelotte M. The use of patient welfare as a criterion measure. *Nurs Res*. 1962;11(1):10–14.

30. Marek KD. Outcome measurement in nursing. *J Nurs Qual Assur*. 1989;4:1–9.

31. Hover J, Zimmer M. Nursing quality assurance: the Wisconsin system. *Nurs Outlook*. 1978;26:242–248.

32. Horn BJ, Swain MA. *Criterion Measures of Nursing Care Quality*. Hyattsville, MD: National Center for Health Services Research; 1978. DHEW Pub. No. PHS78-3187.

33. Daubert E. Patient classification system and outcome criteria. *Nurs Outlook*. 1979;27:450–454.

34. Martin KS. Nursing and patient care processes: nursing care outcomes measurement. In: *Nursing Practice and Outcomes Measurement*. Oakbrook Terrace, IL: Joint Commission on Accreditation of Healthcare Organizations; 1997:17–34.

35. Martin KS, Scheet NJ. *The Omaha System: Applications for Community Health Nursing*. Philadelphia: WB Saunders Co; 1992.

36. Roper WL, Winkenwerder W, Hackbarth GM, et al. Effectiveness in health care: an initiative to improve medical practice. *N Engl J Med*. 1988;319:1197–1202.

37. Nadzam DM. The agenda for change: update on indicator development and possible implications for the nursing profession. *J Nurs Qual Assur*. 1991;5(2):18–22.

38. Lalonde B. *Quality Assurance Manual of the Home Care Association of Washington*. Edmonds, WA: The Home Care Association of Washington; 1986.

39. Rinke L. *Outcomes Measures in Home Care: State of the Art*. Vol. 3. New York: National League for Nursing; 1988.

40. Sorgen LM. The development of a home care quality assurance program in Alberta. *Home Health Care Serv Q*. 1986;7(2):13–28.

41. Marek KD. Measuring the effectiveness of nursing care. *Outcomes Manage Nurs Pract*. 1997;1(1):8–12.

42. Shaughnessy PW, Crisler KS, Schlenker RE, et al. Outcome-based quality improvement in home care. *CARING*. 1995;XV(6):44–49.

43. Research update. Outcome-based quality improvement demonstration. *CARING*. 1996;XV(6):67.

44. Health Care Financing Administration. *Federal Register,* March 10, 1997;62(46):11,004–11,064.

45. Field MJ, Lohr KN, eds. *Clinical Practice Guidelines: Directions for a New Program*. Washington, DC: National Academy Press; 1990.

46. Brown JB, Shye D, McFarland B. The paradox of guideline implementation: how AHCPR's depression guideline was adapted at Kaiser Permanente northwest region. *J Qual Improve*. 1995;21:5–21.

47. Kaegi L. Nurses leading the charge to take national AHCPR guidelines into local settings. *J Qual Improve*. 1995;21:45–49.

48. Dean-Baar SL. Application of the new ANA framework for nursing practice standards and guidelines. *J Nurs Care Qual*. 1993;8:33–42.

49. Montgomery LA, Budreau GK. Implementing a clinical practice guideline to improve pediatric intravenous infiltration outcomes. *AACN Clin Issues*. 1996;7:411–424.

50. Yoos HL, Malone K, McMullen A, et al. Standards and practice guidelines as the foundation for clinical practice. *J Nurs Care Qual*. 1997;11(5):48–54.

51. Joint Commission on Accreditation of Healthcare Organizations. Performance improvement tools for outcomes measurement. In: *Nursing Practice and Outcomes Measurement*. Oakbrook Terrace, IL: Joint Commission; 1997:123–161.

52. Katz JM, Green E. *Managing Quality*. 2nd ed. St. Louis, MO: CV Mosby; 1997.

53. Harris MD, Dugan M. Evaluating the quality of home care services using patient outcome data. *Home Healthcare Nurse*. 1996;14: 463–468.

54. Parisi LL. What is influencing performance improvement in managed care? *J Nurs Care Qual*. 1997;11(4):43–52.

55. National Committee for Quality Assurance (NCQA). *HEDIS 3.0*. Vol. 2. Washington, DC: NCQA; 1997.

56. Nelson EC, Batalden PB, Plume SK, et al. Report cards or instrument panels: who needs what? *J Qual Improve*. 1995;21:155–166.

57. Nugent WC, Schults BA, Plume SK, et al. Designing an instrument panel to monitor and improve coronary artery bypass grafting. *J Clin Outcome Measure*. 1994;1(2):57–64.

58. Schriefer J, Urden LD, Rogers S. Report cards: tools for managing pathways and outcomes. *Outcomes Manage Nurs Pract*. 1997; 1(1):14–19.

59. Martin KS, Scheet NJ, Stegman MR. Home health clients: characteristics, outcomes of care, and nursing interventions. *Am J Public Health*. 1993;83:1730–1734.

60. National League for Nursing (NLN). *Summary of Findings: In Search of Excellence in Home Care*. New York: NLN; 1994.

61. Johnson M, Maas M, eds. *Nursing Outcomes Classification (NOC)*. St. Louis, MO: CV Mosby Co; 1997.

62. Maas ML, Johnson M, Moorhead S. Classifying nursing-sensitive patient outcomes. *Image*. 1996;28:295–301.

63. American Nurses Association (ANA). *Nursing Quality Indicators*. Washington, DC: ANA; 1996.

64. Ellwood PM. Shattuck lecture—outcomes management: a technology of patient experience. *N Engl J Med*. 1988;318:1549–1556.

65. Kania C, Richards R, Sanderson-Austin J, et al. Using clinical and functional data for quality improvement in outcomes measurement consortia. *J Qual Improve*. 1996;22:492–504.

66. Houston S, Miller R. The quality and outcomes management connection. *Crit Care Nurs Q*. 1997;19(4):80–89.

67. Wojner AW. Outcomes management: from theory to practice. *Crit Care Nurs Q*. 1997;19(4):1–13.

68. Hoesing H, Karnegis J. Nursing and patient care processes: interdisciplinary care outcomes management. In: *Nursing Practice and Outcomes Measurement*. Oakbrook Terrace, IL: Joint Commission on Accreditation of Healthcare Organizations; 1997:35–62.

69. Huber D. *Leadership and Nursing Care Management*. Philadelphia: WB Saunders Co; 1996.

70. Simpson R. Case-managed care in tomorrow's information network. *Nurs Manage*. 1993;24(7):14–16.

71. Levknecht L, Schriefer J, Schriefer J, Maconis B. Combining case management, pathways, and report cards for secondary cardiac prevention. *J Qual Improve*. 1997;23(3):162–174.

72. Willoughby C, Budreau G, Livingston D. A framework for integrated quality improvement. *J Nurs Care Qual*. 1997;11(3):44–53.

■ 2 ■

Patient Care Outcomes: A League of Their Own

Dominick L. Flarey

Everyone is talking about them, many are measuring them, and the majority are continuing to define them. Patient care outcomes are a necessary component of the entire process of care delivery. As such, their measurement and management are becoming more and more imperative in this new era of health care. The concept of patient care outcomes is not new. Nursing in particular has been defining and measuring them since the birth of the profession. What is new is their importance in demonstrating the value of our care initiatives in today's managed-care environment.

What exactly are patient care outcomes? Patient care outcomes may be defined as the measurement and assessment of the status of the patient following health care interventions. While this definition may seem simple, the overall concept of measuring outcomes can be challenging. The new adage in managed care is that one must deliver a high-quality service at an affordable price, with good outcomes. Much debate exists as to what is considered high quality, what *affordable* means and by whose standards, and what constitutes good outcomes. Despite the ensuing debate, nursing is emerging as the profession most prepared to address the current issue of patient care outcomes. Other health care professions are looking to us to provide leadership and guidance with this new mandate.

WHY MEASURE PATIENT CARE OUTCOMES?

The first and most important reason to measure patient care outcomes is to provide health care professionals with evaluative feedback regarding the actual care

that is being provided. Each care or treatment modality has a particular goal for the patient. Thus, each activity will also have some potential effect on the patient's overall condition and status. Examining the outcomes of patient care reveals to us whether the patient is responding satisfactorily to the planned interventions and whether the patient is meeting preestablished benchmarks for his or her particular problem at specifically determined times in the course of care delivery.

Health care professionals need to assess patient care outcomes constantly to provide a level of quality care and intervene appropriately when patients are not meeting defined benchmarks for care progression. There are at least nine other important and compelling reasons to measure and manage patient care outcomes:

1. to demonstrate the rationale for choosing specific medical and nursing interventions
2. to provide benchmarks that can be used to evaluate subsequent care outcomes across settings, populations, and other demographics
3. to demonstrate to third-party payers and society as a whole the effectiveness of care delivery
4. to assist health care professionals in adequately defining the concept of "quality" in care delivery
5. to assist the health care team in developing an individualized plan of care based on continual assessments of outcomes
6. to assist health care professionals in developing a research-based practice that is rooted in actual clinical experiences in patient care
7. to develop and test theories of care delivery in the practice setting
8. to assist in placing a monetary value on patient care delivery related to specific outcomes

Source: Reprinted with permission from D.L. Flarey, Patient Care Outcomes: A League of Their Own, *Outcomes Management for Nursing Practice,* Vol. 1, No. 1, pp. 36–40, © 1997, Lippincott-Raven Publishers.

9. to test collaborative practice interventions and evaluate how synergy among disciplines leads to further enhanced care outcomes

While the above reasons for measuring outcomes are not all-inclusive, they do provide a compelling justification for nurses' continuing efforts to define the specific reasons for and benefits of measuring outcomes in patient care delivery. Nursing has carved a niche as a leader in defining and measuring care outcomes, especially those that go beyond the interest of medicine. From the first phases of our educational process to current practice at the bedside, to some extent we have always incorporated outcomes measurement in patient care delivery. Today, the spotlight shines brightly on outcomes measurement and management.

In planning for outcomes measurement, two decisions are what to measure and the "how to" of outcomes measurement and management. Deciding what to measure is based on what we *expect* our interventions to achieve for patients. Outcomes are what we *hope* to achieve for patients.

MEASURE METHODOLOGY

To more fully operationalize the concept of what we measure related to patient care outcomes, this chapter will use the patient condition of heart failure as an example. This particular condition was selected because it currently dominates as an issue in managed patient care due to its complexity, enormous costs, and functional impairment of patients. Heart failure is a frustrating and costly condition to treat; defining outcomes criteria for this population also is challenging.

The first step before actually defining what should be established as outcomes for measurement and management in patients who have heart failure is a methodology that will lead us to making correct choices. The methodology recommended here is based upon the premise of critical needs theory. With each of the major domains—physiologic, psychological, and cognitive—critical questions need to be answered or critical issues defined. Attention to this particular detail will lead to the most obvious outcomes that must be measured for the patient, based on his or her condition.

The starting place for this methodology is viewing the patient physiologically. At this level, a person's needs are related to physiologic functioning or homeostasis and survival. The most critical questions that need to be addressed here are:

- What is disrupting homeostasis?
- What body system or physiologic function is impaired?

- What objective and subjective criteria can be used to evaluate adequately the level of impairment experienced by the patient and the type and degree of response to interventions?
- Which disruption is most damaging to comfort and sustaining life?

On the basis of these questions and their answers, nurses can derive the particular outcomes that should be measured. Exhibit 2–1 demonstrates how this methodology can be used.

WHY WE MEASURE OUTCOMES

From Exhibit 2–1, we can easily identify what outcomes need to be assessed and what outcomes are the most important for evaluating effectiveness of treatment. On the basis of the above analysis, the following patient care outcomes are recommended for assessment from a physiologic perspective:

- degree of dyspnea
- utilization of oxygen
- degree of disability related to activities of daily living
- type of breath sounds
- functional status of the patient
- rate and rhythm of pulse
- color of skin and nail beds
- degree of chest pain
- presence of orthopnea
- degree of jugular venous distention
- degree of abdominal distention
- degree of peripheral edema
- evidence of digitalis toxicity
- adequacy of oxygen saturation

These are the most important outcomes related to patients who experience congestive heart failure. They may be assessed using the format of a nursing diagnosis and an expected outcomes statement. Exhibit 2–2 provides us with a simple example.

Once both a nursing diagnosis and an outcomes statement have been written, it is necessary to provide for assessment and documentation of the defined outcomes. Exhibit 2–3 provides a sample of an assessment related to the nursing diagnosis for congestive heart failure. This assessment should be completed at the time of patient discharge from an acute care hospital.

The assessment of patient care outcomes in Exhibit 2–3 provides a comprehensive, documented assessment of the critical indicators for outcomes measurement at the time of patient discharge. The rationale for providing a full, detailed outcomes assessment is that it

Exhibit 2–1 Patient Care Outcomes: Physiologic Needs

Critical Question	Response	Major Symptoms and/or Dysfunction
What is disrupting homeostasis?	Alteration in the pumping mechanism of the heart, leading to poor pulmonary perfusion, causing hypoxia, which compounds the alteration in the pumping mechanism of the heart	Fatigue Dyspnea Chest pain Adventitious breath sounds
What body system or physiologic function is impaired?	Cardiovascular system, pumping action of the heart, pulmonary system	Cardiac enlargement Pulmonary congestion Cardiac arrhythmia
What subjective/objective criteria can be used to evaluate adequately the level of impairment and the type and degree of response to interventions?	Subjective: feelings of dyspnea, degree of dyspnea, precipitating factors, degree of relief with oxygen, degree of relief from diuretics, degree of relief from cardiovascular drugs	Pulmonary congestion, low arterial P_{O_2}, resulting in inadequate tissue perfusion
Which disruption is most critical to sustaining life?	Poor or failed pumping action of the heart; pulmonary congestion; tissue anoxia, leading to cardiac arrhythmia	Dyspnea, bilateral rales, cyanosis, peripheral edema, pulsus alternans, orthopnea, cough, fatigue, confusion, jugular venous distention, abdominal distention

- provides for a comprehensive assessment of the patient at the time of discharge
- is the patient's profile of progression or lack of progression at the time of discharge from an acute care setting
- provides a means to evaluate treatment effectiveness over the course of the acute episode of care
- provides documentation of the effectiveness of treatment over the course of the acute episode of care
- conveys important patient information to other health care providers who will continue providing patient care services after discharge
- provides a vehicle for the collection and aggregation of data related to patient care
- provides internal benchmarks for analysis by health care providers

- is a means to guide continued treatment of the patient over a longer course of time
- provides critical information for the redesign of processes to improve future patient outcomes
- provides concrete information regarding treatment effectiveness for managed-care networks and regulatory agencies
- documents the outcomes of the nursing process
- provides for defensive documentation in instances of litigation
- provides a profile of the patient's health status at a particular point in time

A LEAGUE OF THEIR OWN

The assessment and documentation of patient care outcomes are more important in health care today than ever before. Consumers want to know the effects of treatment and are demanding a much higher level of quality from the care services they purchase. Patient care outcomes need to be firmly established for all of the diagnostic-related groups for which an organization provides care.

Defined outcomes should be a component of an outcomes pathway and should be documented at least once daily on the pathway. For the documentation of discharge outcomes, a dedicated outcomes assessment form, as provided in Exhibit 2–3, should be created and should be a part of the patient's permanent medical

---

Exhibit 2–2 Nursing Diagnosis and Outcomes Statement

Nursing Diagnosis: Gas exchange, impaired, related to pulmonary congestion and fluid in the alveoli.

Outcomes Statement: "The patient will have improved gas exchange, as evidenced by vital signs within normal limits for patient's age and condition, skin and mucous membranes without cyanosis or pallor, decreased dyspnea, and arterial blood gases within normal limits."[1(p.1175)]

Exhibit 2–3 Assessment of Patient Outcomes at Discharge

Dyspnea	No dyspnea at rest. Mild dyspnea noted after ambulating in hall. Relieved rapidly with rest.
Utilization of O_2	Used O_2 when eating and while giving self a bath. Not in use otherwise. Has O_2 at home.
Activities of daily living	Able to give self a bed bath, able to eat without assistance, able to dress self without signs or symptoms of respiratory or other distress.
Breath sounds	Bronchial, clear to auscultation.
Functional status	Self-care is good for condition. Alert, oriented, answers appropriately.
Pulse	92, slightly irregular. Rhythm strip shows persistent rare premature atrial contractions.
Color	Normal, no cyanosis, no jaundice.
Chest pain	No chest pain. Last episode was 2 days ago. Not using nitroglycerin.
Orthopnea	Sleeping with two pillows as prior to admission.
Jugular venous distention	None today. Last evident upon admission and prior to acute care treatment.
Abdominal distention	Girth is 38, which is preadmission girth. Down from 44 upon admission.
Peripheral edema	Mild, 1+ edema, nonpitting.
Response to digitalis	No evidence of digitalis toxicity. Digitalis level this morning in therapeutic range.
Oxygen saturation	O_2 saturation as of 22 hours ago was 96%. No physiologic evidence of significant change.

record. Information from this discharge outcomes form can be included in a large patient care outcomes database so that outcomes can be aggregated and analyzed over time. This aggregation of outcomes provides the basis for the evaluation of patient care outcomes on an organizationwide or systemwide level.

Critical information can be obtained from this aggregate of outcomes data. Reviewing the outcomes data over time can prompt the development of critical questions that can be posed to improve the quality of outcomes through process improvements. Some of these critical questions for the diagnosis of congestive heart failure are:

- How do our patients compare in their length of stay to patients of other organizations or health care systems?
- What percentage of patients develop digitalis toxicity?
- What percentage of patients can enjoy activities of daily living without significant assistance?
- What percentage of patients demonstrate resolution of adventitious breath sounds by the time of discharge?
- What other medical and nursing interventions or therapies might provide better patient outcomes?

While these questions are not all-inclusive, they do provide an example of how health care providers can use the aggregate information from patient outcomes data to develop newer, more innovative approaches to patient care. Answering these questions is not a easy task. Not all patients are the same; each comes to the acute care episode of care with a different degree of congestive heart failure, with different comorbidities, and with a different level of compliance with treatment. One of our greatest challenges today is adjusting for these discrepancies and problems in adequately assessing outcomes. Though the assessment of patient care outcomes is in its infancy, future applications to help adjust for patient differences will be more commonplace. Methods and models for the aggregation of data related to specific predetermined criteria such as age, sex, and cognitive functioning are being developed and will continue to be a major focus of effort.

Physiologic patient care outcomes are not the only measures that are important. Other individual patient-related outcomes may be equally important and should not be neglected. These outcomes are related to psychological status and cognitive status. With respect to psychological status, some outcomes for congestive heart failure to be measured include

- presence/absence of anxiety
- presence/absence of depression
- ability to cope with a chronic illness
- degree of family support
- effect of disability on mental health
- evidence of suicidal ideations
- ability to plan for the future
- motivation to take responsibility for the illness

Cognitive outcomes also are important because poor cognitive outcomes may prevent patients from achieving levels of wellness or improvements in their condition. A few cognitive outcomes that should be assessed in patients with congestive heart failure are

- the degree of understanding and the depth of knowledge that the patient has of the illness and self-care
- the degree of understanding that the family/support system has of the patient's illness
- the ability of the patient to care for self and plan care for self
- the degree of understanding of needed lifestyle changes

While this is not an exhaustive list of psychological and cognitive outcomes, it does provide a stimulus to think about the psychological and cognitive aspects of the patient's illness and how these can positively or negatively influence the course of illness and ultimate health status outcomes. Quality care means assessing all the dimensions of potential patient outcomes. Patient care outcomes are not exclusive to disease states and physiologic dysfunction. All three domains—physiologic, psychological, and cognitive—interplay to influence the overall health status of individuals. Therefore, parallel outcomes should be assessed in each domain.

CONCLUSION

Defining patient care outcomes is a journey. We are just beginning that journey in health care along the road of patient care outcomes measurement and management. This is an exciting time in health care and one in which nursing has great opportunity to demonstrate its ability to positively affect the overall health of this nation.

As we continue to explore the realm of patient care outcomes, nurses will play more pivotal roles in defining what must be measured and, more important, what needs to be done to influence outcomes positively for patients in the future. We are more than likely to discover that nursing interventions, rather than some new fad in technology, have the most dramatic impact on patient care outcomes. Nurses have always known that nursing assessment and strong nursing care and interventions are what quality care is made of.

REFERENCE

1. Luckman S. *Medical-Surgical Nursing: A Psychophysiologic Approach.* 4th ed. Philadelphia: WB Saunders Co; 1993.

■ Part II ■
Medical Cardiovascular

■ 3 ■

A Rapid Rule Out Process for Myocardial Infarction: Separating Low Risk Cardiac Disease from Uncomplicated MI

Lynne Nemeth, Patricia Calver, Barbara Parlotz, Carol Smith, and Johnese Spisso

Harborview Medical Center (HMC) is a Level I trauma center for a four-state region, serving a mission population, which includes the indigent, underserved, and vulnerable patients, as well as victims of violence and abuse, the mentally ill, and our county prison population. We have a limited scope of cardiology services housed at HMC, as our affiliate hospital, the University of Washington Medical Center (UWMC) specializes in interventional procedures and cardiac surgery, including transplantation.

Harborview's cardiology program has historically focused on out-of-hospital cardiac arrest and ventricular fibrillation, in association with Seattle's Medic One program, which has been nationally known for successful outcomes of survival after sudden cardiac death.[1-3] HMC had developed, with the strength of the cardiology program, a laboratory that focused on conducting electrophysiological studies (EPS) to map areas of aberrant heart conduction. Approximately 9 years ago, this EPS Laboratory was transferred to the UWMC. The arrival of a new cardiology chief at HMC, with strong interest in imaging studies of the heart, including transesophageal echocardiography, has changed the focus of cardiology research being conducted at HMC.

Harborview admits approximately 70% of its patients through the emergency department (ED). The volume of patients seen at Harborview with chest pain or other cardiac symptoms, which prompt rule out myocardial infarction (R/O MI) evaluations, has always been consistent. At this academic medical center, resident physicians rotate throughout the medical service monthly, and a variety of approaches has been taken for this R/O MI process. The

arrival of the new cardiology chief at HMC in 1994 provided a good opportunity for the interdisciplinary team to standardize the evaluation of cardiology patients. We focused on the care processes and outcomes involved in ruling out MI and uncomplicated MI as one of the initial pathways in our new clinical pathways program. This chapter discusses HMC's rapid rule out process for myocardial infarction and clinical pathway.

PATHWAY DEVELOPMENT

An interdisciplinary authoring team was appointed to develop the pathway for this patient population. The team had representation from nursing staff within the ED, cardiac care unit (CCU), cardiology/telemetry monitoring unit, and the cardiology clinic. A cardiologist and a nuclear medicine physician attended the meetings regularly, as did a pharmacist, social worker, occupational and physical therapists (OT and PT), nutritionist, and a representative from laboratory medicine. The literature was reviewed for standards and guidelines that were applicable. The Agency for Health Care Policy and Research[4] had published guidelines for unstable angina, and the American College of Cardiology and the American Heart Association[5] had published guidelines for the early management of patients with acute myocardial infarction. There was an evolving national standard that advocated a "door to decision" guideline regarding the use of thrombolytics with the appropriate candidate, within the first 30 minutes of patient arrival in an ED.[6-10]

The clinical pathway project manager presented a brief overview of the process of pathway development and presented the format that had been developed by an interdisciplinary clinical pathway design team. The cardiology chief described the ideal process for evaluating and

The authors wish to acknowledge the expert assistance of Joan Knecht and Tarek Salaway, MHA, MPH, MA, in the preparation of this manuscript.

managing acute chest pain, as well as the care of the patient who had an uncomplicated myocardial infarction. Team members had the opportunity to discuss each discipline's role in the care of this patient. Much discussion took place about the problems in delivering optimal care, and we began to focus on the development of a pathway, to improve system issues, and provide more coordinated cost-effective care. This prompted efforts between the pharmacy and ED to ensure the delivery of thrombolytics from the pharmacy to the ED in a more timely manner.

The team process facilitated a better understanding of the roles needed to assist patients in the first phase of cardiac rehabilitation after MI. Therapists and nurses on our authoring team networked with others who had developed cardiac rehabilitation programs in other facilities, as we realized we did not have a consistent process in place prior to pathway implementation. The cardiac rehabilitation literature was reviewed and American Heart Association materials were inventoried. This information was used to facilitate patient education, focusing on behavioral changes that would be needed to achieve improved health outcomes. An outgrowth of the initial pathway process was the development of standard patient education packets for patients who ruled in for MI, which would be used by the entire interdisciplinary team. This improved the piecemeal approach to educating patients, where team members brought along their own materials to reinforce their interventions and teaching to the patient.

We limited our patient population to be placed on the uncomplicated myocardial infarction pathway to those patients who were hemodynamically stable and able to meet the goals of an approximately 4-day length of stay, which was the benchmark in our community. However, due to the complex nature of the patient population we see at HMC, we included on the pathway patients who had other comorbid conditions, such as diabetes, hypertension, or chronic obstructive pulmonary disease. We felt there was much to learn about our patient population, and that the clinical pathway would facilitate our care and improve our resource utilization.

Team members needed to learn their new responsibilities as part of a pathway authoring team. It was necessary for the team members to represent the perspectives of their discipline, unit or specialty, and report back to this authoring team. This transition toward a team-based culture was occurring simultaneously with the implementation of a housewide continuous quality improvement education effort.

The cardiology pathway authoring team's beginnings coincided with the recent arrival of the cardiology chief and some changes in nursing management on the CCU, which necessitated transitional reorientation. With perseverance on the process of developing a pathway for our patients, we were able to adapt to the new orientation for the development and implementaton of the Rule Out MI/Uncomplicated MI pathway in June 1995. This was our fourth pathway project at HMC and among the initial group of pathways implemented.

We felt it was essential to develop the concept of a separate ED pathway, to be started when the patient presented with symptoms of a myocardial infarction. This would facilitate the appropriate interventions and evaluations made at the point of entry into our system. This format departed slightly from the format that our clinical pathway design team had developed, but fit well within our overall implementation scheme. It also allowed the ED to track some key indicators regarding timeliness of interventions, which we felt would facilitate our quality improvement goals for this project.

INITIAL PATHWAY AND SUBSEQUENT PATHWAY REVISIONS

We started with a combined pathway for Rule Out MI/ Uncomplicated MI. We made multiple changes after listening to concerns from the clinical staff using the pathways. Most of our pathways at HMC are produced with a standard order set for our resident physicians to use upon admission from the ED, and also for the patients' transfer from the intensive care unit to the telemetry unit as applicable. We adapted the physician order sets to improve the transcription processes for these orders and made logical revisions to the pathway as needed to meet the goals of this evolving documentation system.

The CCU staff had made a transition in the previous year to a computerized medical record procedure. Using a paper-based clinical pathway was felt to be a step backward to many nursing staff members. However, it was not possible for us to have the clinical pathway documentation on the computer. That capability was not available from our computerized clinical information system (CIS) vendor, nor did we have the capability of a computer workstation within all areas of our care continuum for MI patients.

Our implementation strategies included educating the resident physicians upon every rotation about the clinical pathway and standard order sets. Physician resistance was higher with this pathway than with any of the other pathways we had initiated, primarily due to the historical "learning by discovery" mode that dominates the style of academic medical education.

Within a few months of the initial implementation, fine tuning was done to make the pathway easier to use. Within a year of the pathway's use, we were able to analyze our case mix and discover that our patient population who ruled in for a myocardial infarction was about 150 patients a year, and that the majority of the patients needed to be ruled out for MI rather than go through the entire pathway. The original pathway ended if the patient ruled out on day 2, and continued only if the patient had ruled in for MI. The pathway was again revised to reformat it for easier use in June 1996.

In August 1996, a group from the ED and CCU were meeting with cardiology to determine a better way to rule out MI patients even more efficiently. New diagnostic testing modalities were demonstrating effectiveness in evaluation of the impact of chest pain on myocardial perfusion, while decreasing the costs of hospital care in those who have noncardiac causes of chest pain.[11–15]

By October 1996, we implemented the acute cardiac evaluation or acute chest pain (ACE-U) pathway and protocol (Figure 3–1), in which we would more rapidly assess patients with low suspicion of cardiac disease. This would be accomplished by using three CK-MB and myoglobin tests performed, upon arrival in ED, at 3 hours, and at 6 hours after the onset of chest pain. In addition, the qualifying patient would be injected with Technetium-99m-labeled sestamibi (^{99m}Tc-sestamibi) within 1 hour of the onset of symptoms, and SPECT imaging would be conducted within 2 hours of the injection.[15] This imaging technique has facilitated detection of myocardial ischemia and can, with newer gaited techniques, evaluate for defects of left ventricular wall motion. If the resting ^{99m}Tc-sestamibi study is negative, an exercise or pharmacologic stress study can be given to those with nondiagnostic electrocardiogram (ECG) and negative laboratory findings to screen for hemodynamically significant coronary artery disease.[14]

This new approach to evaluating chest pain was exciting to the caregivers involved, however, having two separate protocols and pathways for ruling out MI was causing confusion. Team members liked having a rapid laboratory testing sequence and started to combine aspects of the ACE-U pathway with the R/O MI pathway. After 6 months of using the new ACE-U pathway, the pathways were revised once more in the spring of 1997. Additional data was now available indicating the effective use of Troponin-I as a marker for myocardial injury, especially when there was trauma or multiple organ damage.[16] For our pathway, we adopted the use of Troponin-I as part of the late chest pain panel in our laboratory (drawn 6 hours after the onset of chest pain).

At this time we have two pathways: a combined ACE-U/Rule Out MI pathway and a separate Uncomplicated Myocardial Infarction pathway that is used only when MI has been confirmed. This has vastly clarified the decision making regarding which pathway is appropriate to use and has standardized our care processes to be even more cost-effective.

CARE MANAGEMENT

Care management at HMC is coordinated by an interdisciplinary team. Approaches may vary as needed to meet the individualized outcome goals for the niche population served. Pathways serve as a tool to facilitate care coordination. Case management models are in place in various areas, as well as service lines designed to meet specific care coordination goals along the continuum for target populations. Our case management models in place may utilize nurses, nurse practitioners, and social workers and/or patient care coordinators to best meet the needs of the population served.

The clinical pathway is initiated in the ED by the ED physician, admitting physician, and the ED nurse after evaluation of the patient. Ideally, all patients admitted with suspected cardiac chest pain are placed on the Rule Out MI pathway. Having a very busy emergency department with rotating medical staff makes it difficult, however, to ensure that they are started on the pathway in the ED.

Rule Out MI patients are usually admitted to either the CCU or the cardiology/telemetry unit. If the patient has not been previously placed on a pathway, the nurse on the admitting unit usually discusses this with the admitting resident physician. Often, the resident has written orders that are almost identical to the preprinted pathway orders and does not want to rewrite the orders again. They do not have to rewrite the pathway orders, however, they are asked to indicate that the patient is on the pathway in their notes or orders. On some occasions, residents choose not to place the patient on the pathway. Recently, the attending physicians have spent more time on reinforcing the hospital's goal of using the pathway.

Nursing care is delivered using a total patient care model. We have a predominately registered nurse (RN) staff at HMC, with few assistive personnel. Social workers at HMC are primarily involved in discharge planning. In the case of ACE-U patients, the length of stay has been significantly reduced, therefore, the social worker only gets involved in these cases at the referral of a nurse or physician. As many of the Rule Out MI patients have other medical conditions identified, they are not always discharged. Due to these patients' other medical conditions some may need to continue care in the hospital. When the rule out process has concluded, they are considered to have completed the pathway.

The clinical pathway is kept in the patients' active medical record. In the CCU, where computerized medical records are used, the admitting nurse completes the admission assessment using the computerized form for pa-

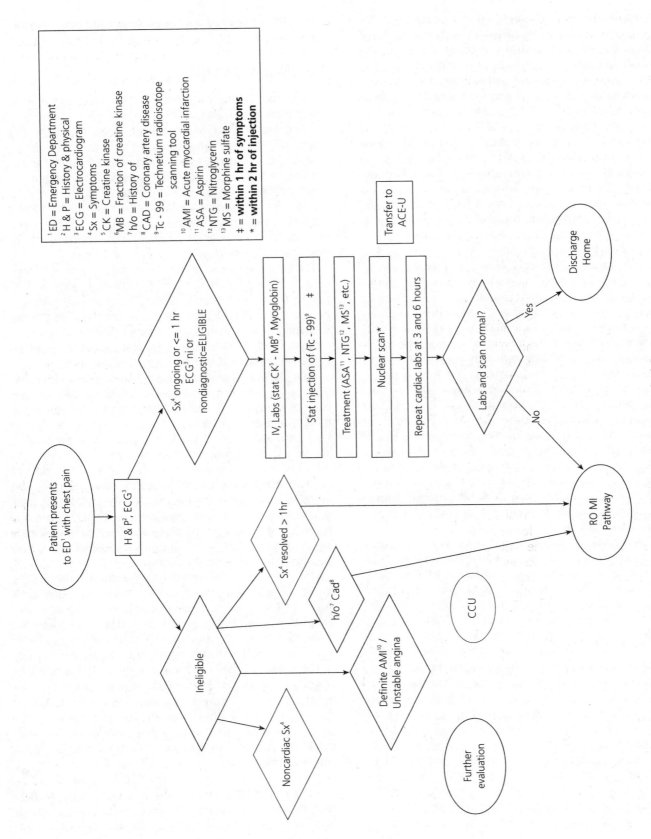

Figure 3–1 ACE-U Chest Pain Evaluation Protocol. *Source:* Copyright © Harborview Medical Center.

tient history. The cardiology/telemetry nursing staff use the admission assessment that correlates with the pathway. Nurses carry out the interventions and evaluate the outcome statements, which are preprinted on the pathway. The outcome statements are printed in a bold italic format (Exhibit 3–A–1, ACE-U/Rule Out MI Pathway in Appendix 3–A).

If the patient rules in for a myocardial infarction, they are placed on the Uncomplicated MI pathway (Exhibit 3–A–2, Uncomplicated MI Pathway), and an additional set of physician orders guides the care in this next phase. In the original MI pathway, physical therapists and occupational therapists assessed and provided specific interventions, provided education, and evaluated patient outcomes on days 3 and 4 of the pathway. The experiences from the earlier pathway implementation indicated that often the patient was independent, prior to the therapists' visit with the patient. In the newly revised pathway, the therapists visit if the patient is not independent in mobility or activities of daily living. Currently, nursing staff evaluate outcomes and make therapy referrals as needed. The optional PT and OT referrals are included in the pathway; however, it is not necessary to have these completed if the patient does not need additional interventions. Nursing staff referral has streamlined the process and decreased frustration of the therapists for visits that were not warranted.

Clinical nutritionists take an active role in providing education to the MI patients regarding dietary modifications needed to live a heart-healthy lifestyle. They provide the patient with specific education aimed toward food and drug interactions, if the patient is placed on Coumadin or other drugs that have been identified as among our high priorities for nutritional counseling. Social work staff evaluate the high-risk patients, screening for support systems, legal issues, alcoholism or substance abuse issues, financial needs, and insurance coverage. Assessment protocols in the social work department drive their in-depth focus on selected issues as needed.

CLINICAL OUTCOMES

Outcome statements are measured regularly on the pathway by evaluating the bold italic remarks during the various phases of the pathway and at discharge. Some examples of specific outcomes in our R/O MI pathway include

- hemodynamically stable
- no arrhythmias
- lungs clear
- $SaO_2 > 92\%$
- pain level is acceptable to the patient or the patient is without pain

In our Uncomplicated MI pathway, numerous other statements are added as the care becomes more focused. Examples of outcomes that appear on this pathway are

- patient experiences no shortness of breath, dizziness, or fatigue after getting up to the chair, or ambulating
- patient verbalizes understanding of any food/drug interactions
- patient/family demonstrate effective coping related to treatment plan
- patient develops plan to control personal risk factors
- patient/family discuss specific plan for physical care and home activity
- patient able to perform basic self-care tasks safely and independently
- patient/family verbalize understanding of concepts related to activities of daily living: precautions, self-monitoring, pacing, work simplification and energy conservation, stress reduction, and sexual activity
- patient regains indoor level of function

Many of these outcome statements were evaluated by OT and PT in earlier versions of the pathway. Nursing is now responsible for evaluating whether these outcomes are met and for making a referral to OT or PT if the patient does not demonstrate physical independence in these activities.

FINANCIAL OUTCOMES

For our Rule Out MI patient population, length of stay for the chest pain evaluation has been reduced from an average of over 2 days to about 1 day.[15] Many of these patients have other underlying medical conditions that may warrant further evaluation and treatment. When evaluating the financial outcomes, it is difficult to use total hospital charges in reviewing this patient population. Many of these patients come in with an admitting diagnosis that is not the same as their discharge diagnosis. Therefore, it is not sound to compare total charges with patients who have ultimately different discharge diagnoses and diagnostic related groups (DRGs).

We do not have a costing methodology in place to evaluate true costs of the current chest pain evaluation pathway, but we are transitioning toward a consolidated decision support system. Through the use of HBO&Company's (Atlanta, Georgia) TRENDSTAR® products, we will be able to model costs in our pathways and monitor specific performance in a more feasible, timely manner.

Our myocardial infarction patients' mean total charges have gone down from $17,598 prepathway implementation, to a mean of $10,754 after 18 months of pathway implementation. Length of inpatient stay has decreased from a mean of 7 days prepathway to 4.4 days after 18 months of pathway utilization.

PATIENT SATISFACTION

We have not compared patient satisfaction specifically within pathway and nonpathway groups. Our medical center participates in the Picker survey for patient satisfaction, and our data have not been analyzed in a manner that facilitates that comparison. We have had a goal of using patient language pathways, which would be written in lay terms using graphic elements and narratives describing the process of care. Development of this patient pathway template is currently underway in a joint effort between the two academic medical centers that comprise the University of Washington's health care delivery system. Clinical pathway design team members, community relations staff, and patient education committee members from both medical centers are guiding this process to ensure its applicability to our diverse, dynamic patient population. We will also be developing mechanisms to further evaluate the impact of the patient pathway and satisfaction after this implementation.

STAFF SATISFACTION

We conducted a staff survey in late 1996 in order to systematically identify staff issues related to the implementation of our pathway program. Most of the interdisciplinary team in cardiology (56% of RNs, 80% of social workers, 100% of nutritionists, 66% of the PT and OT staff, and 100% of the pharmacists) state they know how to use the pathway to document patient outcomes and interventions. The majority (85%) of the nurses know who should be put on the pathway and are working with the physicians to place patients on the appropriate pathway. Patient progress is tracked on the pathway (by 55% of the staff). The pathway is used to monitor patient progress and coordinate care (by 63% of staff), and guide decision making (by 53% of staff). The pathway is not systematically used by the staff in rounds (26%) or for discharge planning (40%).

Our RN staff track pathway variances 46% of the time. No physicians participate in the variance identification process we have established. The other disciplines (PT/OT, nutritionists, and pharmacists) reported 94% to 100% understanding regarding the process of variance tracking and documentation. A new process for variance tracking is being considered as a result of this finding. With each nurse caregiver responsible for variance documentation, and no single individual responsible for this process, we have had difficulty in reaching all of the nu-

merous staff nurses (many of whom are working part-time) to consistently address this component of our process. It is evident through the staff survey that the majority of the nursing staff do not realize the importance of the pathway for patient/family education, quality improvement, or reduction in resource utilization.

Paperwork is perceived as decreased by most of the staff in the cardiology/telemetry unit, however, in the CCU, where there is a computerized medical record, staff view the pathway as an increase in paperwork. One of the goals of our documentation process with the pathways is to decrease redundant narrative entries in the medical record. Using pathways, a narrative note is indicated only if the patient varies from the pathway. This concept has been very difficult for the CCU nurses, who have expressed concern over the thoroughness of their documentation and the legal implications of this approach. We continue to explore ways to handle these concerns until our pathways are fully integrated with our electronic medical record.

The physicians use the orders for these pathways; however, they rarely use the pathway to review patient care or outcomes. How often they read any other documentation by the team has not been systematically studied. In general, physician satisfaction with the pathway is evenly split among the two physicians responding to the survey.

ADVANTAGES/DISADVANTAGES AND ACCEPTANCE OF THE PATHWAY

One of the biggest advantages of the pathway is that a more consistent approach to patient education has been developed and delivered to the patient. This education, implemented through an interdisciplinary process, has allowed a greater focus on activity level after MI and lifestyle changes necessary for reducing risks related to coronary heart disease.

The interdisciplinary team input into the pathway design and revision has provided the opportunity for each discipline to better understand the ideal plan of care for these patients. However, while authoring team members participated and took ownership of the pathway process, these team members were not always able to fully impress this ownership to their counterparts. The staff has not consistently understood and implemented this ideal plan.

Different team members experienced frustration regarding communication and expectations. The OT and PT staff felt they wasted time by seeing inappropriate patients. Delays by the nursing unit in making the appropriate referrals thwarted the optimal delivery of care. This impacted the acceptance of the pathway by many team members who were at the same time making adjustments from the free-text narrative style of documentation used in routine patient care. The abbreviated nature of the pathway left some staff feeling they were not adequately detailing their professional observations, problem-solving skills, and planning processes. These issues are being considered in the revision process for the pathway's format, currently underway by the clinical pathway design team. Staff education around pathway development, implementation, and variance tracking remains essential for success over time.

The lack of an automated clinical information system has been a barrier to smoothly implementing pathways within critical care units. In our work with the clinical information system we use (EMTEK), current expansion of this electronic medical record to all areas of the hospital is taking place. Once that expansion is accomplished, we will begin adding functions including order entry, increased interdisciplinary documentation, and clinical pathways.

VARIANCE AND OUTCOME REPORTING

A lengthy description of our variance tracking and aggregation process, as well as the methodology for our reporting, can be found in the *Health Care Outcomes* book in the chapter on calcaneal fracture in the section on variance as a source of outcome data. Briefly described, variance is defined as any clinical pathway intervention that does not happen or happens later or sooner than defined by the pathway.

An example of the variances for the Uncomplicated MI patient population, trended over time can be seen in Figure 3–2, which shows interventions delayed or dropped from the pathway. Discharge from the hospital delays are the most frequent variance in 1996. The reasons for the delay are summarized on the right side of the chart. An obvious limitation in our variance reporting system is found when looking at the last reason coded to explain the variance: staff did not identify the cause of the discharge delay. This points to the internal system problem we have of not using this variance tracking system as it was intended.

IMPORTANCE OF OUTCOMES ASSESSMENT

There have been numerous benchmarks available to measure against while looking at the cardiovascular patient population in general. The increase in publicly available data regarding cardiovascular surgery outcomes is particularly noteworthy.[17–19] However, much of this data is not relevant for our particular patient population at HMC, as our acuity levels and comorbidities among our inpatient population remain unique in our region.

In measuring the impact of an acute care management process, outcome evaluation is limited by the time frame available for measurement. We are providing episode care management for our current cardiology population, as we have not truly converted to a type of continuum-based care management that would be seen within a health plan. While patients who have confirmed myocardial infarctions become candidates for so-called chronic disease management programs, we have not yet established programs that measure the longitudinal impact of specific disease with population-specific outpatient disease-management strategies.

Outcomes measurement is a field currently in transition. Historical hospital indicators have focused on length of stay, total charges, reimbursement levels, readmission or recidivism rates, infection, and mortality rates. Quality of life, level of functional independence, and prevention of complications become a more important area of focus as we strive to decrease the consumption of resources while providing the highest levels of quality for our designated populations. We need to provide opportunities to get the patients' feedback and input into their plan of care more regularly, as this has a major influence regarding their own health-seeking behavior. We must move from our traditional focus on the acute care episode toward the long-term community-based partnership with our patients. If our goal is to improve their levels of health knowledge and practices, we can ultimately reduce their incidences of chronic disease to achieve the outcome of efficient and effective care.[20]

CONCLUSION

Since the MI pathways began in 1995, ongoing review of the care provided to cardiology patients has taken place. We have identified a method to more rapidly rule out our low-risk chest pain patients, with 86% of this group having a length of stay of less than 1 day. We have been able to systematically study the impact of system issues, such as delays in referrals and appropriate treatment. This has allowed us to refocus our efforts to prioritize interventions to those patients who require individualized treatment, rather than provide standardized treatment that is not needed.

Standardizing the approach we take to care for low-risk patients has created a more cost-effective approach to the evaluation of chest pain. We still have room for many more improvements in the care of cardiology patients. Now we have a systemwide culture that utilizes data to drive decision making regarding the priorities to be further developed. Plans are currently underway to develop a cardiac rehabilitation program at HMC to meet the need that exists for our patient population.

Pathways will be developed at Harborview, within the next year, for a variety of medical problems such as com-

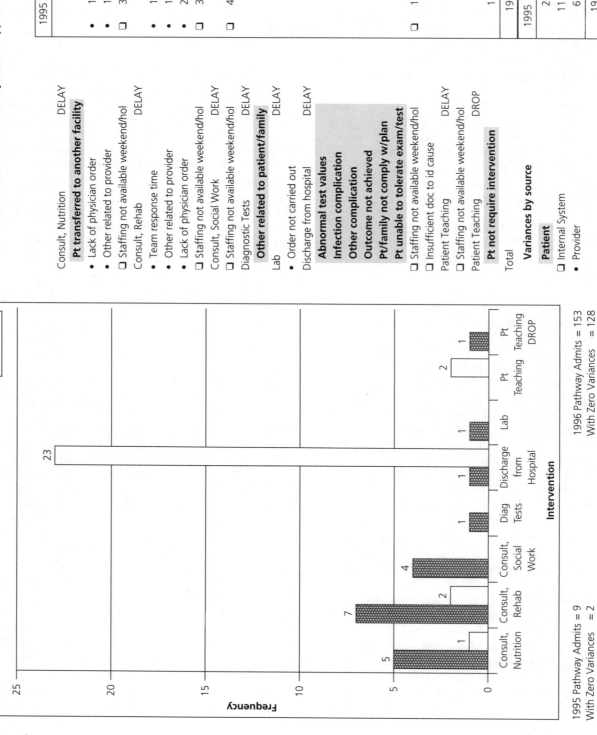

Figure 3-2 Rule In Myocardial Infarction Variance Analysis: Patients who remained on pathway June 1995 to December 1996. *Source:* Copyright © Harborview Medical Center.

munity-acquired pneumonia, deep vein thrombosis, alcohol withdrawal, and several orthopaedic conditions. With a more comprehensive information infrastructure for our pathways, our implementation will be easier, as we facilitate outcome assessment and provide the direct links to decision support systems.

The next steps in our pathway program evolution will be the development of more highly integrated care management models that span the continuum of care for our patients. Hopefully through this effort, we will be able to realize the goals of our collaborative practice: improved health, well-being, and decreased costs for health care.

REFERENCES

1. Cobb L, Weaver WD, Hallstrom AP. Experiences with out-of-hospital resuscitation. In: Lown B, Malliani A, Prosdocimi M, eds. *Fidia Research Series, Vol. 5.* Padua, Italy: Lavinia Press; 1986:433–440.

2. Alvarez H, Miller RH, Cobb L. Medic 1: the Seattle advanced paramedic training program. Procedings of the National Conference on Standards for Cardiopulmonary Resuscitation and Emergency Cardiac Care. May 16–18, 1973.

3. Cobb L, Weaver WD, Fahrenbruch CE, Hallstrom AP, Copass MK. Community-based interventions for sudden cardiac death. *Circulation.* 1992;85[suppl I]:I98–I102.

4. *Clinical Practice Guideline: Diagnosing and Managing Unstable Angina.* Rockville, MD: Agency for Health Care Policy and Research; May 1994. US Dept of Health and Human Services Publication 94-0603.

5. Ryan TJ, Anderson JL, Antman EM, et al. ACC/AHA guidelines for the early management of patients with acute myocardial infarction: a report of the American College of Cardiology/American Heart Association Task Force on Practice Guidelines (Committee on Management of Acute Myocardial Infarction). *J Am Coll Cardiol.* 1996;28:1328–1428.

6. Weaver WD, Eisenberg M, Martin JS, et al. Myocardial infarction triage and intervention project—phase I: patient characteristics and feasibility of prehospital initiation of thrombolytic therapy. *J Am Coll Cardiol.* 1990;15:925–931.

7. Weaver WD, Cerqueira M, Hallstrom AP, et al. Prehospital-initiated vs hospital-initiated thrombolytic therapy. *JAMA.* 1993;270(10):1211–1216.

8. Weaver WD. Time to thrombolytic treatment: factors affecting delay and their influence on outcome. *J Am Coll Cardiol.* 1995;7(suppl):3S–9S.

9. Blank FSJ, Austin M, Bennett A, et al. Decreasing "door to thrombolysis" time at one busy acute care hospital. *J Emerg Nurs.* 1995;21:202–207.

10. MacDonald Hand M, Bracken J, Hartman MB, Shero S. The role of the national heart attack alert program in improving the care of the acute MI patient: focus on the emergency department. *Heartbeat.* 1994;5(2):1–8.

11. Weissman IA, Dickinson CZ, Dworkin HJ, O'Neill WW, Juni JE. Cost-effectiveness of myocardial perfusion imaging with SPECT in the emergency department evaluation of patients with unexplained chest pain. *Radiology.* 1996;199:353–357.

12. Radensky PW, Hilton TC, Fulmer H, McLaughlin BA, Stowers SA. Potential cost effectiveness of initial myocardial perfusion imaging for assessment of emergency department patients with chest pain. *Am J Cardiol.* 1997;79(5):595–599.

13. Tatum JL, Jesse RL, Kontos MC, et al. Comprehensive strategy for the evaluation and triage of the chest pain patient. *Ann Emerg Med.* 1997;29(1):116–125.

14. Selker HP, Zalenski RJ, Antman EM, et al. An evaluation of technologies for identifying acute cardiac ischemia in the emergency department: a report from a national heart attack alert program working group. *Ann Emerg Med.* 1997;29(1):13–87.

15. DeRook FA, Lewis DH, Muzzarelli JR, Reddy RK, Comess KA. Imaging and lab studies for myocardial ischemia and injury. Abstract presented at Society of Nuclear Medicine Western Regional Meeting Sept 20, 1997, Monterey, California.

16. Futterman LG, Lemberg L. SGOT, LDH, HPD, CPK, CK-MB, MB$_1$, MB$_2$, CTNT, CTNC, CTNI. *Amer J Crit Care.* 1997;6(4):333–338.

17. National cardiovascular network (NCN) and the cardiovascular healthcare foundation (CHF). In: *The 1995 Medical Outcomes and Guidelines Sourcebook.* New York: Faulker & Gray Inc; 1994:142–443.

18. New York State Department of Health. Coronary artery bypass surgery in New York State 1990–1992. In: *The 1995 Medical Outcomes and Guidelines Sourcebook.* New York: Faulker & Gray Inc; 1994:466–480.

19. Pennsylvania Health Care Cost Containment Council. A consumer guide to coronary artery bypass graft surgery. In: *The 1995 Medical Outcomes and Guidelines Sourcebook.* New York: Faulker & Gray Inc; 1994:481–525.

20. Bower KA, Falk CD. Case management as a response to quality, cost, and access imperatives. In: Cohen EL, ed. *Nurse Case Management for the 21st Century.* St. Louis, MO: Mosby–Year Book; 1996:161–167.

■ Appendix 3–A ■
Clinical Pathways

Exhibit 3–A–1 ACE-U/Rule Out MI Pathway

		TIME
EMERGENCY DEPARTMENT and CCU/3 EAST	Arrival in ED	_____
	ECG Obtained	_____
DAY ONE	Isotope Injection	_____
	Nuclear medicine scan	_____
	Admission	_____

DATE	ED	Shift	DATE	ED	Shift
specify shift as per admission time in the three boxes					

Assessment and Monitoring

VS: q 15–30 min until stable then per routine
 [Systolic BP ≥90]
 (Hemodynamically stable)

Cardiovascular
- Continuous ECG monitoring
 [No arrythmias]
 [If (+) for MI diagnosis made within 15 min]
 [Acute drug intervention within 30 min]

Neurological
 [Alert and oriented x 3]
 [PERL, moves all extremities equally]
 [Behavior appropriate to situation]

Integumentary
 [Skin warm, dry, intact with good color]
Respiratory
- O₂ per nasal cannula
 [Maintains SaO₂ ≥92%]
 [Lungs clear, able to clear own secretions]
 [Respiratory rate regular, unlabored]

Medications
- Aspirin
- NTG SL
- MSO4 prn

- Consider need for other anti-ischemic therapy:
  - Beta blockers
  - Heparin
  - IV NTG

- Thrombolytics per ST-segment score for uncomplicated MI patient

Pain Management
- Pain location, radiation, assoc. symptoms and relief
 Describe _____
 Time pain started: _____
 Pain scale introduced (0–10) current level _____

 [Pain level is acceptable to patient or pt is without pain]

 Pain reassessed q 15 min until relief

Consults/Diagnostic Studies
- 12 lead ECG within 15 minutes and prn chest pain
- O₂ saturation prn
- CBC, M7, PT, PTT = MI panel
- CXR
- Cardiology notified within 30 minutes of arrival
- **If patient meets ACE-U criteria:**
CK-MB and myoglobin (early chest pain panel) in
ED = Time 0 ❑, at 3° ❑, and 6° ❑

- Technetium injection stat (notify radiology MD or nuclear med if pt meets ACE-U criteria)
- SPECT scan within 2 hours of injection
- **Non ACE-U patients/other Rule Out MI labs:**
Specify three labs ordered (late cp panel = CK-MB & Troponin I).
 ❑ _____ chest pain panel at _____ hrs
 ❑ _____ chest pain panel at _____ hrs
 ❑ _____ chest pain panel at _____ hrs
 or: _____
 _____
 _____

Fluid/Volume
- Peripheral IV site _____ gauge _____
- IV site capped for medication use
- Monitor I/O
Nutrition
- NPO except medications, until nuclear scan (if done), then AHA diet
Elimination GI/GU
 [Abdomen soft]
 [Bowel tones WNL]
 [Stool guaiac negative]
 [No nausea or vomiting]
Activity/Safety
 [No falls or seizure activity]
- Siderails up
Psychosocial/Emotional
 [Appearance and affect appropriate to situation]
- Screen for ETOH/drug abuse, neglect, abuse, prolonged stress or confusion
Patient Education
- Need for diagnostic tests discussed: pt/family
- Pain scale instructions given & request pt to notify RN of increased pain & pressure
- Instructions given re: diet, risk factors, exercise, weight, BP
Discharge Planning
 [Prehospitalization, patient lives independently and/or has adequate support]
- Consider elective cath or stress test as needed
- If (+) for MI, pt placed on Uncomplicated MI pathway

Clinical Pathways provide guidance in the management process for a specified case type. Using Clinical Pathways in actual practice requires consideration of individual patient needs.

continues

Exhibit 3–A–1 continued

LAB and DIAGNOSTIC STUDIES	ED	Time	Ini	RESULTS
If patient meets ACE-U criteria: • CK MB and myoglobin in ED				
• CK MB and myoglobin at 3 hrs				
• CK MB and myoglobin at 6 hrs				
• ECG in ED (all patients)				
• ECG prn chest pain (all patients)				
Non ACE-U patients/other Rule Out MI labs: specified three labs: ❏ _____ chest pain panel at _____ hrs ❏ _____ chest pain panel at _____ hrs ❏ _____ chest pain panel at _____ hrs or: _____ _____ (early chest pain panel = CK-MB and myoglobin) (late chest pain panel = CK-MB and Troponin I)				
Pt qualified for SPECT scan: ❏ Yes ❏ No • Technetium injection within 1 hr of onset of chest pain • SPECT scan within 2 hrs of injection				
ADDITIONAL LABS/DIAGNOSTIC STUDIES *AS NEEDED ON DAY ONE* RN Signature _____ RN Signature _____ RN Signature _____				RN Signature _____ RN Signature _____ RN Signature _____
Source: Copyright © Harborview Medical Center.				

Exhibit 3–A–2 Uncomplicated MI Pathway

	MI (UNCOMPLICATED) STABILIZATION PHASE													
DATE		N	D	E		N	D	E			N	D	E	
FOCUS/PROBLEM ✓ = Normal * = Abnormal NA = Not applicable Ø = Not done	DAY 2 TELE.				DAY 3 TELE.				DAY 4 TELE.					
Assessment and Monitoring *Initial VS:*	• VS per critical care or tele standards *[systolic BP ≥90]*				• VS per tele stds *[systolic BP ≥90]* *[NSR/hemodynamically stable]*				• VS per tele stds *[systolic BP ≥90]* *[NSR/hemodynamically stable]*					
Cardiovascular	• Continuous ECG/tele monitoring *[NSR]* *[Chest pain free]* *[No periph edema or CHF]* *[No new murmur]*				*[Cardiovascular assessment stable]*				*[Cardiovascular assessment stable]*					
Neurological	*[Moving all extremities]* *[Alert, oriented]*				*[Moving all extremities]* *[Alert, oriented]*				*[Moving all extremities]* *[Alert, oriented]*					
Integumentary	*[Skin warm, dry intact, with good color]*				*[Skin warm, dry intact, with good color]*				*[Skin warm, dry intact, with good color]*					
Respiratory/ Airway *Respiratory*	*[lungs clear]* *[able to clear own secretions]* *[SaO₂ ≥92]*				*[lungs clear]* *[able to clear own secretions]* *[SaO₂ ≥92]*				*[lungs clear]* *[able to clear own secretions]* *[SaO₂ ≥92]*					
Medications	• ASA • NTG/MSO$_4$-PRN Consider: • Beta blocker • ACE inhibitor • Coumadin or heparin				• ASA • NTG/MSO$_4$-PRN Consider: • Beta blocker • ACE inhibitor • Coumadin				• ASA • NTG/MSO$_4$-PRN Consider: • Beta blocker • ACE inhibitor • Coumadin					
Pain Management	*[Pt free of chest pain]* • Pain scale q 4°				*[Pt free of chest pain]* • Pain scale q 4°				*[Pt free of chest pain]* • Pain scale q 4°					
	Pain Scale				**Pain Scale**				**Pain Scale**					
	N				N				N					
	D				D				D					
	E				E				E					
Consults/ Diagnostic Studies	• ECG in a.m. • PTT if on heparin, PT if on Coumadin • Fasting lipid profile in a.m. • HCT/WBC/M7 • Echocardiogram if ordered				• ECG in a.m. • O₂ sat prn • PTT if on heparin, or PT if on Coumadin • MD to evaluate for: • Cardiac catherization • Thallium scan • Full-level ETT • Low-level ETT • Pharmacological stress ETT • Referral to outpatient cardiac rehab.				• Schedule outpt. diagnostics f/u • PTT is on heparin, or PT if on Coumadin					

LABS AND X-RAY KARDEX		FREQ	DATE DUE	TIME DUE	RESULTS

RN SIGNATURES	

continues

Exhibit 3–A–2 continued

MI (UNCOMPLICATED) STABILIZATION PHASE													
DATE		N	D	E		N	D	E		N	D	E	
FOCUS/PROBLEM ✓ = Normal * = Abnormal NA = Not applicable Ø = Not done	DAY 2 TELE.				DAY 3 TELE.				DAY 4 TELE.				
Fluid/Volume	• Periph IV site capped Gauge/Site-____/____ Gauge/Site-____/____ **[sites: w/o redness, drainage, tenderness]** **[op site drsg dry & intact]** • Weight_____ • I & O				• Periph IV site capped Gauge/Site-____/____ Gauge/Site-____/____ **[site: w/o redness, drainage, tenderness]** **[op site drsg D & I]** **[No peripheral edema]** • Weight_____ • I & O				• Periph IV d/c'd Gauge/Site-____/____ **[site: w/o redness, drainage, tenderness]** **[op site drsg D & I]** **[No peripheral edema]** • Weight_____ • I & O				
Nutrition	• NPO except meds for fasting lipid profile • AHA diet (advance as tolerated)				**[Eating 75% of AHA meals]** Dietitian to: • Complete assessment & education screen • Assess cholesterol screen • Food/drug interactions for pt on Coumadin **[Pt verbalizes understanding of food/drug interactions]**				**[Eating 75% of AHA meals]** Dietitian to: • Complete education as appropriate • Assign diet(s): ❑ Step I ❑ Step II ❑ 3 gm Na ❑ 2 gm Na ❑ Gen. lowfat (<30% fat) ❑ Basic good nutrition **[Pt verbalizes relationship between prescribed diet and dx]**				
Elimination *Gastrointestinal* *Genitourinary* *Menstrual/Prostate*	**[Bowel function WNL]** **[Stool guaiac negative]** **[No nausea/vomiting]**				**[Bowel function WNL]** **[Urinary output adequate]** **[Reassessed for outpatient referral for GU/GYN issues]**				**[Bowel function WNL]** **[Urinary output adequate]**				
Activity/Safety *Musculoskeletal* *Fall Risk/Safety* *Activities*	• Activity monitored by RN: ❑ chair 15–30 min BP/HR pre _____ BP/HR 2 min _____ **[Pt experiences no SOB, dizziness, fatigue]** • Self-supervised activity • Up to chair TID **[Pt monitors own HR & reports need for return to bed as needed]** • Nursing to contact PT if pt unable to mobilize independently				• Self-supervised activity: up ad lib in room BRP • RN to assist in ambulation down corridor as per guideline (P 5) if pt ambulating independently **[Pt experiences no SOB, dizziness, fatigue]** • Nursing to contact OT if pt unable to perform ADLs independently				• Self-supervised activity: up ad lib in room BRP • RN to assist in ambulation as per guideline (P 5) **[Pt experiences no SOB, dizziness, fatigue]** **[Tolerates activity sufficiently to be scheduled for ETT]**				
Therapy (List any additional treatment/orders)													
RD Signature													

continues

Exhibit 3–A–2 continued

MI (UNCOMPLICATED) STABILIZATION PHASE										
DATE		Time	Ini		Time	Ini			Time	Ini
FOCUS/PROBLEM ✓ = Normal * = Abnormal NA = Not applicable Ø = Not done	DAY 2 TELE.			DAY 3 TELE.			DAY 4 TELE.			

Psychosocial

SW high risk screen completed
- Needs additional assessment for

_____
_____
_____
_____

- No additional SW assessment required at this time

- Contact SW if pt's needs change

Social Work Psychosocial Assessment:

Living Situation: _____
_____

Family, support system, service providers:
❑ Adequate   ❑ Marginal   ❑ Stressed
❑ None   ❑ Unknown
Name   Relationship   Address/Phone
_____
_____
_____

Legal Consent Authority: _____
Guardianship   ❑ Y ❑ N ❑ Needed
Power of Attorney: _____
Other Legal Issues: _____
Psychiatric/Involuntary Treatment Act (ITA)
                    ❑ Y ❑ N
Orientation/Cognitive Issues: _____
_____

Coping Skills: ❑ Pt WNL ❑ Unable to assess
❑ Needs assistance
❑ Family WNL ❑ Needs Assistance ❑ N/A
Issues: _____
Pre-morbid functioning: ❑ Independent
                         ❑ Req. assistance with
_____
_____

ETOH/Substance Abuse:
❑ Assessment completed
❑ Cannot assess   ❑ Does not apply
❑ Medical barrier to Tx ❑ Does not agree to Tx
Additional Assessment: _____
_____

Financial: ❑ Needs funding
❑ Medicaid Pending
❑ Date applicated completed
❑ Has Medicare _____ A & B _____A only
❑ Has Medicaid Active # _____ PIC _____
❑ Insurance: _____
❑ Case manager: _____
Phone: _____
PCP: _____
Phone: _____
Income: ❑ SSD/SSI ❑ Social Security
❑ Other _____

DAY 3 column:

[Pt/family demonstrate effective coping related to treatment plan]
Social Notes Continued:

_____
_____
_____
_____
_____
_____
_____
_____
_____
_____
_____
_____
_____
_____
_____
_____
_____
_____
_____
_____
_____
_____
_____
_____
_____
_____
_____
_____
_____

DAY 4 column:

[Pt/family demonstrate effective coping related to treatment plan]
Social Notes Continued:

_____
_____
_____
_____
_____
_____
_____
_____
_____
_____
_____
_____
_____
_____
_____
_____
_____
_____
_____
_____
_____
_____
_____
_____
_____
_____

SW Signature

_____

continues

Exhibit 3–A–2 continued

MI (UNCOMPLICATED) STABILIZATION PHASE										
DATE		Time	Ini		Time	Ini			Time	Ini
FOCUS/PROBLEM ✓ = Normal * = Abnormal NA = Not applicable Ø = Not done	DAY 2 TELE.			DAY 3 TELE.			DAY 4 TELE.			
Patient/Family Education	• Start pt education using MI education packet			*[Pt develops plan to control personal risk factors]* • Pt taught control of: • smoking • activity • diet • weight • stress • BP • blood sugar • Medication teaching • Discuss when to resume sexual activity			*[Pt/family discuss specific plan for physical care]* *[Pt/family describe purpose, route, dose, schedule, & significant side effects of all d/c meds]* *[Pt/family discuss home activity plan]* *[Pt/family discuss use of SL NTG]* *[Pt/family discuss when to activate EMS]* *[Patient/family verbalize understanding of concepts in relation to ADLs and home activities:* • **precautions** • **self-monitoring/pacing** • **work simplification and energy conservation** • **stress reduction** • **sexual activity]** • referred to outpt cardiac rehab • given next step in activity progression			
Discharge Planning	*[Pt has adequate d/c plan, no further Social Work support services needed]* • Contact Social Work if pt support needs change • If pt needs planning, complete plan below **Date** ___ Discussed d/c options with pt/family ___ D/C support services needed: ❑ Home Health Care Service ❑ Rehab ❑ Sub-Acute ❑ NHP ❑ Other _____ ___ Support services provided by: _____ _____ Anticipated date of d/c: _____ Time d/c: _____ ___ Transportation _____ _____ *If Nursing Home:* ___ FPIP ❑ _____ OBRA ❑ Pending Medicaid Form ❑ _____ Letter ❑ _____ ___ Faxed request to Home and Community Services ❑ ___ DSHS screen completed by: _____ _____ ___ Bed confirmed at: _____ _____ Family contact: _____ Phone: _____ ___ If a contracted facility contact Dr. McCormick/Dr. Grefenson									

continues

Exhibit 3–A–2 continued

DATE		Time	Ini		Time	Ini
	DAY 3/TELEMETRY			DAY 4/TELEMETRY		

Treatments/Therapy:
(If unable to ambulate independently)
Physical Therapy Assessment Day 2
Contraindications to progression, or criteria for termination of Tx:
1. Angina, SOB, dizziness, fatigue
2. Orthostatic BP drop (10 mm Hg systolic BP after 2 min. Tx)
3. HR increases > 20 BPM above baseline

Chart review re: pts history of present illness
Baseline mobility status:
❑ Household mobility
❑ Indoor/outdoor ambulation
❑ Community ambulation
❑ Independent
❑ Requires assistance: _____

Architectural barriers at home:
Stairs   ❑ yes ❑ no
Railing   ❑ yes ❑ no
Other _____

Equipment pt owns/rents:
❑ None
❑ Assistive device
❑ Wheelchair

Equipment ordered/issued from _____
_____
❑ None
❑ Assistive device
❑ Wheelchair

Assistance available at home: ❑ yes ❑ no
Describe _____
_____

• Consult w/RN re: recovery process
• Coordinate activity w/CMT

Initial Mobility Evaluation:
Check vital signs pre, during, post activity. If vital signs are within safe parameters, continue
       BP _____   HR _____
       [AROM WNL]
       AROM WNL except _____
• Supine to sit on EOB
   ❑ Independent
   ❑ Needs assistance
       BP _____   HR _____
• Sit to stand and walk _____ ft and back to sit in armchair
   ❑ Independent
   ❑ Needs assistance
       BP _____   HR _____

• Review with pt/family re:
   • Precautions & need to modified but continued physical activity

PT Signature

_____

Assessment time _____

Physical Therapy

• Consult w/**RN** re: recovery progress and pts tolerance to ADLs
• Coordinate activity with CMT

NOTE PRECAUTIONS

• Consult w/**OT** re: ADL tolerance
• Progress ambulation distance as tolerated
   BP _____/_____ HR _____ (pre walk)
   5–7.5 min ambulation in the hall
   BP _____/_____ HR _____ 2 min into walk
   BP _____/_____ HR _____ (post walk)

• Have staff or family member stand by when testing pt on stairs (rest after every 2 steps)
   BP _____/_____ HR _____ (presteps)
   BP _____/_____ HR _____ (post steps

• Final assessment of F/U needs:
   1) Individualized home activity program
                              ❑ yes ❑ no
   2) VNS   ❑ yes ❑ no
   3) Cardiac rehabilitation outpt program
                              ❑ yes ❑ no

[Regain baseline indoor level of function]
[Pt/family verbalize understanding of instructions]

Treatment Time _____

continues

Exhibit 3–A–2 continued

MI (UNCOMPLICATED) STABILIZATION PHASE						
DATE			Time	Ini		Time	Ini
	DAY 3/TELEMETRY				DAY 4/TELEMETRY		

Treatments/Therapy:
(If unable to perform ADLs independently)
<u>Day 3</u>
<u>Occupational Therapy Assessment</u>
Chart Reviewed _____
• Consult w/RN re: recovery progress
• Pt/family interview re: baseline
❑ ADL function limited by other disease process
❑ Independent w/self-care (bathing, drsg, hygiene)
❑ Needs assist w/self-care
   If yes, how much _____
❑ Independent with home care (cleaning, meal prep, laundry, shopping)
 	
• OT coordinates self-care activity with CMT
NOTE: Contraindications to progression include:
 	
1) SOB, angina, dizziness, fatigue
2) Orthostatic BP drop
   (10 mm Hg systolic BP after 2 min)
3) Heart rate increasing 20 BPM above baseline
• OT performs monitored self-care eval.
Shower/sponge bath
   ❑ Seated
   ❑ Standing
   _____ level of assistance
Dressing
   ❑ Upper body
      ❑ Seated
      ❑ Standing
   ❑ Lower body
      ❑ Seated
      ❑ Standing
   _____ level of assistance
• Vital signs recorded:
      pre-activity
      2 min into activity, and
      post-activity
BP _____/_____ HR _____ pre-resting
BP _____/_____ HR _____ 2 min
BP _____/_____ HR _____ post-activity;
      endurance/
      duration of activity
	• OT will reinforce written/verbal pt/family ed re: ADLs and home activities
	  • precautions
	  • self-monitoring, pacing
	  • work simplification
	  • stress reduction
	  • sexual activity
	• OT informs pt of performance of self-care eval
					• OT consults w/RN re: recovery process
					• OT final assessment of home equip f/u needs
					  ❑ Recommended/arranged home OT eval
					  ❑ Recommended/ordered shower chair/adaptive equipment
					  ❑ Recommended _____ level of assist @ D/C
					
					<u>OT Discharge Goals:</u>
					[Patient able to perform basic self-care tasks safely and independently]
					
					[Patient/family verbalize understanding of concepts in relation to ADLs and home activities:
					  • *precautions*
					  • *self-monitoring/pacing*
					  • *work simplification and energy conservation*
					  • *stress reduction*
					  • *sexual activity]*		

OT Signature _____							

■ 4 ■

Data Driven Critical Pathway Development and Outcomes: The Case of Heparin in Thromboembolic Disease

Carlos A. Estrada, Joan Wynn, and Taffy Klaassen

A hallmark of quality improvement programs is decreasing variation in processes. In hospitals and health care systems, variations in care may be the source of differences in both economic and clinical outcomes.[1,2] To demonstrate efficiency and quality, hospitals must develop methods to outline expectations for care that are based on best practice. Many institutions use optimal care paths or practice guidelines to achieve this outcome.[3,4] Critical pathways are management plans that display goals for patients and provide the corresponding ideal sequence and timing of staff actions to achieve those goals with optimal efficiency.[5] The pathway outlines the major interventions desired for a defined patient population in a defined length of time.[6] It is believed that critical pathways improve outcomes, patient and staff satisfaction, and decrease costs.[2,5]

In our institution, we have implemented optimal care paths for a variety of patient populations in order to decrease variation in practice, to enhance patient outcomes, and to be efficient in utilization of resources. The most successful efforts have strong multidisciplinary representation including physicians, nursing, pharmacy, and administration.

Several barriers to physicians' participation in the implementation and utilization of optimal care paths have been identified. These include fear of "cookbook medicine," liability concerns, resistance to change, time commitment, personal attitudes, and lack of objective data to show that optimal care paths actually improve patient care.[3,5] Time pressures in clinical practice and mismatch between evidence and clinical circumstances are barriers for transferring evidence from research into practice.[7]

We think that a way to ensure participation of health care personnel in this type of quality improvement endeavor is to use objective data to convince them of the need to change or improve a process. This step also provides specific baseline information for benchmarking. With this in mind, we undertook this study of the use of heparin in the treatment of venous thromboembolism at our institution.

VENOUS THROMBOEMBOLISM

It is estimated that approximately 2 million new cases of venous thromboembolism occur each year in the United States.[8] Death can occur when the venous thrombi break off and block pulmonary circulation causing hypoxemia, hemodynamic instability, and arrhythmias. It is estimated that in the United States, 600,000 patients per year develop a pulmonary embolism (PE) and that 60,000 die of this complication. The morbidity associated with thromboembolism is also significant. Deep venous thrombosis (DVT) and PE may recur in up to 30% of patients, and the postphlebitic syndrome may later develop in up to 47% of patients at 5 years.[8,9]

Heparin is the standard treatment for patients with deep venous thrombosis and pulmonary embolism. Our understanding of the use of heparin has increased significantly over the last few years.[8] It is widely accepted that heparin should be used early, that full anticoagulation should be achieved during the first 24 hours, and that oral anticoagulants should be prescribed for at least 3 months.[8,10–18] It has been shown that patients with venous thromboembolism must receive at least 30,000 U/day of unfractionated heparin to achieve full anticoagulation[8,12,13,17,19] and to decrease the risk of recurrence.[10,19] The recommended starting dose is a bolus of

Project funded by the Department of Medicine Research Grant, East Carolina University, Greenville, North Carolina.

39

5000 to 10,000U of IV heparin (or 80U/kg) followed by a continuous infusion rate of 1200 to 1600 U/hr (or 18U/kg/hr).[8,10,12] The infusion is then adjusted to achieve and maintain a therapeutic activated partial thromboplastin time (aPTT) (between 60 sec and 90 sec at our institution). Patients should receive an oral anticoagulant (warfarin) within 24 hours and have an overlap of warfarin and heparin of 4–5 days.[8,12] In patients receiving inadequate treatment, the risk of recurrent thromboembolism is increased fivefold.[8,10]

Previous studies, reporting chart audits, have revealed that heparin dosing is difficult and often there is a delay in achieving a therapeutic aPTT.[13,15,20–24] To address this problem, many institutions have developed guidelines and heparin dosing nomograms to improve outcomes (ie, prevention of recurrence of thromboembolism) while not increasing bleeding rates.[8,10,13,16,17,20] Nomograms are standing orders that define a heparin dose based on a specified aPTT range. The use of nomograms is more effective than "usual care" in achieving a therapeutic aPTT within 24 hours.[10,11,13–16,25] Practice guidelines and optimal care paths for the evaluation or treatment of suspected deep venous thrombosis have recently been published.[26,27] Raschke et al[28] found that a weight-based heparin nomogram was well accepted by clinicians and led to improvements in the time to achieve a therapeutic aPTT (by 12 hours), without increasing bleeding rates.

Our goal was to garner physician support at our institution to develop an optimal care path that would guide the use of heparin in the treatment of thromboembolic disease and improve patient outcomes.

PATHWAY DEVELOPMENT AND OUTCOMES

Continuous quality improvement and optimal care paths are methods that improve outcomes.[6] Optimal care path development may follow these steps prior to implementation (FOCUS):

- Find a process to improve (select a topic and population).
- Organize a team.
- Clarify the current process.
- Understand the current process (analyze and evaluate).
- Select a solution.[5,29–31]

F—Find a Process to Improve

We chose this project because the process of taking care of patients on anticoagulants is relatively straightforward and there are good data that describe best practices and related improvement in outcomes. We also had

anecdotal experience from pharmacists, nurses, and practicing physicians relating that the heparin dosing was problematic.

The medical records database at our hospital was queried to identify patients with a primary or secondary discharge diagnosis of venous thrombosis or pulmonary embolism (Table 4–1). Charts were reviewed to determine eligibility if the billing database showed a charge for intravenous heparin (heparin 25,000 U/500 DW). Secondary diagnoses were also searched because a thromboembolic event may not be the main reason for hospitalization. We also did not want to restrict our review to the Diagnosis Related Group (DRG) specific codes for thromboembolism for the same reason. Between October 1, 1995, and September 30, 1996, 276 patients were

Table 4–1 ICD-9 Codes of Lower Extremity Thromboembolism and Pulmonary Embolism

ICD-9 Code	Description
	Phlebitis and thrombophlebitis
451.11	deep vessels leg, femoral vein
451.19	deep vessels leg, other
451.2	leg, unspecified
451.81	other sites, iliac vein
451.9	other sites, unspecified
	Other venous emoblism and thrombosis
453.2	vena cava
453.8	other specified veins
453.9	unspecified
	Venous complications in pregnancy and puerperium deep venous thrombosis, antepartum
671.30	unspecified
671.31	delivered, with/without mention of antepartum condition
671.32	delivered, with mention of postpartum complication
671.33	antepartum condition or complication
	deep venous thrombosis, postpartum
671.40	unspecified
671.42	delivered, with mention of postpartum complication
671.44	postpartum condition or complication
	unspecified venous complication
671.90	unspecified
671.91	delivered, with/without mention of antepartum condition
671.92	delivered, with mention of postpartum complication
671.93	antepartum condition or complication
671.94	postpartum condition or complication
	Pulmonary emoblism and infarction
415.10	pulmonary embolism and infarction
415.11	iatrogenic
415.19	other

Source: Data from St. Anthony's ICD-9-CM Code Book for Physician Payment, 5th Edition, © 1996, St. Anthony Publishing Inc.

identified through the computer database. Current thromboembolism was not present in 34 patients (wrong coding, 17; prior history of thrombosis, 14; other, 3). We excluded 81 additional patients due to use of anticoagulants within 7 days of admission, n = 43; thrombosis of upper extremity, n = 22; planned caval interruption, n = 9; death prior to the initiation of heparin, n = 3; and catheter associated or other, n = 4. The charts of the first 102 consecutive patients provided data of current practice at our institution.

O—Organize the Team

An interest group was convened to examine the current use of heparin in our hospital. Participants were invited based on expertise, interest, or experience in the use of anticoagulants. Initially, representatives from administration, nursing, pharmacy, and internal medicine participated on the team. The data provided in the first few meetings objectively illustrated the current practice at our institution and was convincing enough to increase interest and participation. The team now consists of representatives from administration; nursing; pharmacy; laboratory; radiology; cardiology; hematology-oncology; surgery; pulmonary and critical care; rehabilitation; and emergency, family, and internal medicine. The team reviewed the initial data that were collected and strategized solutions to improve the process. Maintaining focus is and will be a continuous challenge in part because there are also many other possibilities to explore in this area (prevention, diagnosis, system issues).

C—Clarify the Process

Evaluation of Current Practice

To determine if the current use of heparin in our hospital adhered to recommendations for the treatment of thromboembolism, we reviewed our current practice. These data were used to develop the optimal care path. We performed a retrospective chart review. The inclusion criteria for chart review were

1. diagnosis of pulmonary embolism, suspected clinically and confirmed by high probability lung scan, pulmonary arteriography, or autopsy
2. lower extremity deep venous thrombosis, confirmed by venography or Doppler ultrasonography

Exclusion criteria were

1. anticoagulant or thrombolytic therapy received in 7 days prior to admission
2. active hemorrhage at the time of admission
3. acute major cerebrovascular event

4. history of heparin-induced thrombocytopenia
5. a known allergy to heparin
6. death prior to the initiation of heparin
7. planned inferior vena caval interruption
8. use of low-molecular-weight heparin
9. catheter-associated thromboembolism

We abstracted demographics, clinical information, associated conditions, and calculated the Charlson comorbidity index.[32] The Charlson index is a validated measure of the burden of comorbid disease that assigns a weight to each defined comorbid disease. The diseases and their corresponding weights appear in Table 4–2. The Charlson index is the sum of the weights corresponding to each chronic comorbid condition suffered by the patient. It has been shown to accurately predict the risk of death within 1 year of medical hospitalization.

Risk factors for thromboembolism and bleeding diathesis were abstracted. The risk of bleeding was defined by a history of surgery or a stroke within the last 14 days, peptic ulcer disease, current gastrointestinal or genitourinary bleeding, or platelet count <150,000/ml.[10] Heparin dosing and pertinent laboratory information were also collected. Blood samples obtained within 4 hours of a heparin dose change were not included in the analysis because a steady state of heparin is not achieved during that period.

The data were analyzed using the Statistical Products and Services Solutions™ software (SPSS 7.5 for Windows, Chicago, Illinois). Standard summary statistics were used to describe the data. This project was approved by the Institutional Review Board at East Carolina University and Pitt County Memorial Hospital.

Outcomes

Previous studies have indicated the key processes of care are (1) time to exceed the therapeutic threshold (aPTT $\geq$ 1.5 times control), measured as the time elapsed in hours between initiating heparin therapy and surpassing the therapeutic threshold (aPTT $\geq$ 60 sec at our institution), and (2) time to achieve a therapeutic range (aPTT of 1.5 to 2.5 times control, or 60 sec $\leq$ aPTT $\leq$ 90 sec at our institution). These surrogate outcomes are readily measurable and have been shown to correlate with recurrence of venous thrombosis at 3 months.[10] We elected not to evaluate recurrence of venous thrombosis and the development of the postphlebitic syndrome. Their measurement would have entailed the evaluation of patients for 3–60 months after hospital discharge,[9,10] which is not a viable option for hospital-based quality improvement projects.

Other key processes of care we measured were doses of the initial bolus of heparin, the heparin infusion rate,

Table 4–2 Weights Assigned to Comorbid Conditions by the Charlson Index

Weight	Condition	Observations
1	Cerebrovascular disease	history of cerebrovascular accident or transient ischemic attack
	Congestive heart failure	—
	Connective tissue disease	polymyositis, lupus, rheumatoid, other
	Dementia	chronic cognitive deficit
	Diabetes	with no end organ damage
	Liver disease, mild	chronic hepatitis, cirrhosis
	Myocardial infarction, current	—
	Peptic ulcer disease	current or history of bleeding from it
	Peripheral vascular disease	intermittent claudication, bypass, gangrene, acute arterial insufficiency, aneurysm >6 cm
	Pulmonary disease, chronic	dyspnea, asthma, oxygen use, hypercapnia, $Po_2 < 50$, pneumonia, pulmonary embolism, Chronic Obstructive Pulmonary Disease, pulmonary hypertension
2	Diabetes, with end organ damage	eyes, kidney, nerve
	Hemiplegia	regardless of cause
	Lymphoma	Hodgkin, multiple myeloma
	Leukemia	any
	Renal disease, moderate/severe	dialysis, transplant, uremia, Creatinine >3 mg/dl
	Tumor, any	<5 years, no metastasis
3	Liver, disease, moderate/severe	cirrhosis with varices, or bleed
6	Metastatic solid tumor	any, breast, lung, colon, other
	AIDS	or AIDS-related complex

Source: Data from M.E. Charlson, et al. A New Method of Classifying Prognostic Comorbidity in Longitudinal Studies: Development and Validation, *J Chron Dis,* Vol. 40, pp. 373–383, 1987.

---

time to first dose of warfarin, number of days overlap between warfarin and heparin, the International Normalized Ratio (INR) value at discharge, length of stay, whether or not a heparin nomogram was used by the physician, recommendation of compressive stockings (to prevent postphlebitic syndrome), and blood count determinations during heparin therapy. Physician specialty was also abstracted to assist in targeting future interventions.

Adverse outcomes were defined as (1) major bleeding while on heparin (a decrease of hemoglobin of more than 20 g/L, transfusion of 2 or more units of blood, or location of the hemorrhage in the retroperitoneum, cranium, or a prosthetic joint); (2) minor bleeding (bleeding at any site that did not meet any of the preceding criteria);[10] and (3) the development of heparin-induced thrombocytopenia (platelet count <100,000/mm³).

Benchmarks for Outcomes

It is important to identify specific measures of outcome or key processes of care[5,24,29,33] and establish benchmarks.[33] Benchmarking is a tool to create tension for change, to build awareness of current capability versus best-known capability, and to encourage people to move from a position of inertia to positive action.[33] We chose the benchmarks from randomized controlled trials and expert recommendations (see below). Most of these processes of care are dependent only on physician and health care system actions. Other benchmarks that may

be used are specific targets over a specified period of time. Schoenenberger et al suggested the following targets as benchmarks: 12 hours for the time to exceed the therapeutic threshold, 12 hours to start warfarin after the aPTT exceeds the therapeutic threshold, >4 days for the use of heparin, <5 days for the time to the first INR value of 2–3, or 12 hours for the time from first INR value between 2–3 to discharge.[24]

Laboratory Evaluation and Outcomes

The aPTT therapeutic range should be determined at each institution. This is due to the high variability of the thromboplastin reagents used. The recommended therapeutic range is equivalent to an antifactor Xa heparin level of 0.3–0.7 U/ml or to a heparin level by thrombin/protamine titration of 0.2–0.4 U/ml. This range is not commonly determined using either methodology in US hospitals. In most laboratories this is equivalent to 1.5–3.0 times the mean of the laboratory normal range.[8] The therapeutic range at our institution is determined by a heparin dose response curve equivalent to heparin concentrations of 0.2–0.4 U/ml (which is not the same as the recommended methodology). The implications of not having an accurate and valid estimate of the therapeutic range cannot be underestimated. The dosing changes described on the nomograms rely on aPTT values. Undertreating patients may result in higher recurrences of DVT/PE and overtreating patients may result in higher bleeding rates.

Results

A deep venous thrombosis was diagnosed in 68.6% (n = 70), pulmonary embolism in 21.6% (n = 22), and both conditions in 9.8% (n = 10). Physicians who took care of these patients were general internists (23%), family practitioners (25%), internal medicine subspecialists (25%), surgeons (18%), and others (10.9%). The patient characteristics, Charlson's comorbidity index, and risk factors for thromboembolism or bleeding are presented in Table 4–3.

Measurement of the key processes of care and the associated recommended benchmarks are given in Table 4–4. The mean time to exceed the therapeutic threshold (aPTT ≥ 60 sec) was 25 h (SD 23h) and the average time to achieve a therapeutic range (60 sec ≤ aPTT ≤ 90 sec) was 36 h (SD 31h). The mean initial heparin bolus dose administered was 5599 U (SD 2285 U) or 65.8U/kg (SD 29.7U/kg). The average initial infusion rate was 1009 U/h (SD128 U/h) or 12.1 U/kg/h (SD 3.1 U/kg/h). A platelet count was obtained after starting heparin in 81% and the hemoglobin was measured in 80% during the course of treatment. An explicit goal for the aPTT range was documented in 48%. Major bleeding occurred in 3.9%, minor bleeding in 4.9%, and heparin-induced thrombocytopenia in 2.9%. Other key processes of care are presented in Table 4–4.

Table 4–3 Baseline Characteristics (n = 102)

Patient Characteristic	Value*
Sex, male	52%
Race, white	67%
Smoking	28%
Age, years	58 ± 18.1
Weight, kg	89 ± 24.7
Charlson's comorbidity index	1.8 ± 2.1
Risk factors for thromboembolism†	
Thrombosis, history	40%
Medical conditions‡	20%
Surgery, prior 2 weeks	12%
Cancer, current	13%
History of cancer	9%
Postphlebitic syndrome	9%
Estrogen use	7%
Immobility	7%
Hyper coagulable state§	5%
Other¶	16%
Abnormal bleeding risk† ‖	6%

*Values are expressed as % or mean ± SD
†Each patient may have more than one risk factor
‡Heart failure, sepsis, myocardial infarction, or inflammatory bowel disease
§Protein C or S deficiency, or antiphospholipid antibody
¶Fracture (pelvis, femur, tibia), 1%; postoperative sepsis, 1%; pregnancy, 2.9%; surgery (hip, major knee surgery), 2.9%; surgery, extensive for malignancy, 2%; varicose veins, 6.9%
‖Stroke, last 2 weeks, 2%; gastrointestinal bleed, 2.9%; genitourinary bleed, 1%; current peptic ulcer disease, 1%

U—Understand the Process

In patients with a diagnosis of pulmonary embolism and/or deep venous thrombosis who receive intravenous heparin, we found that the initial intravenous heparin dose that they received was lower than what is usually recommended (bolus of 65.8U/kg, initial infusion 12.1 U/kg/h vs 80U/kg and 18 U/kg/h, respectively). We also found that it took more than 24 hours in half of the cases to achieve and surpass the therapeutic threshold of full anticoagulation. Other aspects in the process of care were also discussed: use of warfarin, overlap of warfarin and heparin, goals for anticoagulation, and optimal determination of the aPTT therapeutic range. The recommended methodology is measuring heparin levels by the antifactor Xa assay or by thrombin/protamine titration.[8]

A great degree of variation in heparin bolus (from 0 to 170U/kg), initial heparin infusion (from 5.5 to 20 U/kg/h), mean time to exceed the therapeutic threshold (from 4.5 to 104 hours), and the average time to achieve a therapeutic range (from 4.5 to 191 hours) exists at our institution. Some physicians never obtained a blood count during therapy. We explored the reasons for variability. The lack of using the patient's weight to guide dosing is explained in part by the reliance on a standard dose for every patient (standard bolus of 5000U and initial infusion of 1000U/h). Communication barriers and systems issues were also noted during the chart abstraction and group discussions (patient volume, timeliness of intravenous infusion starting, bed availability, delays in adjusting medication dose). Each physician wrote orders in the way that he or she usually practices, and no coordinated efforts were discovered among disciplines to achieve a measurable goal. These observations will be included in the second phase of the path development to explore and understand sources of variation.

In summary, heparin use at our hospital was not consistent with best practice as defined by the literature. Based on our results, an opportunity to improve the dosing, titration, and therapeutic use of heparin exists in our medical center.

S—Select Solutions

The team discussed options to deal with this problem. Several issues were raised at different levels. The team discussed other published reports that have shown the use of higher initial doses of heparin (bolus and infusion) and the use of a prespecified nomogram to adjust the heparin infusion rate result in a shorter time to achieve a therapeutic aPTT.[8,10,13,16,17,20] It has been previously shown that the implementation of a heparin nomogram is possible.[11,13,16,25,28] The development of an optimal care

Table 4–4 Key Processes of Care (n = 102)

	Value*	Recommended benchmark	References
Time to exceed the therapeutic threshold[†] (aPTT ≥ 60 sec), hours	24 ± 23	within 24 h	10
Time to achieve a therapeutic range[†] (60 sec ≤ aPTT ≤ 90 sec), hours	36 ± 31	within 24 h	10
Initial heparin bolus, U, or	5599 ± 2285	5000	8, 10, 12, 13
weight-based, U/kg	65.8 ± 29.7	80	10
Initial heparin, U/h, or	1,009 ± 128	1250 to 1660	8, 10, 12, 13
weight-based, U/kg/h	12.1 ± 3.1	18	10
Time to start of warfarin, days	1.64 ± 1.6	0 to 1	8, 12
Overlap heparin and warfarin, days	3.86 ± 3.5	4 to 5	8, 12
International Normalized Ratio at discharge (INR)	2.36 ± 0.88	2 to 3	8, 12
Length of stay, days[‡]	8.3 ± 8.1	na	—
Nomogram use	10%	100%	8, 12
Compressive stockings	10%	100%	8, 9
Platelet count during heparin therapy	81%	100%	8, 12
Hemoglobin level during heparin therapy	80%	100%	—

*Values are expressed as % or mean ± SD, aPTT = activated partial thromboplastin time, na = not applicable.

[†]The therapeutic range at our institution is determined by a heparin dose response curve equivalent to heparin concentrations of 0.2–0.4 U/ml (the recommended methodology is measuring heparin levels by the antifactor Xa assay or by thrombin/protamine titration).

[‡]The length of stay in patients with thromboembolism may relate to the underlying condition.

path for the use of heparin in the treatment of deep venous thrombosis is the solution the team has chosen. How to diagnose and prevent venous thromboembolism were examples of issues discussed for future quality improvement projects. The recommendations to hospital administrators were: request for funding to determine the therapeutic range of aPTT (by using the antifactor Xa assay) and continued administrative support for hospitalwide dissemination of the optimal care path.

Optimal Care Path Development

The organization is committed to the development and use of optimal care paths as a means to improve quality and efficiency of care while optimizing use of resources. A model has been established for development of care paths that depends on physician-led interdisciplinary teams. The team is responsible for reviewing literature, analyzing available data, and obtaining feedback from appropriate disciplines. The organization provided clerical support to the team. A standard housewide path format was used to expedite consistency and education once the path was completed. A variance tool will be used to track variations in care from the established interventions outlined on the care path.

The optimal care path shown in Exhibit 4–1 was constructed after careful review of the steps described above. The literature was reviewed giving special emphasis to well-designed randomized controlled trials. We also reviewed observational studies and expert opinion. Data on current practice at our institution were also exten-

sively discussed. The specific outcomes recommended in the path are supported by randomized controlled trials or by expert opinion. The recommended benchmarks for these outcomes were also supported by available literature. No studies are available that compare the nomograms to adjust heparin; they all seem to work the same. The weight-based nomogram was chosen because it has been shown to improve outcomes (Exhibit 4–2), the dosing is tailored to body habitus, and it was recommended by local clinical pharmacists and thrombosis specialists. The path was developed by the team via consensus and brainstorming using available data. In terms of practical issues, we were faced by constraints in physicians' and other team members' availability.

Clinical and Financial Outcomes

What are the outcomes and financial implications of our findings? In a cost-effectiveness analysis, we explored the impact of using the lower dose described above versus the recommended higher dose.[34] We found that in caring for 100 patients, 24 cases of recurrent DVT would be seen at 3 months using the lower dose described above (5000U bolus followed by an initial infusion rate of 1000U/h) as compared to 11 cases of recurrent DVT using the recommended higher doses; ie, 13 additional cases are prevented at a net cost savings of $41,000/100 subjects (or a saving of $3200 per each additional case prevented). Implementing a path in which a higher dose of heparin is given would prevent more cases, at the same time saving resources to the health care system.

Exhibit 4–1 Optimal Care Path for Proximal Deep Venous Thrombosis and Pulmonary Embolism

	Day of admission	Day 2	Day 3	Day 4	Day 5 or beyond
Assessment	Color, T°, pulse (LE) Symptoms of PE, bleed, risk of fall	→	→	→	→
Test/ Procedures	STAT aPTT q6h after heparin bolus/after any dose change* Radiology report	When 2 consecutive aPTT within 60–90 sec then qd*	INR qd*		
Consults	per MD				
Treatments	Local heat†, elevate LE for comfort O₂ per MD	Fit for graduated compression stockings	→	→	→
Activity/ Mobility	Bed rest†	Ambulate prn after 24 hrs of therapeutic heparin†	→	→	→
Medications	Heparin (weight-based nomogram*) Acetaminophen prn Warfarin, 24 hrs after initiation of heparin*	→ →	→ Adjust warfarin*	→ →	→ →
Patient/Family Teaching†	DVT complications Heparin risks Need for regulation of heparin	Rationale for warfarin Need for daily labs Explain INR target Explain activity level	Written instructions Warfarin video Medications to avoid Dietary restrictions	Reinforce symptoms and signs of DVT, PE, bleeding Teach how to use stockings	Specify how patient will communicate with MD
Discharge Planning			Identify primary MD responsible for f/u Specify how patient will communicate with MD	→	Arrange for next INR within 1–3 days† Fax INR, warfarin dose to MD or to anticoagulation clinic†
Outcomes	Heparin bolus within 4 hours of diagnosis‡ aPTT within 60–90 sec within first 24h Heparin ≥ 5 days* Overlap warfarin*-heparin ≥ 3 days		INR 2–3 Knowledge about DVT, PE, risks†	INR 2–3 for 2 consecutive days*	Symptoms improved No bleed No fall Stockings use 2 years* LOS target ≥ 5 days* (above knee DVT) f/u within 3 days

Suggested processes and outcomes supported by randomized controlled trials (*) or by expert opinion (†). aPTT = activated partial thromboplastin time, DVT = deep venous thrombosis, f/u = follow-up, INR = International Normalized Ratio, LE = lower extremity, LOS = length of stay, O₂ = oxygen, PE = pulmonary embolism
Source: Copyright © Pitt County Memorial Hospital.

CONCLUSION

Since we began sharing the initial data, additional physicians have joined our interest group. The use of objective data to describe the state of the problem is most beneficial in garnering physician and other staff support. Objective data are also crucial to support recommendations to hospital administrators to allocate resources. We

intend to evaluate the optimal care path for the use of heparin in the treatment of venous thromboembolism. We plan to study both clinical and financial outcomes to ensure that implementation of the path has had the desired effect.

Our mission is to provide the best care, at the lowest cost, with the least side effects or complications of therapy. Designing systems to overcome the barriers in

Exhibit 4–2   Heparin Order Sheet

(All blanks must be filled in by physician, ✓ ❑ as necessary

1. Make calculations using total body weight: _____ kg.

2. Indication for heparin: _____

3. Bolus heparin, 80 U/kg = _____ Units IV.

4. IV heparin infusion, 18 U/kg/h = _____ Units/h
   (25,000 U in 500 cc/ D5W, 50U/ml)

5. Warfarin _____ mg PO QD for 2 days, start 24 hrs after heparin was initiated (further orders to be based on INR results).

6. Laboratory ❑ aPTT, PT/INR, CBC now
   ❑ CBC with platelet count Q 3 days
   ❑ STAT aPTT q 6 hours after heparin bolus
   ❑ Stool for occult blood
   ❑ PT/INR Q day (start on second day of heparin)

7. Radiology: _____ (see Radiology orders)

8. Adjust heparin infusion based on sliding scale below:
   aPTT < 35 sec   give bolus of 80 U/kg = _____ units, and
                      increase drip by 4 U/kg/h = _____ units/h
   aPTT 35 to 59 sec give bolus of 40 U/kg = _____ units, and
                      increase drip by 2 U/kg/h = _____ units/h
   aPTT 60 to 90 sec No change
   aPTT 91–110 sec   Reduce drip by 2 U/kg/h = _____ units/h
   aPTT > 110 sec   Hold heparin for 1 h, reduce drip 3 U/kg/h = _____ units/h

9. When 2 consecutive aPTTs are therapeutic (60 to 90 seconds), order aPTT every 24 hours. Order aPTT 6 hours after any dosage change, adjusting heparin infusion by the sliding scale until aPTT is therapeutic.

10. No IM injections are to be administered during therapeutic heparin.

Please make changes as promptly as possible and round off doses to the nearest unit/h (1 ml of infusion contains 50 units of heparin).

Signed: _____   Date: _____

Source: Adapted with permission from Raschke et al., The Weight-Based Heparin Dosing Nomogram Compared with a "Standard Care" Nomogram, *Annals of Internal Medicine,* Vol. 119, pp. 874–881, © 1993, American College of Physicians.

the delivery of care go beyond the individual patient, nurse, and physician interaction. We believe the approach we used is an excellent means of exploring the scope of the problem as well as a method to gain objective data for path design and implementation.

We conclude that physicians at our institution use lower doses of heparin than recommended in the treatment of thromboembolism. An opportunity exists to implement an optimal care path. Physician and other allied health personnel participation seems enhanced because of the use of objective data to state the significance of the problem and the need to decrease variations in care.

REFERENCES

1. Prager L. Groups embrace guidelines as part of managed care contract. *Am Med Assoc News.* 1996;39(34):1–3.

2. Lumsdon K, Hagland M. Mapping care. *Hosp Health Netw.* 1993;67(20):34–40.

3. Spath P. *Clinical Paths: Tools for Outcomes Management.* Chicago: American Hospital Publications; 1994.

4. Lord JT. Architects of care. *Hosp Health Netw.* 1994;68(6):20–21.

5. Pearson SD, Goulart-Fisher D, Lee TH. Critical pathways as a strategy for improving care: problems and potential. *Ann Intern Med.* 1995;123:941–948.

6. Peters DA. Outcomes: the mainstay of a framework for quality care. *J Nurs Care Qual.* 1995;10:61–69.

7. Haynes RB, Sackett DL, Guyatt GH, Cook DJ, Gray JAM. Transferring evidence from research into practice, 4: overcoming barriers to application. *ACP J Club.* 1997;126(1):A14–A15.

8. Hirsh J, Hoak J. Management of deep vein thrombosis in pulmonary embolism. *Circulation.* 1996;93:2212–2245.

9. Brandjes DPM, Buller HR, Heijboer H, et al. Randomized trial of effect of compression stockings in patients with symptomatic proximal-vein thrombosis. *Lancet.* 1997;349:759–762.

10. Raschke RA, Reilly BM, Guidry JR, Fontana JR, Srinivas S. The weight-based heparin dosing nomogram compared with a "standard care" nomogram: a randomized controlled trial. *Ann Intern Med*. 1993;119:874–881.

11. Elliott CG, Hiltunen SJ, Suchyta M, et al. Physician-guided treatment compared with a heparin protocol for deep vein thrombosis. *Arch Intern Med*. 1994;154:999–1004.

12. Hyers TM, Hull RD, Weg JG. Antithrombotic therapy for venous thromboembolic disease. *Chest*. 1995;108:335S–351S.

13. Cruickshank MK, Levine MN, Hirsh J, Roberts R, Siguenza M. A standard heparin nomogram for the management of heparin therapy. *Arch Intern Med*. 1991;151:333–337.

14. Hollingsworth JA, Rowe BH, Brisebois FJ, Thompsom PR, Fabris LM. The successful application of a heparin nomogram in a community hospital. *Arch Intern Med*. 1995;155:2095–2100.

15. Petitta A, Kaatz S, Estrada CA, Effendi A, Anandan JV. The transition to medication system performance indicators. *Top Hosp Pharm Manage*. 1995;14:20–26.

16. Dartnell JG, Allen B, McGrath KM, Moulds RF. Prescriber guidelines improve initiation of anticoagulation. *Med J Aust*. 1995;162:197.

17. Hyers TM. Integrated management of venous thromboembolism. *South Med J*. 1996;89:20–26.

18. Litin SC, Gastineau DA. Current concepts in anticoagulant therapy. *Mayo Clin Proc*. 1995;70:266–272.

19. Hull RD, Raskob GE, Rosenbloom D, et al. Heparin for 5 days as compared with 10 days in the initial treatment of proximal venous thrombosis. *N Engl J Med*. 1990;322:1260–1264.

20. Corbett NE, Peterson GM. Review of the initiation of anticoagulant therapy. *J Clin Pharm Ther*. 1995;20:221–224.

21. Fennerty AG, Thomas P, Backhouse G, Bentley P, Campbell IA, Routledge PA. Audit of control of heparin treatment. *Br Med J Clin Res*. 1985;290(6461):27–28.

22. Lee HN, Cook DJ, Sarabia A, et al. Inadequacy of intravenous heparin therapy in the initial management of venous thromboembolism. *J Gen Intern Med*. 1995;10:342–345.

23. Wheeler AP, Jaquiss RD, Newman JH. Physician practices in the treatment of pulmonary embolism and deep venous thrombosis. *Arch Intern Med*. 1988;148:1321–1325.

24. Schoenenberger RA, Pearson SD, Goldhaber SZ, Lee TH. Variation in the management of deep vein thrombosis: implications for the potential impact of a critical pathway. *Am J Med*. 1996;100:278–282.

25. Fennerty AG, Renowden S, Scolding N, Bentley DP, Campbell IA, Routledge PA. Guidelines to control heparin treatment. *Br Med J Clin Res*. 1986;292(6520):579–580.

26. Pearson SD, Polak JL, Cartwright S, et al. A critical pathway to evaluate suspected deep vein thrombosis. *Arch Intern Med*. 1995; 155:1773–1778.

27. Pearson SD, Lee TH, McCabe-Hassan S, et al. A critical pathway to treat proximal lower-extremity deep vein thrombisis. *Am J Med*. 1996;100:283–289.

28. Raschke RA, Gollihare B, Peirce JC. The effectiveness of implementing the weight-based heparin nomogram as a practice guideline. *Arch Intern Med*. 1996;156:1645–1649.

29. Nelson EC, Batalden PB, Plume SK, Mohr JJ. Improving health care, part 2: a clinical improvement worksheet and user's manual. *Jt Comm J Qual Improve*. 1996;22:531–548.

30. Eckhart J, Gilbert P. Improved Coumadin therapy using a continuous quality improvement process. *Clin Lab Manage Rev*. 1996;10: 153–156.

31. Ramirez O, Lawhon J. Quality improvement team uses FOCUS-PDCA method to reduce laboratory STAT volume and turnaround time. *Clin Lab Manage Rev*. 1994;8:130–141.

32. Charlson ME, Pompei P, Ales KL, MacKenzie CR. A new method of classifying prognostic comorbidity in longitudinal studies: development and validation. *J Chron Dis*. 1987;40:373–383.

33. Mohr JJ, Mahoney CC, Nelson EC, Batalden PB, Plume SK. Improving health care, part 3: clinical benchmarking for best patient care. *Jt Comm J Qual Improve*. 1996;22:599–616.

34. Estrada CA, Mansfield C, Wynn J. Deep venous thrombosis: which heparin should I use? *J Gen Intern Med*. 1997;12:60. Abstract.

■ 5 ■

Heart Failure: Managing Care and Outcomes Across the Continuum

Linda D. Urden, Susan Casamento, Mary Mitus, Marsha Terry, and E. Thomas Arne Jr.

Heart failure (HF) is a rapidly increasing cardiovascular condition and one of the leading causes of morbidity and mortality in the United States.[1,2] HF affects an estimated 2 million Americans with average mortality rates of 10% at 1 year, and 50% after 5 years.[3] Because the incidence of HF rises after 65 years of age, the prevalence of the condition is expected to increase as the population ages.[3] In addition to multiple clinical manifestations, HF can greatly affect quality of life and functional abilities.[2,4] Factors such as social isolation and noncompliance with medications and diet also contribute to early hospital readmissions. It has been shown that coordinated collaborative care across the continuum that addresses all problems associated with HF has positively impacted the outcomes of HF patients.[5,6]

Issues related to the complexity and scope of treating HF patients described at the national level are consistent with our local and regional setting. Our team identified the opportunity to meet the challenges of HF management and to work toward quality improvements in the areas of medical, nursing, pharmacologic, dietary, and lifestyle change management. Because of our linkage with home care, there was also an opportunity to work toward seamless provision and continuation of services. This chapter presents the inpatient and home care HF pathways, variance tracking, and quality improvement activities related to our outcomes management program.

Butterworth Hospital (BH), a 529-bed teaching community tertiary care hospital in western Michigan, is the acute care facility for the Butterworth Health System (BHS). The Visiting Nurses Association (VNA) is the BHS agency that provides home care. The VNA is a nonprofit, community-based agency that provides a wide range of home health and hospice services, providing 135,000 home visits per year, with an average daily census of 1100 patients. Clinical pathways were identified by our system as a tool for interdisciplinary quality improvement that would enhance outcomes for our patients. The development of pathways is facilitated by an outcomes management director who works with interdisciplinary teams to develop, implement, and monitor pathways and outcomes. The BHS program began in 1993 with the introduction of an inpatient total joint replacement pathway, followed by the implementation of nine other pathways the following year. After the successful implementation and enhancement of outcomes in these inpatient populations, the program was expanded to other BHS entities and included both inpatient and outpatient diagnoses.[7]

Clinical pathway teams consist of appropriate disciplines, such as medicine, nursing, pharmacy, respiratory, physical therapy, social work, and other specialty clinicians. Teams are cofacilitated by a physician and an advanced practice nurse (in areas where there is not an advanced practice nurse, a nurse with specialty expertise in the area serves in this role under the leadership of the outcomes management director). Other services that are important to the success of the pathway are represented: community agencies, physician office staff, health maintenance organizations (HMOs), and other external specialists. Since pathways are frequently modified based on new literature and/or outcomes, pathways are not part of the permanent medical record and are used as guidelines for the practitioner in the management of care. All practitioners are highly encouraged to use the pathways. Since there was initial controversy among the medical staff regarding the use of pathways, it is not mandatory for physicians to use pathways and actual use varies among various specialties. The pathways and outcomes management program has recently been integrated into our newly organized corporate Care Management Program.

HEART FAILURE PATHWAY

Inpatient Pathway

The inpatient heart failure pathway (DRG 127), developed by an interdisciplinary team, was cofacilitated by a cardiologist and cardiology nurse practitioner. Members of the pathway team consisted of emergency medicine, family practice, and internal medicine physicians; inpatient cardiovascular clinical nurse specialist; cardiac rehabilitation nurse specialist; pharmacist; VNA cardiac clinical coordinator, HMO representative; nursing cardiovascular services director; dietitian; medical social worker; and medical and nursing quality improvement staff. The entire team held six 1-hour meetings over a 3-month period. In addition, there was work done in small groups or independently by specialists between meeting times. Clinical practice patterns of all involved clinicians prior to the pathway were reviewed so that all team members would understand the current treatment methodologies. The literature was reviewed for the latest trends and research regarding heart failure treatment. The pathway was formulated based on research, previous successful treatments and outcomes, and feasibility and acceptability into our setting.

The pathway is illustrated in Exhibit 5–A–1 in Appendix 5–A. Essential treatment/care categories are found in the first column, with the specific function listed on the appropriate day in columns 2 through 6. The pathway is initiated in the emergency department and is based on an expected length of stay of 4 days. Expected outcomes are delineated in the last column. Criteria for dietary, pharmacy, social work, nurse practitioner, and cardiac rehabilitation consultations are listed at the back of the pathway. Variances from the pathway and explanation codes are also described at the end of Exhibit 5–A–1.

Home Care Pathway

The home care heart failure pathway (ICD-9 code 428) was also developed with a multidisciplinary approach, with nursing, therapy, social work, and physician input. Development and refinement solicited feedback from the BH inpatient Congestive Heart Failure (CHF) Service medical director, cardiology nurse practitioner, cardiac rehabilitation staff, and VNA home care CHF team. Communication with these persons ensured that the home care pathway would interface with other treatment practices and provide continuity. The pathway design was modeled from the BH format so that is easily recognizable by all and can provide a smooth transition from hospital to home (see Exhibit 5–A–2). Columns 2 through 4 are described in weeks versus the inpatient days. The last column is the same as the inpatient pathway in that it delineates expected outcomes. A different variance tracking method is used at VNA, and therefore the coded variance tracking system seen in the inpatient pathway is not used for the home care pathway.

CASE MANAGEMENT

Inpatient Management

The pathway was implemented after finalization by the team and comprehensive inservice education for all practitioners, including nursing, medicine, physician assistants, secretarial staff, pharmacy, physical therapy, dietary, and social work. The heart failure pathway "package" consists of the clinical pathway, nursing guidelines of care, discharge instructions, education record, and imprinted order sheet. The imprinted order sheet serves as a summary and reminder of medical treatments and diagnostics that are delineated on the pathway. Pathways are kept on all of the patient care units and are added to the admission packet by the secretary for all heart failure admissions. Since patient and family education is crucial for the management of heart failure, a special CHF Kit was designed for the patients as an educational component of the pathway. The kit is a canvas bag with velcro closure and carrying handle that contains the CHF education book and any special instructions regarding medications and diet that they may receive prior to discharge from the hospital. Patients are asked to keep all of their medications in the bag and to bring the bag to physician appointments, emergency department visits, or subsequent hospitalizations.

The facilitators of the heart failure pathway team are also the codirectors of the BH CHF Service. Referrals from the patient's primary care medical practitioner indicate the CHF Service consultation and heart failure treatment and education plan. They follow the patient for the duration of the inpatient stay, make referrals to appropriate support services, such as VNA for home care, and continue to follow the patients in an outpatient setting in the BH CHF Clinic. The patients are also monitored closely through phone follow-up.

Although the pathways are not part of the permanent record, they are placed in the medical record during the episode of care as a guide and method of communication among practitioners for a quick reference to see where the patient is on the pathway. All clinicians have accountability for implementing the interventions specific to their practice and share accountability for patient outcomes. There is not one specific person designated to be the case manager for the inpatient care. Staff nurses coordinate the care of patients and consult with the appropriate disciplines so that the pathway is maintained and patient outcomes are maximized.

Variances are noted and the cardiology nurse practitioner is consulted for assistance in addressing the conditions causing the variance. As clinicians carry out interventions, care is documented as appropriate (Exhibit 5–A–1). Any variance from the pathway is documented with reason why the activity did not take place.

The cardiology nurse practitioner is accountable for reviewing all pathways, completing quarterly audits and conducting quarterly heart failure team meetings. The purpose of these meetings is to review the quarterly quality reports and to make recommendations for changes in the pathway and associated tools. Additionally, any systems issues are reviewed that may require intervention and follow-up. The pathway is revised as appropriate and changes, if any, are communicated to the other practitioners.

Home Care Management

The system of home care delivery is a multidisciplinary team approach with a primary care nurse case manager. The nurse case manager is responsible for managing the pathway and initiates it with the opening paperwork on new heart failure patients. Most case managers file the pathway in the patient's home care folder in the patient's home, checking off interventions as they are completed. This ensures continuity of care if a nurse other than the case manager sees the patient. It also encourages patient participation and serves as a resource for the patient between nursing visits. Other disciplines routinely refer to the home care folder in the patient's home to review the patient's progress. Patients also take the home care folder to their physician's office and to hospital emergency room visits, should that occur. The paper pathway is not part of the permanent medical record. Documentation is done on the nursing care plan in the computerized medical record by the case manager. The pathway is used as a guide when giving progress reports from one nurse to another, or to physicians.

Pathways are reviewed routinely every 6 months as problems or new procedures or protocols are introduced. The cardiac program manager, cardiac clinical coordinator, director of quality management, and vice-president of clinical services are responsible for the review. Additional feedback is solicited from other clinicians, as appropriate.

OUTCOMES

Inpatient

Outcomes that are essential for the management of heart failure patients are established by the pathway team members. The initial listing of outcomes is dis-

cussed and formulated prior to initiation of the pathway. As changes in care occur, consistent attainment of meeting thresholds of care, new research findings, and modifications in expected outcomes may occur. The pathway is therefore considered a more fluid document and not one that is etched in stone. It is constantly reviewed and evaluated for currency and appropriateness.

Outcome Indicators

Outcomes are listed in the last column of the pathway and relate to the major care/treatment categories: consults, assessment and interventions, medications, diet, activity, tests, education, and discharge planning. Refer to Exhibit 5–A–1. Since the pathway displays a succinct summary of expected interventions and outcomes, other outcomes important to this patient population are also analyzed. BHS uses a report card format for our outcomes management program. This tool provides a quick reference and balanced view of all outcomes relevant to the specific pathway, and tracks improvements in outcomes over several reporting periods. It also serves as an excellent way to share pathway quality improvement efforts with many disciplines and audiences.[8]

As the report card illustrates (see Exhibit 5–A–3 in Appendix), there are four major categories of outcomes assessment: clinical, patient satisfaction, cost and utilization (fiscal), and functional health. This provides a comprehensive approach for outcomes assessment.

Clinical Outcomes

Clinical outcomes that are measured consist of interventions demonstrated to impact the outcome of heart failure patients: daily weights, cardiac functional class, and discharge planning/education. Other parameters reflect diagnostic tests that are essential in managing heart failure: use of diuretics, serum potassium, cholesterol and creatinine, echocardiogram, and cardiac ejection fraction. Additional indicators listed are helpful in monitoring clinician practices regarding the pathways: consultations, documentation of care and teaching provided, and completed diagnostic testing.

Patient Satisfaction

The quality of care and services perceived by the patient is assessed in the areas of overall care received, nursing care, medical care, various services provided by hospital departments; discharge instructions, specific condition information, readiness for discharge from the hospital; and likelihood of returning to BH or referring a friend or family member to the hospital.

Functional Outcomes

For patients who are admitted to the CHF Service, quality of life is an outcome that is initially assessed with ongoing monitoring during subsequent outpatient visits. Quality of life is assessed in both physical and emotional dimensions.

Cost/Utilization

The fiscal parameters of cost and utilization are examined in the following areas: length of stay (LOS), total charges, readmissions within 30 days, average charges per case (intensive care unit [ICU], non-ICU, laboratory, respiratory therapy, pharmacy, and radiology).

Outcome Measurement and Data Collection

The cardiology nurse practitioner collects data on the clinical indicators from medical record reviews using a specially designed audit tool. Utilization review staff, medical staff, quality improvement staff, and other quality or research assistants assist with data collection based on additional indicators that the pathway team may want collected. Our market research department conducts quarterly phone interviews in which the patient satisfaction survey (PSS) is administered. The PSS is a structured questionnaire developed and refined by BHS that takes approximately 20 minutes to administer. It asks patients to respond with their answer on a 0–5 Likert scale, with 5 being the maximum score. Patients are also encouraged to give additional feedback and comments regarding their care and services received.

The functional outcome quality of life is measured by the *Minnesota Living with Heart Failure (LlhFE)* questionnaire. The *LlhFE* is a 21-item, self-administered questionnaire that assesses the quality of life of those with heart failure. A six-point response format, ranging from 0 to 5 is used to quantify the patient's perception of disability related to the questions. The instrument has established validity and reliability and has been used in multiple studies to examine quality of life in heart failure patients.[9] Patients are administered the questionnaire upon admission to the CHF Service, and at 3, 6, 9, and 12 months.

Cost and utilization indicators are obtained using a variety of reports from the finance and medical records departments. Standard queries are built and reports are sent quarterly to the outcomes management department for integration into the report card and other reports that may be requested. Additional fiscal parameters can be added to the report and queried to the appropriate data source as necessary.

Outcomes Reports and Integration with Quality Program

The cofacilitators of the pathway team conduct quarterly meetings in which outcomes and pathways status to date are reviewed. After all of the data have been received from the various sources and entered into the computer by the outcomes management department

staff, data are reviewed and analyzed by the cardiology nurse practitioner. The patient case mix for the current report period is documented on the report card (see Exhibit 5–A–3) so that the team can have an accurate description of the patient population for the quarter being reviewed. Current quality improvement activities in progress are also listed for reference. Outcomes are reported for the current quarter in the appropriate box, with previous quarters listed for comparison. Due to space restrictions on the report card, a maximum of four quarters can be listed.

The pathway team reviews the report card and discusses variances and any identified trends in quality from the previous quarter. Any outstanding issues related to systems or processes are also discussed. Action plans to address any problem areas are formulated and assigned to the appropriate person(s) for follow-up. Since all disciplines involved in the pathway are members of the team, it is expected that they report the quality issues related to their departments at their quality meeting and assume accountability for any specific follow-up that may need to be done. The pathway outcomes also are reported to the hospital quality coordinating council and to the board of trustees.

Pathway Variances

Variances from the pathway are tracked on the back of the pathway (see Exhibit 5–A–1). A coding schema has been formulated for variances from pathways in the outcomes management program so that computer entry and retrieval is easy. Common variances for each pathway have been identified and are listed for reference for the user to select and enter the appropriate code number. There is also space for any comments that the clinician would like to communicate to the team. As variances become more or less common, coding may be added or deleted. We also encourage comments and suggestions for improving care and pathway format. The variance portion is not part of the permanent medical record and is used for quality improvement purposes only. Variances are reported in the quarterly pathway team meetings. The cardiology nurse practitioner reviews all variances and formulates appropriate actions to address unexpected variances from the pathways.

Quality Improvements in the Care of Heart Failure Patients

Since the pathways and outcomes are components of our quality improvement program, we cannot share specific outcomes outside of our system. However, we are pleased that we have had improvements in the several areas. The pathway was implemented in August 1996, and outcomes will be described year-to-date (current date July 1997). Clinical improvements include the following increases: use of diuretics upon admission 10–20%; dietary consult 10%; social work consult 23%; ejection fraction done/documented 30%; cardiology consult 10%; and discharge functional class by 0.5 points. Overall patient satisfaction increased by 0.20 points. The quality of life scores (baseline, 3 months, 6 months, 12 months) are in process at the current time.

We have also positively impacted length of stay from 6.84 days for fiscal year 1996–1997 to a current 4.96 days LOS year-to-date for fiscal year 1996–1997. Year-end costs are not calculated for the fiscal year, but are expected to also be decreased due to the decrease in LOS and hospital emphasis on cost containment.

Home Care

Outcome Indicators

Outcomes are determined prior to implementation of the pathway, similar to the inpatient pathway team approach. The report card format is also used to display outcome data. Home care outcomes for the heart failure pathways are evaluated in five areas: patient case mix, patient satisfaction, clinical outcomes, functional outcomes, and cost and utilization. See Exhibit 5–1 for a complete listing of all indicators.

- *Patient/Case mix.* Three indicators are collected to describe the population: cases per year, patient age, and patient gender. Future report cards will also include significant comorbidities such as diabetes.
- *Patient Satisfaction.* Patient satisfaction with care and services is evaluated in the following areas: overall care received, access and efficiency of services, reliability, how well patient needs were met, understanding of payment for services, patient and family teaching, and the likelihood of returning to VNA or referring a friend or family member to the agency.
- *Clinical Outcomes.* Clinical outcomes consist of the patient's overall condition at discharge from home care as compared to admission status; discharge reason and disposition; and pathway-specific outcomes, as listed in the last column of the home care pathway and individualized for each patient (see Exhibit 5–A–2 in Appendix).
- *Functional Outcomes.* Functional outcomes are assessed in six areas: general health, pain, mental health, ability to walk one block, ability to climb stairs, and ability to perform personal care.
- *Cost/Utilization.* Six areas are evaluated regarding cost and utilization: LOS, visits per patient by discipline, rehospitalizations while on service, extended care facility (ECF) placement, transfers to hospice,

Exhibit 5–1 Home Care Heart Failure Outcomes

Patient/Case Mix
  Cases per year
  Age
  Gender
Patient Satisfaction
  Needs met
  Reliability
  Would use again
Clinical Outcomes
  Discharge disposition
  Reason for discharge
  Discharge condition
Functional Outcomes
  General health
  Pain
  Mental health
  Walking one block
  Climbing flight of stairs
  Personal care
Cost and Utilization
  Length of stay
  Visits per patient by discipline
  Rehospitalizations
  ECF placement
  Transfer to hospice
  Expired

and patients expired. The information is entered concurrently into the automated information system and reported for the specific diagnostic categories every 6 months.

Outcome Measurement and Data Collection

The cardiac clinical coordinator and the cardiac program manager collect data on the clinical indicators from the medical record reviews and reports generated from the automated system. Quality management staff assist with data collection as needed. The community relations department coordinates the patient satisfaction surveys. A survey is mailed to past patients every 6 months. Results are tabulated and reported to the program managers and the quality management committee.

The functional outcomes are measured by the SF-36,[10] which is administered by the nurse during the discharge visit. The survey is easy to administer and takes approximately 10 minutes for most patients to complete. The survey is scored and analyzed by the BHS market research department.

Cost and utilization indicators are obtained using reports from the automated information system. Standard queries are built and reports are generated every 6 months and sent to the cardiac program manager for inclusion into the report card.

Outcomes Reports and Integration with Quality Program

The VNA pathway and outcomes steering committee meets every other month to monitor pathway development and review outcome data. The vice-president of clinical services and the director of quality management serve as links between this committee and the quality management committee.

Outcomes are displayed on the home care CHF report card (see Exhibit 5–A–4 in Appendix). Outcome report cards are updated every 6 months and submitted to the pathway and outcomes steering committee. The cardiac program manager, cardiac clinical coordinator, and multidisciplinary team review the report card, discuss variances and trends and work on significant issues related to the care of heart failure patients. Action plans are developed and implemented to address unresolved problems.

The cardiac program manager continuously reviews results of the heart failure program and collaborates with quality management to incorporate improvement activities for the program with overall agency quality improvement initiatives.

Pathway Variance

A formal variance tracking program is in development at the VNA. Variances from the heart failure pathway are monitored by the cardiac program manager and the cardiac clinical coordinator by reviewing the outcome data and information shared by the cardiac team. Follow-up actions consist of working with the cardiac team and consulting with the cardiologist and cardiology nurse practitioner to adjust the pathways and patient care interventions. Suggestions for improving care and the pathway are solicited from all members of the inpatient and home care teams.

Quality Improvement in the Care of Heart Failure Patients

Since the home care heart failure pathway was implemented less than 1 year ago, outcome data reflect baseline measures only. It is anticipated the care delivery will be more efficient and that continuity of care will be enhanced.

OVERALL PATHWAY PROGRAM ASSESSMENT AND FUTURE PLANS

Inpatient

Our outcomes management program has been in place since 1993; however, the heart failure pathway has been in existence just short of a 1-year period. One of the biggest challenges to our team in implementing this pathway was that there are multiple physician specialties admitting patients with heart failure. Although the internal medicine and family practice team members are very involved in the pathway development and ongoing refinements, their colleagues have accepted the pathway in varying degrees. That pathways are used as guidelines for practice and are not mandatory contributes to the less than 100% use of the pathway for all patients admitted with heart failure.

Another concern is that there has been varied use by practitioners of the imprinted order sheet. While some of our medical staff faculty are supportive and encourage their medical students and house staff to use them, others discourage them from using the imprinted order sheet as it does not allow for the student to "critically think" through what orders are appropriate for each patient. Both of these issues are being addressed by our chief of medical staff, who is supportive of the pathway program and the codirector of our newly designed Care Management Program.

Nursing staff and other disciplines have accepted the pathways and see them as a quick reference and overview of the patient's progress to outcomes. Some nurses use the pathway in their exchange reports, others do not. Since the pathways are not part of the permanent record, some consider the documentation on the pathway to be redundant with their required documentation. We are currently implementing a computerized documentation system in the hospital. There are plans in the future to incorporate the pathways into the medical record, which will eliminate duplication of documentation in various parts of the medical record.

Data collection and interface of data has been a long process since we currently have many data sources that are not integrated. All departments have been extremely helpful by running their specific reports and sending them to the outcomes management staff for entry into the outcomes database. Future plans include interfaces between the various databases so that the compilation of data into one report will be much more efficient.

One of the greatest advantages to the implementation of the pathway is the interdisciplinary teamwork that developed or was enhanced. Each discipline and service or entity (inpatient, home care, rehabilitation, HMO) recognizes how each contributes to the outcomes of the patient and strives to provide comprehensive care. All outcomes are examined in total, versus singularly, by each discipline or entity. The challenges each discipline makes to the others has made a difference in treating this chronic disease state. Pathways, outcomes, and report card are shared with HMOs and payers upon request.

Home Care

The VNA outcomes management program has been in place for just over 1 year. During this time, pathways have been developed for high volume and high-risk populations. The pathways have been well supported by the staff due to their involvement in the development of the pathways and the benefits realized by the staff and patients.

Currently, pathways are used as guidelines for care and are not part of the automated clinical documentation record. Staff use care plans to document patient care on laptop computers. Future plans include automating pathways and using them as the basis for clinical documentation. Variances will be easily identified, tracked, and reported. At that point, we will be able to efficiently use more than one pathway when the patient needs care for both primary and secondary diagnoses.

Outcome measurement will soon be enhanced when the Outcome and Assessment Information Set (OASIS)[11] is implemented. OASIS is a data collection tool consisting of 79 questions targeted at patient assessment of health and functional status. The assessment is completed at admission to home care, every 57 to 62 days, and again at discharge from home care. It is anticipated that HCFA will require this assessment as condition of participation for home care agencies to be certified for Medicare/Medicaid reimbursement. The VNA has begun to pilot OASIS using the 10-item minimum data set with selected diagnoses.

CONCLUSION

In summary, this chapter has presented a description of our unique outcomes management program that links inpatient care with home care for HF patients. We have provided a brief overview of the program structure and delineated the multidisciplinary pathway team approach for clinical pathway development and management of outcomes. The program continues to evolve to meet the changing needs of our patients, clinicians, and partner providers. It is truly a continuous quality improvement effort that we are pleased to be able to share so that components may be used or modified to meet the needs of others in their unique health care settings.

Clinical pathways can be used as tools in an overall quality improvement plan to meet the needs of specific patient populations in all settings. It is important that pathways are developed by all health care providers who will be involved in the care of that population. Collaboration in planning the care and designing the appropriate outcomes to monitor will lead to shared accountability for the outcomes of the patients. Targeting specific outcomes in predetermined priority areas will focus the team to work toward those quality areas. Using an organized framework such as a report card will provide a balanced approach to outcome measurement and provide reports that are easy to use by all health care professions and providers, and serve as a succinct summary of the quality indicators for the population. Improvement and maintenance of appropriate outcomes will positively impact efficiencies in health care.

REFERENCES

1. Garg R, Packer M, Pitt B, Yusuf S. Heart failure in the 1990's: evolution of major public health problem in cardiovascular medicine. *J Am Coll Cardiol.* 1993;22(suppl A):3A–5A.

2. English M, Mastream M. Congestive heart failure: public and private burden. *Crit Care Nurse Q.* 1995;18(1):1–6.

3. Konstam MA, Dracup K, Baker DW, et al. *Heart Failure: Evaluation and Care of Patients with Left-Ventricular Systolic Dysfunction.* Rockville, MD: Agency for Health Care Policy and Research; US Dept of Health and Human Services; June 1994. Clinical Practice Guideline No. 11, AHCPR Publication No. 94-0612.

4. Funk M, Krumholz H. Epidemiologic and economic impact of advanced heart failure. *J Cardiovasc Nurs.* 1996;10(2):1–10.

5. Venner GH, Seelbinder JS. Team management of congestive heart failure across the continuum. *J Cardiovasc Nurs.* 1996;10(2):71–84.

6. Welsh C, McCafferty M. Congestive heart failure: a continuum of care. *J Nurs Care Qual.* 1996;10(4):24–32.

7. Levknecht L, Schriefer John, Schriefer Janice, Maconis B. Combining case management, pathways, and report cards for secondary cardiac prevention. *J Qual Improve.* 1997;23(3):162–174.

8. Schriefer J, Urden L, Rogers S. Report cards: tools for managing pathways and outcomes. *Outcomes Manage Nurs Pract.* 1997;1(1):16–21.

9. Rector TS, Cohn JN. Assessment of patient outcomes with the Minnesota Living with Heart Failure questionnaire: reliability and validity during a randomized, double-blind, placebo-controlled trial of pimbendan. *Am Heart J.* 1992;124:1017–1025.

10. Stewart AL, Hays RD, Ware JE. Short-form general health survey: reliability and validity in a patient population. *Med Care.* 1988;26:724–735.

11. Shaughnessy PW, Crisier KS. *Outcome-Based Quality Improvement.* Washington, DC: National Association for Home Care; 1995.

■ Appendix 5–A ■
Clinical Pathways and Report Cards

Exhibit 5–A–1 Inpatient CHF Clinical Pathway

CONGESTIVE HEART FAILURE CLINICAL PATH (DRG 127)

Ejection Fraction: _____
NYHA Functional Class:
Admit _____ Discharge _____
I—no limits II—limit w/activity
III—SOB w/ADLs IV—symptoms @ rest

This pathway is only a guideline. Patient care will vary on their individual needs. Please refer to CPM Guidelines of Care and Education Record for the Congestive Heart Failure Patient.
Please fill in dates and check all boxes that apply. Write NA over boxes that do not apply.
Length of stay will vary for each patient. Pathway is based on average stay. Less complicated and younger patients may have shorter than average length of stays. Some patients may stay longer than the average if their condition is more severe.

Stamp with addressograph

Dates	ER	Day 1	Day 2	Day 3	Day 4	Outcomes
Consults (See criteria on back)	☐ Consider cardiology NP ☐ Consider cardiology ☐ Consider MSW	☐ Consider Cardiac Rehab ☐ Consider cardiology NP ☐ Consider cardiology ☐ Consider Dietary ☐ Consider Pharmacy				☐ Follow up established with PCP/cardiology
Assessment & Interventions	☐ VS/strict I&O ☐ Consider BiPAP/CPAP ☐ Assess oxygen/O₂ protocol ☐ Bronchodilator protocol ☐ Old records/echos ☐ Cardiac monitor ☐ Foley/Inter. Infusion Device	☐ VS/strict I & O ☐ Daily weight ☐ Continue O₂ assessment	☐ VS/strict I & O ☐ Patient weighs self and logs their weight ☐ Use same scale if possible ☐ Consider d/c monitor, Foley	☐ VS/strict I & O ☐ Patient weighs self and logs their weight ☐ Use same scale if possible	☐ VS/strict I & O ☐ Patient weighs self and logs their weight ☐ Use same scale if possible	☐ Patient keeps log of their weight daily and notifies MD/NP with weight gain as per education record teaching. ☐ F & E status WDL for this patient
Medications	☐ Pre-admission meds ☐ IV diuretics ☐ Consider digoxin, ACE inhibitors, nitrates, inotropes	☐ Adjusting of meds ☐ Consider ACE inhibitor if not on coumadin ☐ Sq heparin if not on coumadin ☐ Consider pneumovax vaccine	☐ Change to po meds ☐ Consider adjusting ACE inhibitor ☐ D/C sq heparin if OOB	☐ Stabilization of meds ☐ Consider adjusting ACE inhibitor ☐ D/C sq heparin if OOB		☐ Patient verbalizes purpose of medications, doses, side effects, etc., per ed record
Diet	☐ Fluid restriction ☐ NPO except ice chips	☐ Fluid restriction ☐ Sodium restriction	☐ Fluid restriction ☐ Sodium restriction	☐ Fluid restriction ☐ Sodium restriction	☐ Fluid restriction ☐ Sodium restriction	☐ Patient verbalizes understanding of cardiac diet and fluid restriction
Activity	☐ Bedrest with BRP	☐ Bedrest with BRP ☐ Physical therapy prn	☐ OOB	☐ OOB		☐ Cardiac rehab or exercise program in place prn
Tests	☐ CXR, EKG, U/A ☐ CPK/MB, SMA + lytes, Mg++, plts, CBC, TSH—if a-fib/new onset CHF ☐ PT/PTT if on coumadin/hep ☐ Digoxin level if on digoxin	☐ Chem 7	☐ Consider repeat CXR ☐ Chem 7	☐ Chem 7	☐ Consider CXR prior to d/c ☐ Chem 7	☐ Diagnostic studies WDL for pt
Education (see Ed Record)	☐ See CHF Ed Record	☐ Reinforce CHF Ed Record ☐ Smoking counseling/ **Smoking cessation program initiated prn** ☐ **Provide copy of CHF book/ CHF kit—review w/pt**	☐ See Ed Record ☐ Smoking counseling ☐ **Review CHF book w/pt**	☐ See Ed Record ☐ Smoking counseling ☐ **Review CHF book w/pt**	☐ See Ed Record ☐ Smoking counseling ☐ **Review CHF book w/pt**	☐ Patient d/c with copy of the CHF book/kit and verbalizes understanding as per ed record ☐ **Smoking cessation program initiated prn**
Discharge Planning	☐ Assess support systems/ home environment ☐ Identify pot. barriers to d/c	☐ Assess support systems/ home environment ☐ Identify pot. barriers to d/c ☐ Assess need for cardiac rehab after d/c	☐ Assess need for home care ☐ Assess need for assistive devices (i.e. O2)	☐ RN review discharge plan. ☐ Assess need for assistive devices (i.e. O2)	☐ Discharge plans per pre-printed d/c instruction sheet ☐ Support services in place (i.e. O2, rehab, home nursing)	☐ Patient has home care and assistive devices in place prn ☐ Able to verbalize d/c plan ☐ Document d/c NYHA Functional class

continues

Exhibit 5–A–1 continued

Criteria for Medical Social Work Referral
Increased anxiety
Depressed mood/affect
Caregiver of a dependent spouse
Socially Isolated/limited support system
Inadequate financial resources to meet needs

Criteria for Cardiology NP Consult
New onset CHF
Readmission within 30 days
NYHA function class III or IV
Needs CHF education

Criteria for Cardiac Rehab Referral/PT
Decreased mobility/lack of ability to perform ADLs
Lack of knowledge of ways/activities to decrease tissue oxygen demands
New diagnosis
History of smoking/smoking cessation referrals
Secondary diagnosis of Ischemic Heart Disease

Criteria for Dietary Consult
Serum albumin < 3.0
Diet for discharge is new to patient or different from diet prior to admission
Admission weight gain of ten pounds or greater over last discharge weight

Criteria for Pharmacy Consult
Patient on 7 or more medications
Patient on two medications of same class
Serum Creatinine > 2.5 mg/dl excluding admission creatinine
Serum potassium > 5.6 or < 3.0 mEq/liter
Medication related admission (eg: Digoxin toxicity)

Instructions: Please indicate any significant variances that may prolong the stay in the boxes to the right. Write in the variance code, the date the variance occurred, comments and your initials. Your comments for improving the care for this population or the pathways format are appreciated.

CODES FOR COMMON VARIANCES

Code	Issue	Code	Issue	Code	Issue
427.9	Arrhythmia	P1	Lives alone/lack of home support	S1	Pathway documentation incomplete
427.5	Cardiac arrest	P2	Pt education not complete	S2	Imprinted orders not used
786.50	Chest pain	P4	Anxiety	S3	Discharge instructions not used
298.9	Confusion	P6	Unexpected transfer to ICU	S5	Discharge facility delay
780.6	Fever/FUO	P8	Abnormal lab values	S6	Delay in PT consult
401.9	HTN	P16	Concurrent complex med card	S8	Delay in MSW consult
458.9	Hypotension	P21	Patient noncompliance	S9	Delay in dietitian consult
277.8	Metabolic instability	P25	Activity intolerance	S10	Other
787.02	Nausea	P39	Patient expired	S18	Physician discretion
997.3	Pulmonary complication	P50	Patient taken to OR	S19	Late consult or test
787.01	Vomiting			S45	Outcomes not recorded

Pathway Variance/ Exception Code	Date	Comments	Initials

Ideas for improving care for this population:

Ideas for improving format of this pathway:

Source: Copyright © Butterworth Hospital.

Exhibit 5–A–2 Outpatient CHF Clinical Pathway

VISITING NURSE ASSOCIATION OF WESTERN MICHIGAN
CONGESTIVE HEART FAILURE

This pathway is only a guideline. Patient care will vary on their individual needs.

MSW Frequency _____ HHA _____ PT _____ OT _____

	Adm Visit/Week 1 4–5x/week	Weeks 2–3 3–4x/week	Weeks 4–9 1–2x/week	Outcomes
Consults	— Consider PT — Consider HHA — Consult with CHF clinic if applicable — Refer to MSW			
Assessment & Interventions	— RN if assess: financial problems, problems with support system, depression, anxiety, need for resources, refer to MSW — Safety, home environment — Skilled nursing assessment — 12-lead EKG/Pulse Oximetry if appropriate — Review discharge instructions — Make appt. with PCP — Daily wt-record	— MSW: Assess psychological adjustment to need for resources, financial need, support need, support system, make referrals, begin counseling RN: — Skilled nursing assessment — EKG/Pulse Ox prn — Daily wt-record — Respirex Q4H	— MSW: Counseling regarding adjustment, relaxation, follow up on resource referrals RN: — Skilled nursing assessment — EKG/Pulse Ox prn — Med/Tx changes form physician's appt. — Daily wt-record — Assess new problems as they occur	— MSW: Psychological factors related to rate of recovery are addressed, pt. is maintained in a safe environment, linkage with community resources RN: — Will follow up with PCP, cardiologist, and/or CHF Clinic
Medications	— Teach med regime — Teach one med purpose/side effects each visit — Administer IV medications as ordered	— Eval effectiveness of med teaching; reinstruct prn — Teach one med purpose/side effects each HV — Administer IV medications as ordered	— Eval effectiveness of med teaching and reinstruct prn — Administer IV medications as ordered	— Verbalize med regime, purpose, and potential side effects and appropriate response
Diet	— Review meal planning — Teach low fat, low cholesterol diet, 2gm Sodium diet — Teach 2 liter/day fluid restriction	— Eval effectiveness of previous diet teaching; reinstruct prn — Review at least 3 meals each home visit — Give AHA handout, "Eating with Less Salt"	— Review meal planning and principles of prescribed diet	— Verbalize understanding of cardiac diet — Can quote "Rule of Two's"
Activity	— Begin cardiac rehab/home exercise program per hospital discharge plan — Eval response to minimum activity requirements — Increase activity to level of tolerance — Instruct re: avoidance of overexertion; frequent rest periods	— Eval response to home exercise/cardiac rehab — Increase activity to level of tolerance	— Eval response to home exercise/cardiac rehab — Increase activity to level of tolerance	— Enrolled in outpatient cardiac rehab program or participating in home cardiac rehab
Tests	— Lab work as ordered	— Lab work as ordered	— Lab work as ordered	— Receives lab work per PCP order
Education	— Review VNS booklet — Review CHF booklet — How to take pulse before, during, and after exercise — Fluid intake allowance — CHF disease process — S/S electrolyte imbalance — S/S disease exacerbation	— Progressive cardiac rehab/home exercise program — Stress reduction — Smoking cessation — Weight reduction — Energy conservation	— Review lifestyle changes—diet, exercise, stress, smoking; reinstruct prn	— Verbalizes understanding of permanent lifestyle changes and recognizes S/S of CHF exacerbation and F/U with physician
Discharge Planning	— Assess need for home vs. outpatient cardiac rehab program	— Schedule appointment for orientation to outpatient cardiac rehab	— Administer SF-36 — Enroll in outpatient cardiac rehab program	— Verbalizes understanding of discharge plan and instructions
Signatures				

continues

Exhibit 5–A–2 continued

Criteria for Medical Social Work Referral	Criteria for Physical Therapy Referral	Criteria for HHA Referral
Increasd anxiety Depressed mood/affect Problems with support system Need for resources Financial problems	Very little or no progress made by end of week one Unable to perform ADLs due to deconditioning	No caregiver/unable to perform ADLs

Source: Copyright © Butterworth Hospital.

Exhibit 5–A–3 Report Card for Inpatient CHF Pathway

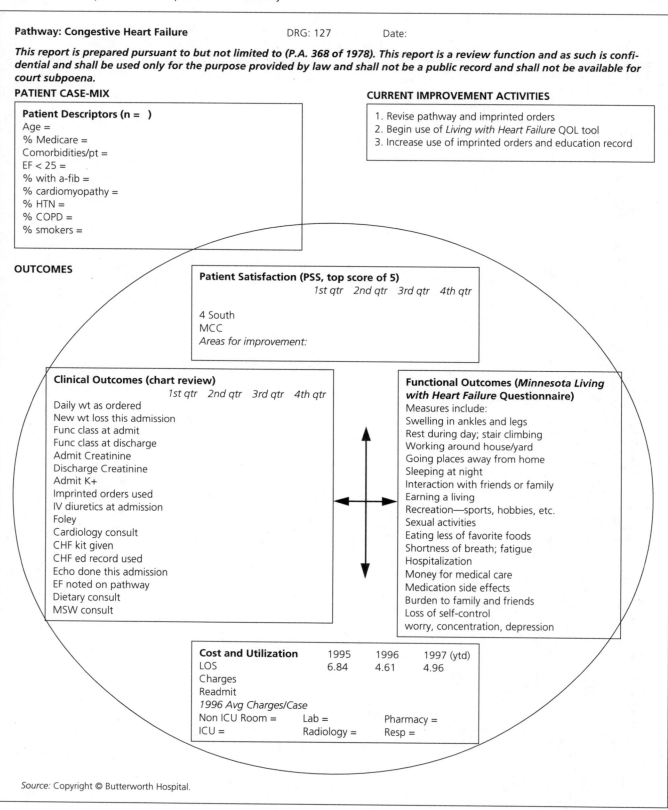

Pathway: Congestive Heart Failure   DRG: 127   Date:

This report is prepared pursuant to but not limited to (P.A. 368 of 1978). This report is a review function and as such is confidential and shall be used only for the purpose provided by law and shall not be a public record and shall not be available for court subpoena.

PATIENT CASE-MIX

Patient Descriptors (n =)
Age =
% Medicare =
Comorbidities/pt =
EF < 25 =
% with a-fib =
% cardiomyopathy =
% HTN =
% COPD =
% smokers =

CURRENT IMPROVEMENT ACTIVITIES

1. Revise pathway and imprinted orders
2. Begin use of *Living with Heart Failure* QOL tool
3. Increase use of imprinted orders and education record

OUTCOMES

Patient Satisfaction (PSS, top score of 5)
                              1st qtr   2nd qtr   3rd qtr   4th qtr

4 South
MCC
Areas for improvement:

Clinical Outcomes (chart review)
                    1st qtr   2nd qtr   3rd qtr   4th qtr
Daily wt as ordered
New wt loss this admission
Func class at admit
Func class at discharge
Admit Creatinine
Discharge Creatinine
Admit K+
Imprinted orders used
IV diuretics at admission
Foley
Cardiology consult
CHF kit given
CHF ed record used
Echo done this admission
EF noted on pathway
Dietary consult
MSW consult

Functional Outcomes (*Minnesota Living with Heart Failure* Questionnaire)
Measures include:
Swelling in ankles and legs
Rest during day; stair climbing
Working around house/yard
Going places away from home
Sleeping at night
Interaction with friends or family
Earning a living
Recreation—sports, hobbies, etc.
Sexual activities
Eating less of favorite foods
Shortness of breath; fatigue
Hospitalization
Money for medical care
Medication side effects
Burden to family and friends
Loss of self-control
worry, concentration, depression

Cost and Utilization	1995	1996	1997 (ytd)
LOS	6.84	4.61	4.96
Charges			
Readmit			

1996 Avg Charges/Case
Non ICU Room =   Lab =   Pharmacy =
ICU =   Radiology =   Resp =

Source: Copyright © Butterworth Hospital.

Exhibit 5–A–4 Report Card for Outpatient CHF Pathway

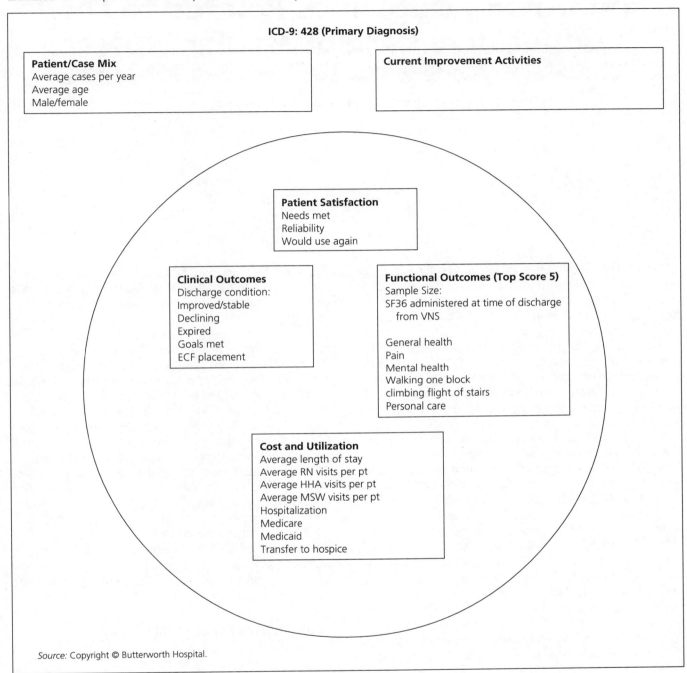

ICD-9: 428 (Primary Diagnosis)

Patient/Case Mix
Average cases per year
Average age
Male/female

Current Improvement Activities

Patient Satisfaction
Needs met
Reliability
Would use again

Clinical Outcomes
Discharge condition:
Improved/stable
Declining
Expired
Goals met
ECF placement

Functional Outcomes (Top Score 5)
Sample Size:
SF36 administered at time of discharge
    from VNS

General health
Pain
Mental health
Walking one block
climbing flight of stairs
Personal care

Cost and Utilization
Average length of stay
Average RN visits per pt
Average HHA visits per pt
Average MSW visits per pt
Hospitalization
Medicare
Medicaid
Transfer to hospice

Source: Copyright © Butterworth Hospital.

■ 6 ■

Improving Outcomes Related to the Cooperative Cardiovascular Project

Bette Keeling, Joycelyn Weaver, and Emily Murph

Methodist Medical Center (MMC) is a regional referral hospital for South Dallas. It is the larger of the two-hospital Methodist Hospitals of Dallas (Texas) system, with its highly complex services tending to be concentrated at the 467-bed trauma hospital. Referrals for these services are encouraged from the smaller community hospital further south in Duncanville. In the early 1990s the MMC established several centers of excellence, one of them being a Heart Center. The Heart Center's programs were developed with the intent of serving the high-volume, high-risk needs of patients in our primary and secondary service areas. Existing services include a cardiac catheterization laboratory, open heart surgery services, a coronary care unit, a medical telemetry unit, and a cardiac rehabilitation program. The vision is to view these departments as a seamless service, reporting through one chain of command. This design enhances our ability to provide comprehensive cardiovascular care to our customers.

In 1994, MMC formed a Clinical Path Task Force for the purpose of recommending a case management system that included developing, implementing, and evaluating clinical pathways. About 6 months earlier, the Heart Center established a Practice Improvement Committee (PIC) to design an overall project strategy to develop, implement, and evaluate a case management approach to provide care for the cardiovascular patient population. The Heart Center Practice Improvement Committee worked to create a clinical pathway grid and a variance analysis grid that could be standardized across the organization. The Heart Center PIC was composed of members representing the following areas: cardiovascular and emergency medicine physicians, Heart Center Leadership, cardiac rehabilitation, social services, nutrition services, respiratory therapy, physical medicine, and staff nurses from the intensive care unit (ICU), cardiac

care unit (CCU), telemetry units, emergency/trauma unit, and the cardiac catheterization laboratory. There were consulting members from the sister hospital from medical records and nursing.

In the fall of 1995, MMC was requested to submit 80 medical records of acute myocardial infarction (AMI) discharges that had occurred between March and November of 1994 to the Texas Medical Foundation (TMF) for review for the Cooperative Cardiovascular Project (CCP). The CCP is the first phase of the first national effort of the Health Care Quality Improvement Program.[1] The Health Care Financing Administration (HCFA) initiated the CCP as a way of reviewing the care provided to patients with AMIs. The results from the analysis of the performance data were shared with MMC staff and physicians during a feedback presentation by the TMF in November of 1995. The existence of these performance data, compiled by an outside benchmarking organization, led to the selection of DRG 122 (acute myocardial infarction, without cardiovascular complications, discharged alive) as the first clinical pathway to be implemented in the Heart Center at MMC.

HEART CENTER PRACTICE IMPROVEMENT COMMITTEE

The early meetings of the Heart Center PIC were used to educate the group about case management and the use of clinical pathways as a planning and communication tool for patient care. The education of the group included an extensive literature review, review of sample pathways from other institutions, and educational sessions with key individuals from the managed care office, utilization review, and medical records. A timeline was established by mid-June of 1994, and the core group began chart reviews to develop the first clinical pathway, acute myocardial infarction, DRG 122.

By mid-July, objectives for the PIC were established and the chart review was ongoing. By August, the housewide clinical pathway grid was complete and tailored to meet the needs of the cardiovascular patient population. By mid-August, the variance report grid was formulated and accepted by the core group. Between mid-August and mid-September, the results of the chart review were applied to the clinical path grid to best represent the current cardiology practice and adhere to the DRG guidelines for desirable length of stay (LOS). During this time, regular one-on-one sessions with the cardiologists were conducted to inform them of the progress of the group and obtain their input on the content of the clinical path. The physician group was supportive and willing to share their expertise. The completed draft for the AMI clinical path was presented to the CCU/MICU Committee in August and was approved for pilot.

The first clinical pathway was piloted for 3 months beginning October 3, 1994. There was a periodic monthly scheduled time to meet with the cardiology group to maintain communication with the progress of the clinical path component of the Case Management Initiative in the Heart Center. Concurrently, the development of the next clinical pathways continued as planned.

During the last week of November 1995, the TMF, along with a local cardiologist, presented the results of the acute myocardial infarction chart review for the 10 AMI Indicators. These results came from the medical records that had been requested earlier. Using the results of the chart audit report on the CCP's AMI Indicators, a focused educational effort was made during the first quarter of 1996 to share these findings with the Heart Center personnel, cardiology section, emergency trauma services, PIC, and inpatient reengineering teams. It was the goal of the Heart Center to actively pursue quality improvement by using this external benchmarking data to assess our ongoing performance in complying with the indicators.

DEVELOPMENT OF THE CLINICAL PATHWAY

In the absence of a cost analysis system, medical records and patient bills for these records were manually reviewed to determine trends in practice and trends in charging. The analysis of these documents rendered a baseline for delivering care to a patient with AMI and for the average costs associated with that care.

The greatest driving factors in developing the AMI clinical pathway were to incorporate practices to improve compliance with the 10 CCP indicators and lower costs of care delivery while maintaining or improving the quality of care that was being delivered. The multidisciplinary meetings were critical in ensuring that all the disciplines that were involved in delivering care were represented and that their standards of practice were incorporated into the clinical pathway. These meetings had a secondary advantage of improving communication between the disciplines. For example, many of the staff nurses were unaware of the magnitude of work required for the social workers to arrange for durable medical equipment to be delivered to a patient's home.

The MMC is a conservative organization by nature, and PIC members decided to incorporate existing practice guidelines, preprinted order sets, and programs into the AMI pathway. MMC's respiratory therapy department had sponsored a smoking cessation program for many years, which received a limited number of patient referrals. Incorporating an informal respiratory therapy assessment for smoking cessation into the AMI clinical pathway would offer an opportunity to reinforce with nursing staff that such a program existed and offer the respiratory therapy department the opportunity to achieve better utilization of an existing program.

Preprinted order sets existed for many different physicians. All of the AMI sets were reviewed and trends and outliers were identified. Physicians and nurses reviewed the trends and outliers and determined which ones needed to be incorporated into the clinical pathway.

The respiratory therapy department had developed and implemented a therapist-assisted protocol (TAP), that had been developed in conjunction with the pulmonologists. With preestablished parameters for oxygenation and ventilation, these protocols provide the respiratory therapists with greater autonomy and flexibility in delivering care. Since the therapist is no longer dependent upon the physician for changing flow rates, based on the parameters, necessary changes to care delivery can occur much more rapidly than before the implementation of the clinical pathways with the TAPs incorporated.

Including pastoral care, social workers, and nutritionists in the multidisciplinary team provided an opportunity for improved communication regarding the variety of services that each specialty offers. In addition, they got to know each other as coworkers, working together rather than working around each other in the clinical setting. For all of the disciplines represented, educating each other about work processes proved to enhance the development of the pathways.

STANDARDIZATION OF THE CLINICAL PATHWAY AND VARIANCE REPORT GRIDS

Since the clinical pathway and variance report grids were intended to be used on both campuses of the hospital system, there had to be agreement on certain items. The first area of agreement was to call them "clinical pathways" instead of "critical pathways." There was a concern that labeling them as "critical" might be construed as being "critical to the patient's care" if litigation involving a pathway was to arise. Then how would variances be defended in a court of law? There was no case law in which pathways had been involved at the time the pathways were being developed to guide us. So we went for the more conservative title. In another conservative decision, the pathway was determined to be a tool for quality improvement and, as such, would not remain a permanent part of the patient's chart, nor would it require a physician's order for implementation.

After reviewing the literature and samples of pathways from other organizations, trends in the types of categories that were being included on the pathways emerged. The PIC agreed that there should be 11 categories of interventions, which would be consistent on all clinical pathways. The categories of interventions, which appear in the left-most column of all clinical pathways at MMC, are listed and defined in Exhibit 6–1.

The members of the PIC also recognized the need for a consistent appearance for the pathways. If the general presentation of all of the pathways was similar, the learning time that staff would need when new pathways were introduced would be minimized. The consistent presentation that was selected for the clinical pathway grid and the variance report grid are presented in Exhibits 6–A–1 and 6–A–2 (in Appendix 6–A), respectively.

The PIC recognized that there were diagnoses that would not fit well within a day 1, day 2, day 3 format (such as the phases of labor), and that even today's LOSs may be shortened. In that spirit, the columns to the right of the categories of interventions are simply labeled with Date and Unit (see Exhibit 6–A–1). In this way, phases of labor or congestive heart failure will also fit within the standardized format. The dashed lines in the teaching section of the grid simply serve to separate the documentation that will be done by the different disciplines, with regards to patient/family education. Solid lines are used to separate major categories on the clinical pathway.

The Clinical Path Variance Report (Exhibit 6–A–2) includes a legend that guides staff in deciding what constitutes a variance. If any of those items occur, they are to write them down, as well as what caused the variance and what action was taken, and the individual who is documenting the variance signs the appropriate area. The signature is necessary if follow-up is needed for clarification at a later time. Coding the variances into groups enables us to see where we might make further improvements in delivering care.

CASE MANAGEMENT

Most of the clinical areas at MMC have clinical nurse specialists (CNS) or clinicians assigned to them. These employees were designated with the responsibility for reviewing existing case management systems and recommending a system for MMC. After several months of lit-

Exhibit 6–1 Methodist Medical Center's Clinical Pathway Components and Definitions

Component	Definition or Example
Outcomes	Desirable outcomes for each phase are listed and the critical outcomes (those items that are determined necessary to progress) are noted with an asterisk.
Assessment/monitoring	Intake and output, weight, vital signs, height, calorie count, cardiac monitor, and pulse oximetry.
Consults	Physician, social work, nutrition, pharmacy, pastoral care, physical therapy, occupational therapy, cardiac rehabilitation. Notification of ancillary team members is for communication purposes and does not require a physician order.
Tests	X-rays, laboratory tests, echocardiograms, bone marrow biopsy, and electrocardiograms.
Treatment	Respiratory therapy, physical therapy, occupational therapy, oxygen, heparin lock, central line placement, tube placement, Foley catheter insertions.
Meds	Intravenous (IV) fluids, oral, intramuscular, and IV meds.
Nutrition	Diet, snacks, supplements, and tube feedings.
Activity	As tolerated, bed rest, and bathroom privileges.
Teaching	Self-injection, diet, medications.
Discharge Planning	Home health, hospice, skilled care, spiritual needs.
Other	May include needs such as pastoral care, Mended Hearts Volunteer visit.

Source: Copyright © Methodist Hospitals of Dallas.

erature review and networking with peers in other organizations, a recommendation was made to create a model for case management at MMC.

The model designates two levels of cases: patternable and unpatternable. Patternable (predictable) cases consist of patients with diagnoses for which a specific set of guidelines has been established (such as the AMI clinical pathway). Unpatternable (unpredictable) cases consist of patients with diagnoses for which a set of guidelines either has not been or cannot be established (such as a multiple trauma victim). The staff nurses were designated to be the "case managers" for patternable cases. In this role, the bedside nurse becomes responsible for the day-to-day management of the delivery of care and monitoring of outcomes and variances. The CNSs and clinicians were designated as consultants to the staff for unpatternable cases. In addition, the CNSs and clinicians were designated with the responsibility for retrospectively analyzing outcomes and variances and presenting these findings to the appropriate individuals or committees.

At the time of admission, the admitting nurse assesses the patient and determines eligibility for being placed on a clinical pathway. To facilitate this process, a criteria page for including the patient in the pathway or excluding the patient from the pathway was initiated in early 1997 based on staff feedback. The average staff nurse was not aware of the criteria that would exclude a patient from being placed on a clinical pathway. Exhibit 6–2 presents the criteria page for the AMI clinical pathway.

Once the patient is assessed and determined to meet the inclusion criteria, the staff nurse is responsible for initiating the clinical pathway and notifying the appropriate disciplines if informal assessments are needed. The clinical pathway is placed within the patient's bedside chart in most areas of the hospital. This way, it is available for all disciplines to document and initial as care is being delivered. Each discipline is responsible to document their interventions on the clinical pathway grid and to note variances on the variance report grid. Because the clinical pathway and variance report grids are not a permanent part of the chart, the staff do have to document their interventions both in the chart and on the clinical pathway.

The clinical pathway is reviewed during change of shift report and at the time a patient is transferred between units. The staff nurse is responsible for reviewing and documenting the progression of nursing care as it occurs. The CNSs make rounds periodically to assess for any problems with which the staff may need assistance and for the ongoing monitoring of the unpatternable cases. Physicians refer to the clinical pathways to assess progress, but seldom document on them.

Clinical pathways are reviewed for revisions approximately 3 months after implementation (what we con-

Exhibit 6–2 Inclusion/Exclusion Criteria Page for AMI Clinical Pathway

---

**Methodist Medical Center
Acute Myocardial Infarction, w/o Cardiovascular
Complications, Discharged Alive Clinical Pathway
DRG 122
Target LOS 5.0 Days**

Criteria for Initiation of Pathway:
- Any patient received to CCU/ICU with a confirmed diagnosis of AMI
- CCU/ICU—any patient with an admitting diagnosis of "chest pain, R/O MI"
- Telemetry—do not place patient on AMI pathway until diagnosis of AMI is confirmed

Exclusion Criteria:
- Any patient with cardiopulmonary complications including:
  - cardiopulmonary arrest
  - severe uncontrolled hypertension
  - congestive heart failure (severe)
  - cardiogenic shock
  - multiple pharmacologic drips
  - IABP
  - pulmonary artery catheter
  - ventilator

Criteria for Discontinuation of Pathway:
- Comorbidity becomes predominant disease process

Discharge Indicators:
- Meets expected outcomes for discharge

This form is not a permanent part of the medical record.

This Clinical Path is *not* intended to set or reflect the prevailing standard of care; rather it is intended as a tool to be used for data collection in an effort to improve efficiency and consistency. It does not replace the independent judgment of experienced medical practitioners applying their knowledge and expertise to a specific case. This path is *not* intended to take the place of the physician's orders and may be modified according to individual patient needs.

Source: Copyright © Methodist Hospitals of Dallas.

---

sider a pilot) and then every 6 or 12 months thereafter, depending upon the review cycle of the department. Minimally, they are reviewed annually along with practice guidelines for revisions. The CNSs and clinicians are responsible for initiating and completing the reviews. The initial revisions to the clinical pathways were occasionally as simple as providing a box for employees to check that the task had been completed. The logistics of using the clinical pathway have continued to be the driving force behind revisions. There have been few revisions for content. The one exception was to clearly reflect the 10 CCP indicators in the AMI clinical pathway.

Exhibit 6–A–3 in the Appendix presents the AMI clinical pathway. The "Discharged to" portion has recently been added so we can begin tracking where our patients go when they leave us and determine whether there is a way to begin tracking their wellness progress after they are discharged. MMC is in a preferred provider agreement for home care service with the Visiting Nurse Association (VNA). We now have an in-house contact person for VNA, so we can easily track any discharged patients who are being followed by VNA.

Education of the staff, physicians, and residents has been key to the successes that have been achieved with the clinical pathways. One of the concerns of the physicians was that if a procedure was on the pathway someone would call them and ask whether or not they wanted the procedure ordered. In an attempt to prevent this type of knee-jerk response, the idea was reinforced among staff that the actions and indicators within the clinical pathways do not necessarily mean that it is done on all patients with that diagnosis, rather they are interventions to be considered in the care of those patients. If the patient is not eligible, then it should not be done. The development of the inclusion and exclusion criteria furthered this effort as well.

OUTCOMES

The Heart Center PIC assumed the responsibility of tracking and measuring patient outcomes. Monitoring every item in the clinical pathway would be too involved. As a result, the PIC selected outcomes that would be most meaningful to MMC and to clinical practice. These outcomes include the ten CCP indicators, LOS, cost per case, and Phase I Cardiac Rehabilitation referrals. Patient, staff, and physician satisfaction were not selected to be monitored during the implementation of the DRG 122 clinical pathway.

Outcomes Related to the CCP Indicators

Monitoring outcomes on the AMI pathway centers around the CCP indicators. This is our primary focus because we are required to respond to the initial report. Failure to be in compliance in the future may affect our reimbursement. Every 6 months, the cardiovascular services manager reviews all of the AMI cases for the period and prepares and submits a report to TMF, where the initial audit on the CCP indicators indicated acceptable compliance to the proposed benchmark on only one indicator, which was administration of aspirin during the hospital stay (Exhibit 6–3). That left eight areas of opportunities for improvement. Staff were inserviced regarding the indicators and the first content revision of the AMI clinical pathway was completed. These actions led to a significant improvement in compliance with the other indicators.

Even though there continued to be changes to MMC's compliance into the first half of 1996, we remained higher than the national average on all indicators except one (administration of thrombolytics in 1 hour). We continue to educate and reinforce with staff and physicians the significance of compliance with these indicators to patient outcomes and to the financial livelihood of the organization.

Outcomes Related to Length of Stay

One of the natural components of clinical pathway implementation is to strive for consistency in LOS in patients with the same diagnosis. This is one of the cost-reducing benefits of clinical pathways and one of the areas where we can demonstrate that we have achieved a high level of success. Our benchmark LOS in 1994 was 6.2 days for our AMI patients. Each year since implementation of the AMI clinical pathway, the LOS has been reduced until we reached the current 4.4 days. See Figure 6–1 for the reductions that have been achieved in LOS with the implementation of the AMI clincal pathway. If LOS is a reflection of the organization's efficiency in providing care to a specific population of patients, then MMC's system has gained efficiency in providing care for AMI patients.[2]

Outcomes Related to Charges per Case

Charges per case have also been tracked as outcomes for the AMI clinical pathway. Since the clinical pathway was implemented in late 1994, we have seen variation in our average charges per case. We noted an increase in charges in 1995. During chart review, we noted that many patients had received a new thrombolytic, Riopro. Upon investigation and consultation with the hospital's pharmacy department, we learned that physicians who were continuing to use Ticlid and aspirin combination were having similar outcomes as were the physicians who were using Riopro, but at significant savings. These findings were presented to the PIC as well as to the medical staff section meeting. Charges per case have declined since 1995, along with a reduction in use of Riopro. Figure 6–2 demonstrates the reduction in average charges per case that we have experienced since implementing the AMI clinical pathway.

Outcomes Related to Cardiac Rehabilitation Referrals

Early in the implementation of the AMI clinical pathway, we noted consistent 48-hour to 72-hour delays in getting the AMI patient population referred to cardiac

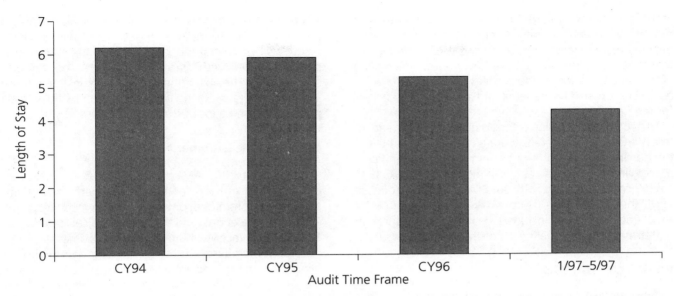

Figure 6–1 Changes in Average Length of Stay per Case with Implementation of Clinical Pathways. *Source:* Copyright © Methodist Hospitals of Dallas.

rehabilitation. Upon exploration of this variation, we found that staff nurses were waiting until the physician wrote orders for the referral before it was made. Cardiac rehabilitation orders were seldom written within the first 2 days of hospitalization. The nursing staff explored the idea of nursing referrals to cardiac rehabilitation with the physician staff. The intent would be to have the cardiac rehabilitation staff interview and assess the patient's readiness for Phase I Cardiac Rehabilitation and then con-

tact the physician with the findings and recommendation of the assessment. Orders for initiation of cardiac rehabilitation could then be received. Within 6 months of the implementation of nursing referrals to cardiac rehabilitation, there was a 200% increase in the number of referrals to Phase I Cardiac Rehabilitation, as compared to the prior 6 months' number of referrals. In addition, there were only 12 different physicians referring to cardiac rehabilitation during the 6 months preceding the imple-

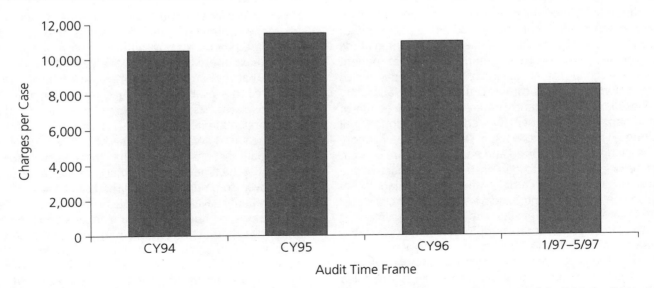

Figure 6–2 Changes in Average Charges per Case with Implementation of Clinical Pathways. *Source:* Copyright © Methodist Hospitals of Dallas.

mentation of nursing referrals. In the subsequent 6 months, the number of different physicians referring AMI patients to cardiac rehabilitation rose to 20. We believe that this represents an increased awareness among the physician population of the services that are available to rehabilitate their AMI patients and also improved access for the patients.

The concept of nursing referrals gained widespread interest and additional areas were added. We also believe the nursing referrals brought about the most significant process improvement to aid us in moving from a 43% to a 63% compliance rate with smoking cessation counseling (Exhibit 6–3). Nursing referrals to the pastoral care department were also included to create a mechanism to better meet the spiritual needs of the AMI patients.

VARIANCE ANALYSIS

Opportunities to improve in the area of variance analysis exist. We are finding that with a manual charting format, charting variances is a low priority when delivering patient care. Compliance with documenting variances has been poor. MMC is pursuing an opportunity to develop an on-line documentation of variances through its clinical information system. One of the successes is the realization that we need to monitor our variations and explore things that cause an increase in the length of stay. Variances are not routinely monitored at this time, but are extremely helpful when they are documented.

For example, the utilization review staff reported an increase in middle of stay denials for payment. Upon investigating the cases of two patients, who seemed to have similar presenting criteria and treatment courses, yet had significantly different length of stays, research into their cases demonstrated that one was admitted at the end of the week whereas the second patient was admitted at the beginning of the week. The end of the week patient needed a cardiac catheterization on the second day, which was Saturday. MMC's cardiac catheterization laboratory is closed on Saturday and Sunday with the exception of emergencies. The patient waited in the hospital until Monday or Tuesday to have a cardiac catheterization. This variance was reported and enabled the cardiovascular services manager to demonstrate the impact of the decision to have the cardiac catheterization laboratory closed on Saturdays and Sundays to this patient, the cost of his care, his length of stay, and our reimbursement.

The most significant process improvement noted in the area of variance analysis is a change in physician behavior. Physicians will now call the cardiovascular services manager when they cannot get their patients scheduled into the cardiac catheterization laboratory on day 2, as recommended by the clinical pathway. This is a win be-

cause we are able to schedule additional staff or rearrange schedules to be able to get their patients scheduled. This demonstrates a willingness on the part of physicians to work with us to lower lengths of stay and reduce costs. The absence of a system to deal with these impediments prior to the implementation of the clinical pathway caused a lot of frustration for our physicians.

Outcomes Measurement

Outcomes for the AMI clinical pathway are measured and reported every 6 months to the Performance Improvement Committee, physician section meetings, nursing staffing meetings, and the Texas Medical Foundation. These groups receive and review the outcomes data. Occasionally, members of these committees assist in identifying, designing, and implementing process improvements that will help us continue to improve our delivery of care systems.

The outcomes that were available for the AMI clinical pathway have been presented in Exhibit 6–3 and Figures 6–1 and 6–2. Calendar year 1994 is considered our internal benchmark, even though we piloted in October of 1994. During the last 3 months of 1994, we were much more focused on teaching the staff how to use the clinical pathways than we were on the outcomes that occurred during that time. The Texas Medication Foundation audit of the CCP Indicators provides us with an external benchmark that we do not have with any of the other outcomes that we monitor.

Advantages to Patient Care Delivery as a Result of Clinical Pathway Implementation

The increased awareness of the bedside caregiver to initiate the informal consults has been the biggest advantage to the patient. As a result of the informal consults, we decrease the wait time for the patients to receive services that they are entitled to receive. The inclusion of references to available patient teaching materials in the clinical pathway has made us more efficient in providing educational materials to patients.

The increased awareness of the bedside caregivers to the special functions that other health care professionals carry out at the time of the informal consults has resulted in improved communication between caregivers and patients. Patients have anecdotally expressed that they feel that all of the members of the team are working together to improve their health situation. There is an improved collegiality and rapport between the professionals at the bedside as well. The use of the clinical pathways during nurse-to-nurse report at transfer or change of shift has resulted in improved continuity of care, which is demonstrated by the reductions in LOS and charges per case.

Exhibit 6–3 MMC's Compliance with the Cooperative Cardiovascular Project AMI Indicators Across Time

CCP's AMI Indicators	Hospital % 4/1–11/30/94 (TMF Audit)	Hospital % 7/1–12/31/95 (MMC Audit)	Hospital % 1/1–6/30/96 (MMC Audit)	National %	Benchmark %
Aspirin during stay	100%	100%	100%	84%	100%
Aspiring on day 1	62%	96%	90%	57%	100%
Reperfusion attempted	75%	100%	100%	58%	100%
Thrombolytic in 1 hour	33%	83%	57%	67%	70%
ACE inhibitors for low LVEF	75%	84%	81%	55%	100%
Aspirin at discharge	85%	96%	100%	78%	100%
Beta blockers at discharge	33%	76%	100%	57%	100%
Calcium channel blockers withheld for low LVEF	0%	83%	87%	82%	100%
Smoking cessation counseling	43%	75%	63%	39%	100%

Legend: LVEF = Left Ventricular Ejection Fraction; TMF = Texas Medical Foundation; MMC = Methodist Medical Center

Source: Copyright © Methodist Hospitals of Dallas.

Disadvantages to Patient Care Delivery as a Result of Clinical Pathway Implementation

Staff currently place a low priority on the clinical pathways. It is one more form that needs to be completed—and it is not a permanent part of the medical record. Staff report that it is inconvenient to double document onto two different manual forms. Another area for improvement is providing timely feedback to staff regarding patient outcomes. Placing story boards on the units that are involved with clinical pathways is one way that we are attempting to provide earlier and more consistent feedback to staff and physicians.

Acceptance of the Clinical Pathways

Acceptance has been consistent with what we expected. Even though they participated in the development of the AMI clinical pathway, the physicians do not use it to drive patient care. They are very interested in the results of the AMI clinical pathway when the outcomes are shared. Similarly, our staff responded as we would expect. Several staff initially resisted the change. A small core group of staff gave it their best and continue to give it their best. Then there are some who continue to resist. For them the pathway does not hold value because they do not directly benefit from the documentation or monitoring. There is no reward or penalty for compliance with use of the clinical

pathway at this time. Efforts are directed toward giving immediate and positive feedback regarding use of the clinical pathways. Over time, these efforts pay off. All of a sudden, documentation will appear on the clinical pathway from someone who may have been a resister earlier.

Automation of Clinical Pathways

MMC is involved in the development and implementation of a comprehensive clinical information system. The first components to be developed will revolve around making improvements in the process of order entry. Modules that will follow will include assessment, medication administration, and, lastly, clinical pathways. It will be several years before clinical pathways are fully automated at Methodist Medical Center. By that time, pathways will have been developed for most diagnoses that are patternable, and the on-line database of clinical pathways that staff will be able to choose from will be large.

SHARING OF CLINICAL PATHWAYS WITH PATIENTS

At this time, sharing of clinical pathways with patients is left to the judgment of the nurse. We have not yet developed patient versions of clinical pathways. Patient versions, written in lay language and easier to understand than the current versions of the pathways, are being developed for the near future.

PLANS FOR THE FUTURE

We will continue to monitor the CCP indicators, Phase I Cardiac Rehabilitation Referrals, LOS, and charges per case. We are planning to monitor the number of days from admission to cardiac catheterization laboratory and then the number of days from cardiac catheterization laboratory to discharge. The intent is to monitor the efficiency with which we make the diagnosis, to determine whether the patient needs a cardiac catheterization, and to get the patient into the cardiac catheterization laboratory. This will help us determine if it takes the same length of time for a patient who is admitted by a family practice or internal medicine physician instead of a cardiologist to get diagnosed and into the cardiac catheterization laboratory—and if it does not, what can we do to improve the situation?

Other types of analysis we would like to conduct concern the variances that are captured. However, we have to improve the documentation of variances in order to do that. If we were documenting variances, could we make even greater gains in patient care delivery? We simply do not know the answer to this question yet.

RESPONSES TO UNACCEPTABLE OUTCOMES OR VARIANCES

The first challenge is to define "unacceptable." At MMC, we have routinely defined it as "a trend going in an undesirable direction." The best example of this would be previous situation shared under the discussion of average charges per case in which an increase in average charges per case resulted when Riopro was being used as the choice pharmacologic agent in 1995. In this case, we did not make physicians change their practice. We simply provided them with information that enabled them to make better, more cost-effective choices for their patients.

As we improve with our documentation of variances, we may find that a different definition of "unacceptable" may be needed. Not all variances are unacceptable. In the case of the AMI patient who is not a candidate for an echocardiogram, the absence of an echocardiogram is an acceptable variance.

FEEDBACK REGARDING THE CLINICAL PATHWAYS

Physicians and staff continue to voice an interest in the results of the AMI clinical pathway outcomes. From the improvements in the CCP Indicators, Phase I Cardiac Rehabilitation Referrals, LOS, and charges per case that we have seen over time, we know that they are listening to

this information, assimilating it, and incorporating it into their patient care delivery at their own pace. The administrative team at MMC has discussed with staff the advantages of having quantifiable information that can be measured repeatedly to monitor for improvements when changes in services are made.

The managed care department has reported that the information from the clinical pathways is used in contract negotiation. When negotiating with a prominent managed care company, the vice-president of managed care shared copies of the clinical pathways to demonstrate to potential reimbursers that we do have a mechanism in place to lower our costs and improve patient outcomes. This improves our ability to compete with other health care organizations.

The Texas Medical Foundation has clarified the definitions of the CCP indicators based on the feedback from individuals who are abstracting charts. In 1996, they developed a mechanism for automating the abstraction tool, which will hopefully increase the efficiency of reporting compliance with the indicators.

Outcomes assessment and management is important to tell us where we've been and where we need to go. With this sense of direction comes a mechanism for providing feedback against the plan. This provides consistent reinforcement that we are headed in the right direction. Quantification of outcomes has provided a universal dialogue between clinical teams of care providers and executive teams and managed care companies.

LESSONS LEARNED: WHAT WOULD WE HAVE DONE DIFFERENTLY?

We have learned that we should have educated the educators a little more thoroughly. The individuals involved in the development and implementation of the clinical pathways were clinicians. They were not well-educated regarding the health care market and managed care. The growth these individuals experienced as they have learned more about the market, the differences in managed care plans, and the available financial and charging mechanisms has occurred slowly. There is a feeling by some that this type of education would have enhanced the implementation of clinical pathways, resulting in earlier and larger improvements.

Also, we would have explored incorporating incentives for compliance into staff performance appraisals and physician credentialing. Even though we have achieved great outcomes and we have compliance in many areas, we feel that it could be better if there were incentives for compliance.

CONCLUSION

With the monumental changes that have occurred in the way health care is being provided, implementing and monitoring clinical pathways provides a mechanism for measuring outcomes from irrefutable sources. Quantifying the efficiencies of a care delivery system increases the likelihood that the decision makers will listen.

The CCP Indicators are intended to provide a standard of care for AMI patients across the country. All AMI patients should receive the same quality of care regardless of the geographic location in which the AMI is experienced. The CCP Indicators attempt to assure patients of the consistency with which care will be delivered for patients admitted to hospitals with AMI. The CCP Indicators also provide a mechanism to determine how the organization is performing against itself and against other organizations. This nationwide benchmarking becomes critical as capitation becomes more common.

The overall purpose of clinical pathways is to provide a mechanism to coordinate care and to reduce fragmentation and, ultimately, cost. We have demonstrated that we have achieved gains in these areas as a result of implementing the AMI clinical pathway.

REFERENCES

1. Vogel R. HCFA's cooperative cardiovascular project: a nationwide quality assessment of acute myocardial infarction. *Clin Cardiol.* 1994;17:354–356.

2. Flarey DL, Blancett SS, eds. *Handbook of Nursing Case Management: Health Care Delivery in a World of Managed Care.* Gaithersburg, MD: Aspen Publishers Inc; 1996.

■ Appendix 6–A ■
Clinical Pathways and Variance Reports

Exhibit 6–A–1 Methodist Medical Center's Standardized Clinical Pathway Format

Admitting Physician Service:	Consult(s):	METHODIST MEDICAL CENTER
Cardiology ____	Cardiology ____	CLINICAL PATH
Int. Med. ____	Int. Med. ____	
Family Practice ____	Renal ____	DRG ____
Teaching ____	Other ____	
Clinics ____		Target LOS: ____

Addressograph

Admit Date ____　D/C Date ____

23 Hour Observation　☐ Yes　☐ No

	Date: ____ Unit: ____	Date: ____ Unit: ____	Date: ____ Unit: ____	Date: ____ Unit: ____
Outcomes				
Assessment/Monitoring				
Consults as Needed				
Tests				
Treatments				
Medications				
Diet				
Activity				
Teaching (Nursing)				
Social Work				

continues

Exhibit 6–A–1 continued

	Date: _____ Unit: _____	Date: _____ Unit: _____	Date: _____ Unit: _____
Nutrition			
Respiratory Therapy			
Other			
Discharge Planning			
Signatures			

Source: Copyright © Methodist Hospitals of Dallas.

Exhibit 6–A–2 Methodist Medical Center's Standardized Variance Report

(Addressograph)

**METHODIST MEDICAL CENTER
CLINICAL PATH VARIANCE REPORT**

DRG _____

DATE	UNIT	VARIANCE CODE	CAUSE	ACTION	SIGNATURE	CODE VARIANCE	CODE CAUSE	DEPT

continues

Exhibit 6–A–2 continued

DATE	UNIT	VARIANCE CODE	CAUSE	ACTION	SIGNATURE	CODE VARIANCE	CODE CAUSE	DEPT

For Review Use Only

VARIANCE CODE LEGEND

1. Internal (Institution)
   a) Appropriate bed not available
   b) Department overbooked
   c) System delays: dept. closed, transportation not available
   d) Equipment broken
   e) Turnaround time for equipment or test results
   f) Staffing
   g) Other

2. Clinician/Physician
   a) Caregiver omission
   b) Delay in or inadequate teaching or DC planning
   c) Health care team decision
   d) Physician availability
   e) Other

3. Patient/Family
   a) Clinical condition—list cause, ie, arrest, arrhythmias, infection, abnormal labs, etc.
   b) Lack of family/social support (Adult Protective Services—APS)
   c) Cultural/ethical considerations
   d) Family not available and/or delayed for transportation
   e) Other

4. External (Community)
   a) Reimbursement not available—Inadequate funding for services needed
   b) System delays: ambulance, personnel, nursing home/ECF not available
   c) Inadequate social support system
   d) Extended care not available
   e) Lack of family caretaker at home
   f) Other

Source: Copyright © Methodist Hospitals of Dallas.

Exhibit 6–A–3 Acute Myocardial Infarction Clinical Pathway

Admitting Physician Service: **Consult(s):**

Cardiology _____   Cardiology _____

Int. Med. _____   Int. Med. _____

Family Practice _____   Renal _____

Teaching _____   Other _____

Clinics _____

METHODIST MEDICAL CENTER
CLINICAL PATH

Addressograph

DRG 122 AMI, w/o C.V. Complications, Discharged Alive

Target LOS: 5.0 _____

Admit Date _____   D/C Date _____   **23 Hour Observation** ☐ Yes ☐ No

* = Critical Outcome Criteria for CCP Indicators

Page 1	Date: _____ Unit: _____	Date: _____ Unit: _____	Date: _____ Unit: _____
Outcomes	☐ Life-threatening arrhythmias controlled ☐ V/S stabilized	☐ Patient tolerating diet ☐ CPK trending down ☐ Patient free of chest pain or angina controlled ☐ Expected clinical progress (pathway) explained to patient/family	☐ Decision for medical management vs. invasive procedures made ☐ CPK trending down ☐ ST segments returning or back to baseline ☐ Progress to monitored step-down unit
Assessment/Monitoring	☐ Advanced directives on chart _____ ☐ V/S q _____ ☐ Admission Ht & Wt _____ ☐ Cardio-respiratory assessment: Is patient a smoker? ☐ Yes ☐ No *If yes, notify respiratory therapy to assess for smoke-stoppers program* ☐ I & O	☐ Daily wt _____ ☐ Assessment q _____ ☐ I & O	☐ Daily weight _____ ☐ Assessment q _____ ☐ I & O ☐ Transfers to telemetry _____ date ☐ Clinical pathway reviewed at time of transfer report _____ initial
Consults as Needed	☐ Cardiologist: Initial visit _____ date _____ time ☐ Case Manager notified ☐ Social Work notified ☐ Cardiac Rehab (for pathway only) ☐ Nutrition services notified ☐ Pastoral care as appropriate	☐ Assessed by Respiratory Therapy for smoke stoppers program _____ Patient Visited/Evaluated by: • Case Manager _____ initial _____ date • Social Work _____ initial _____ date • Cardiac Rehab _____ initial _____ date • Nutrition _____ initial _____ date • Pastoral Care _____ initial _____ date	
Tests	☐ CPK ☐ LDH ☐ PT ☐ SGOT ☐ PTT ☐ FBP ☐ TSH ☐ CXR ☐ **CBC** ☐ EKG ☐ Other _____ ☐ Consider Cardiac Cath _____ Date: _____ ☐ **Echocardiogram**	☐ Complete CPK series ☐ EKG ☐ Daily PTT or q _____ hours. *Confirmation of dx. of AMI documented by: ☐ ↑ Enzymes ☐ AMI on EKG ☐ Unresolved pain ☐ Echocardiogram	☐ Daily PTT (if on heparin) or q _____ hours
Treatments	☐ O₂ at _____ ☐ Pulse oximetry initial: _____ ☐ Other: _____ _____ initial:		

continues

Exhibit 6-A-3 continued

Page 2	Date: _____ Unit: _____	Date: _____ Unit: _____	Date: _____ Unit: _____

Medications	☐ IV Fluids as ordered: _____
	  Assess need for:
	  *Thrombolytic therapy _____
	  *Time initiated: _____
	☐ Cardiac cath _____
	☐ *ASA _____
	☐ *Anticoagulants (heparin) _____
	☐ Nitrates _____
	☐ Cardiac meds _____
	☐ Beta blockers _____
	☐ Analgesics _____
	☐ Antiarrhythmics _____
	☐ *Ca channel blockers ordered.
	  ☐ Yes ☐ No
	  (avoid in pts w/low LVEF (40%))
		☐ IV Fluids as ordered. _____
		☐ ASA _____
		☐ Anticoagulants _____
		☐ Nitrates _____
		☐ Cardiac meds _____
		☐ Beta blockers _____
		☐ Analgesics _____
		☐ Antiarrhythmics _____
		☐ Ca channel blockers _____
			☐ Heparin Lock _____
			☐ D/C IV _____
			☐ Anticoagulants _____
			☐ Topical nitrates _____
			☐ Antiarrhythmics _____
			☐ Cardiac meds _____
			☐ Other _____

Diet	☐ NPO ☐ Liquids	☐ Low cholesterol/no added salt OR ☐ Diet as appropriate (specify) _____	
Activity	☐ Complete bedrest ☐ Space out ADLs	☐ Assist with ADLs	☐ ↑ Activity as tolerated ☐ Cardiac Rehab per protocol ☐ Cardiac Rehab not ordered

Teaching (Nursing)	☐ Orient patient/family to:
	  ☐ Staff ☐ Room ☐ Equipment
	☐ Education regarding treatment plan (including thrombolytic therapy)
	  Init. _____
	☐ Complete needs assessment of patient/family/caregiver
		☐ MI booklet reviewed with patient/family
		  Init. _____
			☐ Orient to telemetry staff/unit
			☐ Cath/PTCA booklet given if applicable

Social Work	As appropriate: ☐ Caregiver support ☐ Financial ☐ Psychosocial		
Nutrition	☐ To be screened or assessed per nutritional services guidelines Init. _____		
Respiratory Therapy	☐ As appropriate.	*Smoker stopper if applicable _____	
Other	☐ Psychosocial: Pastoral care as appropriate.		
Discharge Planning	☐ Problem list reviewed—needs assessed ☐ Appropriate referrals for high risk patients ☐ Discharge needs discussed with family (*note on DPR)		
Signatures			

continues

Exhibit 6–A–3 continued

Admit Date _____ D/C Date _____

Page 3	Date: _____ Unit: _____	Date: _____ Unit: _____	Date: _____ Unit: _____
Outcomes	☐ Patient/family verbalize understanding of disease process, treatment, and therapy ☐ Cardiac rhythm stabilized ☐ Patient/family state understanding of medication management	☐ Patient/family verbalize understanding of discharge instructions and follow-up care ☐ CCP indicators addressed in medical record documentation	
Assessment/Monitoring	☐ Daily weight ☐ Assessment q _____	☐ Daily weight	
Consults as Needed			
Tests	*Possible Stress Test Date of stress test _____ (especially if no Cardiac Cath)		
Treatments			
Medications	☐ Heparin lock OR ☐ IV Fluids ☐ Other: _____		☐ D/C Heparin lock Date _____ Initial _____
Diet			
Activity	☐ Ambulating	☐ Ambulating	
Teaching (Nursing)	☐ Reinforce previous medication teaching with patient/family _____ initial ☐ Review home meds _____ initial ☐ Follow-up appointment ☐ Understood ☐ Scheduled ☐ Discharge instructions	☐ Reinforce MI/Cardiac Rehab teaching _____ ☐ Review home meds/prescriptions • ASA included ☐ Yes ☐ No • Beta blockers included ☐ Yes ☐ No ☐ *ACE Inhibitors (for pts w/low LVEF)	
Social Work	☐ Follow up as appropriate ☐ D/C teaching ☐ Transportation needs ☐ Community resources		
Nutrition		☐ Nutrition education completed or scheduled _____ Initial	
Respiratory Therapy			

continues

Exhibit 6–A–3 continued

Page 4	Date: _____ Unit: _____	Date: _____ Unit: _____	Date: _____ Unit: _____
Other			

Discharge Planning	Cardiac Rehab:
	❑ Referral To:
	  • Phase II _____
	  • CMH _____
	  • Home Program _____
	❑ D/C needs discussed with patient/family
			❑ Discharge To:
			  ❑ Home
			  ❑ Home with outpatient cardiac rehabilitation referral
			  ❑ Home with home exercise program
			  ❑ Home with HH
			  ❑ Nursing Home
			  ❑ TCU
			  ❑ Other
			  *Refer to Alternative Care Placement Tool

Signatures			

Source: Copyright © Methodist Hospitals of Dallas.

■ 7 ■

Pathways and Outcomes Across the Continuum: A Community Medical Center's Experience with Congestive Heart Failure

Judy Conarty, Rita Zenna, Harriet V. Werkman, and Kathleen A. Shafer

IMPETUS FOR ACTION

The honor bestowed upon a health care facility of having "national recognition" is, one would assume, enviable. One caveat, however, is the reputation for having the second highest Medicare admission and readmission rate nationally for patients diagnosed with congestive heart failure (CHF). In the May 1997 issue of Consumers Digest, Rapaport illustrates the most common diagnoses for hospital inpatients with hospital rankings by volume for some diagnoses. The first most common diagnosis related group (DRG) for hospitalization in the United States is normal vaginal delivery with normal newborn. The second most common reason for hospitalization is CHF (DRG 127). In this article, Community Medical Center is second in ranking compared to 19 other hospitals throughout the United States for 1994. Differences may be due to variances in the severity of patients' illness, method and extent of treatment, length of hospital stay, local cost of living, type of hospital, and other factors.[1]

At Community Medical Center, there were approximately 1154 patients admitted with a diagnosis of CHF in 1996. Of those patients, 1066 of them (92%), were 65 years of age and older and had an average length of stay (LOS) of 5.9 days. In 1994, 1004 patients admitted with CHF were 65 years of age or older, with an average LOS of 10.2 days. This chapter discusses our efforts to address our CHF population utilizing disease management strategies across the continuum of care.

Community Medical Center (CMC) is a 596-bed, nonprofit, acute care facility located in Toms River, New Jersey. Since 1996, the hospital has been an affiliate of the Saint Barnabas Health Care System (SBHCS), which includes nine hospitals, six extended care facilities, numerous ambulatory care facilities, a statewide behavioral health network, its own Medicare Certified Home Health/Hospice Program, and more than 4000 affiliated physicians. CMC has a full range of services for cardiac patients with the exception of a cardiac surgery program, planned for the near future.

Local and National Demographics

Since the 1950s, Ocean County has been the fastest growing county in New Jersey. During the 1970s, the overall population grew by 66% from 208,470 to 346,038 residents. The Office on Demographic and Economic Analysis (ODEA) reports that not only is Ocean County's population growing, the population is living longer. It is the third most popular retirement location for older adults in the United States, just behind Florida and Arizona.

National CHF Demographics and Statistics

People 65 years and older at any given time occupy 40% of acute care hospitals in the United States. In 1996, an estimated 4.8 million Americans were diagnosed with CHF. As CHF increases in prevalence, hospitalizations and deaths from this disease have made it a major chronic condition in the United States. It is often the end stage of cardiac disease. Half of the patients diagnosed with CHF will not survive beyond 5 years, with 400,000 new cases diagnosed each year. The annual number of deaths attributed to CHF increased from 10,000 in 1968 to 42,000 in 1993 with another 219,000 deaths related to the condition (National Heart, Lung and Blood Institute, September 1996). CHF is listed as the primary diagnosis in 875,000 hospitalizations, and the most common diagnosis for hospitalized patients ages 65 years and older. In this age group, an additional one fifth of all patients have a secondary diagnosis of heart failure. The magnitude of the CHF problem is expected to grow:

- As more and more cardiac patients (myocardial infarction, atherosclerotic heart disease, cardiomyopathy) are able to survive and live longer with disease, the opportunity for developing CHF increases.
- Future growth in the elderly population will likely result in increasing numbers of persons with this condition regardless of trends in coronary disease morbidity and mortality.

Spending Health Care Dollars

Visits to physicians' offices for CHF increased from 1.7 million in 1980 to 2.9 million in 1993. More than 65,000 persons with CHF receive home care each year. In 1993, an estimated $17.8 billion was spent for the care of CHF patients in hospitals, physicians' offices, home care, and nursing homes, as well as for medication expenses. Financial and other losses of caregivers for these patients are burdensome. The rates of hospitalization from heart failure increased more than three times between 1970–1994 at age 45–64 and age 65 and older, with a large absolute increase in the older group. In 1994, CHF was the primary discharge diagnosis in 874,000 hospital discharges (alive or dead) and the secondary diagnosis in another 1.8 million discharges. One in five of all discharged patients age 65 and older had CHF as a primary or secondary diagnosis. At the same time, CHF patients' mortality decreased from 11.3 percent in 1981 to 6.1 percent in 1993. This trend is observed for all CHF patients.[2]

PATHWAYS DEVELOPMENT

A number of factors influence the selection of diagnoses for pathway development. However, it has not been disputed that the decision is often based on which DRG, International Diagnostic Code—9th edition (ICD-9 Code), diagnostic test, or a surgical procedure that is high volume, high cost, high risk, or problem prone for the organization.[3] In analyzing the facts about CHF and our organization's admission rates, we came to speedy consensus that we would begin with this disease.

Pathway design efforts at CMC began in 1991. The medical staff's resistance was not surprising. The phrase "cookbook medicine," familiar to most of us by now, was voiced, along with considerable trepidation in reducing complex and dynamic medical decisions to a few sheets of paper. Our team, though modest in participation, had clear vision and perseverance, and eventually the first patient care pathways were completed. A physician and a registered nurse drafted the first pathway

documents, following a literature search and retrospective record review. Then, appropriate disciplines, (dietitian, respiratory therapist, etc), critiqued the draft. However, initial efforts lacking concession rarely succeed in health care and utilization of this early prototype was nonexistent to minimal at best, despite the high quality of the document. A new plan was required to improve physician acceptance and utilization. We utilized our "Partners in Care" collaborative practice group for early pathway development. This group, consisting of five physicians, along with social workers and registered nurses, was formed in November 1991. The purpose of the group was to better manage the hospital care and transfer of patients who resided in nursing homes. These initial five physicians grew to seven, with leadership shifting over time from the director of social work to the vice-president for patient care services. The group is now co-led by our vice-president for medical quality and director of case management. Ultimately, this team became a precursor for case management. Through "Partners in Care," we gained a larger physician base and our pathway initiative moved forward.

The results (Figure 7–1), reflected improvement. However, the need to address the high readmission rate for CHF became a priority. Through interdisciplinary discussions and analysis of posthospital follow-up, it became evident that improved and planned utilization of home health services and interventions was needed to prevent disease exacerbation and subsequent readmissions. At this time a decision was made to reconvene the cardiology pathway team to provide needed clinical expertise in this area.

Initial Steps

In 1994, the cardiology services team was organized as a part of the ongoing process improvement initiative throughout the organization. The membership (Exhibit 7–1), represented stakeholders in the division of cardiology services and serves as the model for the design of any pathway within our organization. The team's initial mission was to develop pathways for CHF, myocardial infarction (MI), angina pectoris, and others for utilization in the acute care setting. After 6 months, a representative from the hospital's home care department was invited.

With the addition of home health representation, the team focused on strategies to break the paradigm of patient dependence on acute care services. Efforts were directed toward the development of outpatient management strategies that help identify problems before they become acute and keep patients in relatively stable health and at home. Ideally, as much as half of inpatient care for CHF should be provided outside of the acute hospital setting, if needed at all.[4]

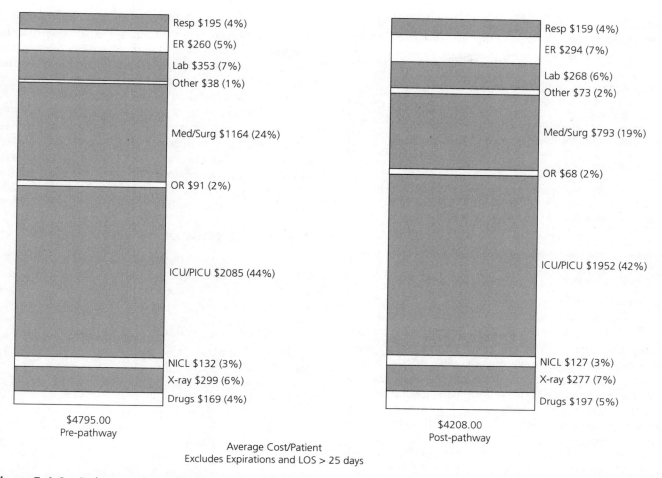

Resp $195 (4%)
ER $260 (5%)
Lab $353 (7%)
Other $38 (1%)
Med/Surg $1164 (24%)
OR $91 (2%)
ICU/PICU $2085 (44%)
NICL $132 (3%)
X-ray $299 (6%)
Drugs $169 (4%)

$4795.00
Pre-pathway

Resp $159 (4%)
ER $294 (7%)
Lab $268 (6%)
Other $73 (2%)
Med/Surg $793 (19%)
OR $68 (2%)
ICU/PICU $1952 (42%)
NICL $127 (3%)
X-ray $277 (7%)
Drugs $197 (5%)

$4208.00
Post-pathway

Average Cost/Patient
Excludes Expirations and LOS > 25 days

Figure 7–1 Pre-Pathway and Post-Pathway Consumption (CHF-DRG 127) Implemented 5/5/95. *Source:* Copyright © Community Medical Center.

Through the collaborative efforts of the team, the CHF pathway was expanded to include a home care pathway. We targeted end stage CHF patients, who require frequent hospitalizations—usually on an emergency basis. This home care initiative was piloted with four cardiology physician practices and goals were as follows:

1. Improve patient's quality of life.
2. Decrease hospital readmissions.
3. Manage CHF cost-effectively.
4. Follow CHF patients for 6 months through home care.

One advantage we had is that the close working relationship between the hospital and home care department would facilitate increased continuity and ultimately help to achieve high patient satisfaction and optimal outcomes.

Pilot Study

During the pilot study, January through June 1995, there were 58 patients referred to the home care conges-

tive heart failure assessment team (CHAT). These patients were followed for 6 months. Nurses trained and experienced in cardiac rehabilitation and/or intensive care or coronary care nursing were selected as CHAT nurses. (Exhibit 7–2 outlines the Home Health CHAT basic plan of care. See also Figure 7–2.)

Of 58 patients referred, 7 were rehospitalized within 6 months; 2 patients elected hospice care, with 88% of the initial group successfully managed at home for 6 months. Encouraged by these early results, the CHAT program was opened to physicians from two hospitals in the SBHCS network: Community Medical Center and Kimball Medical Center, a 354-bed acute care facility, located in Lakewood, NJ.

The CHAT Program

Patients were educated in the process of monitoring and reporting symptoms as well as in the availability of the 24-hour on-call system. Patient/family education regarding self-care and symptom management was rein-

Exhibit 7–1 Community Medical Center Cardiology Service Team Membership

Cardiology Subsection Chair and Coleader
RN Case Management, Critical Care and Coleader
Medical Director, Emergency Department
Vice-President, Medical Quality
Manager, Outcomes and Pathways, Team Facilitator
Partners in Care Representative
Pharmacist
Director, Patient Care, Critical Care Telemetry Unit
Director, Patient Care, Critical Care Post–Cardiac Care Unit
Clinical Liaison, Information Services and Technology
Clinical Nurse Specialist, Critical Care
Cardiologists
Cardiac Rehabilitation Nurse
Coordinator, Quality Resource Services
Administrative Director, Home Health/Hospice
Disease Management Coordinator
Critical Care Nurse

Source: Copyright © Community Medical Center.

forced throughout their stay on the program. Patients were seen under the regulatory guidelines of their health insurance and provided with visits beyond the usual discharge guidelines for Medicare or private insurance for a 6-month period. Follow-up consisted of weekly nursing visits or phone monitoring. The cost of these nonbilled visits was logged and compared against the projected readmission rate prior to the program's onset. Of the 113 patients referred to the CHAT Program in 1996, there was only a 10% hospital readmission rate. (The overall CHF readmission rate for CMC was 8.6% in 1994, 8.5% in 1995, and 7% in 1996. Excluded from data were patient expirations and those with a length of stay greater than 25 days.) At CMC we define a CHF readmission as any patient discharged more than once with the final billed DRG 127 within 30 days.

During the second quarter of 1996, we became aware of an opportunity to improve home health services for patients newly diagnosed with CHF, and a Non-CHAT pathway was formulated. The same premise of monitoring, symptom management, teaching, and follow-up was utilized, with the addition of basic disease management content designed for patients newly diagnosed with CHF.

The criteria for referral to Non-CHAT is at least one hospital admission for CHF while the CHAT criteria is two or more hospital admissions within a 6-month period. Readmission rates for this program are still being collected, as utilization of this program was not what we had anticipated.

A nine-visit plan of care that spans over 9 weeks was developed following a literature search and a 49-patient

record review (49%) of home health CHF total available sample for 6 months (see Exhibit 7–3 for this plan of care). Home Health Nurses with cardiology experience developed this pathway in collaboration with the hospital's cardiology services team.

With several successes now well documented, another gap was beginning to emerge. The cardiology services team, with the initial partners in care group, had reached or exceeded goals of decreasing LOS, readmission rates, and resource utilization. However, the acute care pathway was in need of review and revision. Utilization of the CHF pathway began to decline, and it was evident that a broader physician usage base needed to be established.

Changes in Best Practice Standards

During this time, the literature seemed to explode with data related to pathways and outcomes across all health care disciplines. Numerous sources held that, if developed and implemented properly, critical paths can lead to desirable outcomes for the patient and improved operating effectiveness/efficiency for the health care facility.[5] By aligning the patient care pathway goals with the physicians' own goals of improving patient care and outcomes, interest in participating in the development of pathways markedly increased.

The CHF pathway team was restructured in mid-1996, into the cardiology services team, following a product line philosophy. The cardiology subsection chairperson now co-leads this team with the RN case manager responsible for cardiac care. The group addresses a comprehensive array of services for patients diagnosed with all cardiac diagnoses including CHF. In addition to pathway revision and development, this multidisciplinary team (Exhibit 7–1) will assess (1) clinical outcome measures, (2) financial resource consumption, and (3) patient satisfaction data. Action plans and performance improvement activities will be based on a complete picture of CHF, providing a balanced view of the outcomes of care across the continuum.

Steps to Patient Care Pathway Construction

Patient care pathway development, implementation, and follow-up adhere to a structured continuous quality improvement FOCUS PDCA methodology. This format was developed by Walter Shewart, as PDSA, and later expanded by Dr. Deming as PDCA.[6]

F — find a process to improve
O — organize a team
C — clarify current knowledge of process
U — understand causes of process variation
S — select a process improvement

Exhibit 7–2 Home Health Certification and Plan of Care

1. Patient's HI Claim No.	2. Start Of Care Date 010101	3. Certification Period From: 070196 To: 090196	4. Medical Record No. 0052349	5. Provider No. 317067

6. Patient's Name and Address TEST, CHF 000-000-0000	7. Provider's Name, Address and Telephone Number COMM MED CTR HOME HEALTH 99 HWY 37 W CN 2002 TOMS RIVER, NJ 08755 732-818-6800

8. Date of Birth 010101	9. Sex ☒ M ☐ F	10. Medications: Dose/Frequency/Route (N)ew (C)hanged

11. ICD-9-CM 08889	Principal Diagnosis OTHER	Date 000000	
12. ICD-9-CM N/A	Surgical Procedure N/A	Date N/A	
13. ICD-9-CM N/A N/A N/A N/A	Other Pertinent Diagnoses N/A N/A N/A N/A	Date N/A N/A N/A N/A	

14. DME and Supplies N/A	15. Safety Measures N/A

16. Nutritional Req. N/A	17. Allergies: N/A

18.A. Functional Limitations

1 ☐ Amputation	5 ☐ Paralysis	9 ☐ Legally Blind
2 ☐ Bowel/Bladder (Incontinence)	6 ☐ Endurance	A ☐ Dyspnea With Minimal Exertion
3 ☐ Contracture	7 ☐ Ambulation	B ☒ Other (Specify)
4 ☐ Hearing	8 ☐ Speech	

18.B. Activities Permitted

1 ☐ Complete Bedrest	6 ☐ Partial Weight Bearing	A ☐ Wheelchair
2 ☐ Bedrest BRP	7 ☐ Independent At Home	B ☐ Walker
3 ☐ Up As Tolerated	8 ☐ Crutches	C ☐ No restrictions
4 ☐ Transfer Bed/Chair	9 ☐ Cane	D ☒ Other (Specify)
5 ☐ Exercises Prescribed		

19. Mental Status:	1 ☒ Oriented	3 ☐ Forgetful	5 ☐ Disoriented	7 ☐ Agitated
	2 ☐ Comatose	4 ☐ Depressed	6 ☐ Lethargic	8 ☐ Other

20. Prognosis:	1 ☐ Poor	2 ☐ Guarded	3 ☒ Fair	4 ☐ Good	5 ☐ Excellent

21. Orders for Discipline and Treatments (Specify Amount/Frequency/Duration) *

SN: 1–3 W 3–4 plus 3 prn visits to evaluate: D/spnea, Nocturnal, Dyspnea, increased orthopnea, chest pain or reported wt. gain; 2 lbs/day or 5 lbs/wk
( ) Assess cardio-pulmonary status: persistent cough, rapid, labored, or irregular respirations; S&S to report to RN immediately
( ) I/A S/S disease process; CHF & management
( ) I/A increased weakness; fatigue, decreased activity tolerance, any restlessness confusion, syncope
( ) I/A in cardiovascular risk factors that can affect c/v status & methods of modification; S&S to report to RN
( ) Instruct planned rest periods, balance w/activity
( ) Instruct in taking daily radial pulse & record; to report if rate 50 or below; 110 or above
( ) I/A medication management; diuretic therapy; actions adverse side effects food/drug interactions
( ) I/A prescribed diet; food low in sodium; meal planning

22. Goals/Rehabilitation Potential/Discharge Plans *
( ) In _____ vs recalls proper medications & administration
( ) In _____ vs verbalizes the difference in medical emergency vs nursing emergency & access to 24 hr on-call system or activation of 911
( ) In _____ vs recalls importance of rest periods balanced with activity

23. Nurse's Signature and Date of Verbal SOC Where Applicable:	25. Date HHA Received Signed POT

24. Physician's Name and Address 000-000-0000 DOCTOR, OTHER 99 ROUTE 37 W TOMS RIVER, NJ 08755	26. I certify/recertify that this patient is confined to his/her home and needs intermittent skilled nursing care, physical therapy and/or speech therapy or continues to need occupational therapy. The patient is under my care, and I have authorized the services on this plan of care and will periodically review the plan.

27. Attending Physician's Signature and Date Signed	28. Anyone who misrepresents, falsifies, or conceals essential information required for payment of Federal funds may be subject to fine, imprisonment, or civil penalty under applicable Federal laws.

continues

Exhibit 7–2 continued

ADDENDUM TO:	☒ PLAN OF TREATMENT	❏ MEDICAL UPDATE		Page: 2

1. Patient's HI Claim No.	2. SOC Date 010101	3. Certification Period From: 070196 To: 090196	4. Medical Record No. 0052349	5. Provider No. 317067

6. Patient's Name TEST, CHF	7. Provider Name COMM MED CTR HOME HEALTH

8. Item No.	
21	**ORDERS FOR DISCIPLINE/TREATMENTS CONTINUED:** Fluid Management ( ) Instruct daily weights; record daily log ( ) Assess skin integument; edema; fluid retention; lower extremities; assess temp U/LE appearance ( ) Instruct avoidance of temperature extremes, restrictive clothing; over exertion ( ) Instruct coughing, deep breathing exercises ( ) Pulse oximetry PRN increased dyspnea; VN may order O2 @ 2 liters via nasal cannula for pulse oximetry 88 or below ( ) Instruct bowel management; avoid straining; VN may digitally disimpact if laxatives ineffective ( ) Instruct O2 use, safety precautions, how to access DME
22	**GOALS/REHABILITATION DISCH/PLANS CONTINUED:** ( ) In _____ vs demonstrates 24 hr meal planning according to RX diet ( ) In _____ vs independent with ADLs with or without community resources ( ) In _____ vs verbalizes 3 S&S of _____ reportable to MD ( ) In _____ vs demonstrates coughing w/deep breathing ( ) In _____ vs demonstrates safe use of DME _____ ( ) In _____ vs demonstrates wound care independently ( ) In _____ vs independent insulin administration ( ) In _____ vs demonstrates independent Foley catheter management GOALS TO BE WRITTEN BY VN: ( ) In _____ vs verbalizes 3 foods low in Na+ ( ) In _____ vs demonstrates safe use of O2 and precautions ( ) In _____ vs verbalizes importance of rest periods, tolerates current activity level without s/s CHF ( ) In _____ vs or D/C manages activities of daily living without symptoms ( ) In _____ vs verbalizes 24h meal planning according to rx diet ( ) In _____ vs verbalizes 6 foods high in K+
99	**MISCELLANEOUS CONTINUATION TEXT CONTINUED:** HX—

INSURANCE INFORMATION

PRIMARY _____ SUBSCRIBER _____

CONTACT/CASE MANAGER _____ PHONE _____

INITIAL PREAUTH COMPLETED: ( ) YES ( ) NO

NUMBER OF VISITS AUTHORIZED: _____

INSURANCE CHECK: INITIATED _____ ATTACHED _____

9. Signature of Physician	10. Date

11. Optional Name/Signature of Nurse/Therapist	12. Date

Source: This form has been reprinted from Health Care Financing Administration. The content is from Community Medical Center.

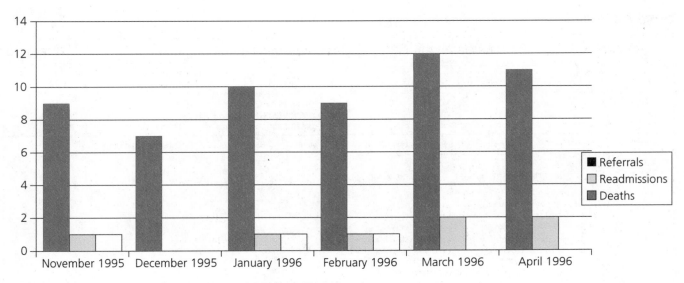

Figure 7–2 Congestive Heart Failure 6-Month Pilot Study Home Health Program

P — plan
D — do
C — check
A — act

A physician-led multidisciplinary team is formed (Exhibit 7–4). This team is considered an organizational process improvement team, with regular reports to the organization's performance improvement council (PIC), and board of trustees. As "just in time" team training is completed, which includes education on continuous quality improvement methodology, the patient care pathway process steps are implemented (Exhibit 7–5 and Figure 7–3).

The first step is to develop an effective mission statement, which is not a simple process. This statement must balance detail with generality and provide the team with enough clarity to focus time and resources expeditiously. A mission that is too detailed may bias the team's objectivity and prevent the discovery of breakthrough solutions, while a mission that is too general will cause the group to lose focus.[7] The appointment of a skilled team facilitator can ensure that a solid clear mission statement is formulated, and for us, this has often been critical to the team's success.

The second step is data collection, and it is essential that appropriate baseline data be collected initially. This will allow for comparative analysis that is vital to the evaluation of the team's progress. Reports should include baseline LOS analysis, resource consumption for the DRG, baseline care delivery practices, based on chart review, and current literature review to provide best current practices.

The third step is to flowchart the current steps in caring for a patient with that particular DRG. This step often leads to the identification of actual and potential barriers—system, patient, or caregiver that will impede the smooth transition of care. These barriers must be addressed using quality improvement techniques if the team's efforts are to be successful.

The fourth step is to develop the actual patient care pathway content, based on identified categories. To do so, the following questions need to be answered:

1. What treatment modalities produce the most desirable patient outcomes for the identified DRG?
2. What are the most cost-effective medications, laboratory, and other testing options, within the identified "best practice"?
3. What is the most appropriate utilization of the services of the multidisciplinary team?
4. What patient/significant other education needs to be addressed?
5. What discharge planning needs exist that are unique to this population?
6. What patient outcomes, measured on a daily basis, will provide us with evidence that our care is producing the desired effects *and* guide future care decisions?

The fifth step, upon completion of a draft, is to invite input from all customers who will execute the pathway. Although cross-representation within the multidisciplinary team assists in the development of the pathway, the survey of more clinicians is needed during this phase.

Exhibit 7–3 Home Health Certification and Plan of Care

1. Patient's HI Claim No.	2. Start Of Care Date 010101	3. Certification Period From: 100196 To: 100296	4. Medical Record No. 0054064	5. Provider No. 317067

6. Patient's Name and Address TEST, NON CHAT 000-000-0000	7. Provider's Name, Address and Telephone Number COMM MED CTR HOME HEALTH 99 HWY 37 W CN 2002 TOMS RIVER, NJ 08755 732-818-6800

8. Date of Birth 010101	9. Sex ❑ M ☒ F	10. Medications: Dose/Frequency/Route (N)ew (C)hanged

11. ICD-9-CM 4280	Principal Diagnosis CONGESTIVE HEART FAILURE	Date 000000
12. ICD-9-CM N/A	Surgical Procedure N/A	Date N/A
13. ICD-9-CM N/A N/A N/A N/A	Other Pertinent Diagnoses N/A N/A N/A N/A	Date N/A N/A N/A N/A

14. DME and Supplies N/A	15. Safety Measures . N/A
16. Nutritional Req. N/A	17. Allergies: N/A

18.A. Functional Limitations

1 ❑ Amputation	5 ❑ Paralysis	9 ❑ Legally Blind
2 ❑ Bowel/Bladder (Incontinence)	6 ❑ Endurance 7 ❑ Ambulation	A ❑ Dyspnea With Minimal Exertion
3 ❑ Contracture	8 ❑ Speech	B ☒ Other (Specify)
4 ❑ Hearing		

18.B. Activities Permitted

1 ❑ Complete Bedrest	6 ❑ Partial Weight Bearing	A ❑ Wheelchair
2 ❑ Bedrest BRP	7 ❑ Independent At Home	B ❑ Walker
3 ❑ Up As Tolerated	8 ❑ Crutches	C ❑ No restrictions
4 ❑ Transfer Bed/Chair	9 ❑ Cane	D ☒ Other (Specify)
5 ❑ Exercises Prescribed		

19. Mental Status:

1 ☒ Oriented	3 ❑ Forgetful	5 ❑ Disoriented	7 ❑ Agitated
2 ❑ Comatose	4 ❑ Depressed	6 ❑ Lethargic	8 ❑ Other

20. Prognosis: 1 ❑ Poor   2 ❑ Guarded   3 ☒ Fair   4 ❑ Good   5 ❑ Excellent

21. Orders for Discipline and Treatments (Specify Amount/Frequency/Duration) *

SN: (Non-CHAT) 3W2; 2W1; 1Q2–4W; 1Q4 WKS +2 PRN for assessment of symptom exacerbation and/or venipuncture
Cardiopulmonary assessment, establish medication schedule and instruct on management, side effects, precautions, food/drug interactions.
   Instruct frequent rest periods and progress activity 5–15' Q visit beginning with week two as tolerated.
I   Elevation of extremities when sitting, slow position changes energy conservation, work simplification. I in keeping a weight/symptom log and review every visit.
I   Patient to report a weight gain of _3 lbs/24 hr or _5 lbs in 1 week to MD.
   VN to facilitate/arrange MD appointment within 2 wks of hospital discharge.
I   Dietary mgt to include RX diet, foods low Na, high K+, 24h meal planning and eating out. I in coughing and deep breathing exercises. I importance of compliance with treatment, consequences of noncompliance & emergency management. VN to draw electrolytes 3–4 days prior to next MD visit

22. Goals/Rehabilitation Potential/Discharge Plans *

   Goals: Adequate cardiac output as evidence by no obvious jugular vein distention; lungs clear; stable weight; warm/dry
      skin or VN will notify MD.
   Compliance with medication schedule.
   State own fluid/dietary restrictions.

23. Nurse's Signature and Date of Verbal SOC Where Applicable:	25. Date HHA Received Signed POT

24. Physician's Name and Address 000-000-0000 DOCTOR, OTHER 99 ROUTE 37 W TOMS RIVER, NJ 08755	26. I certify/recertify that this patient is confined to his/her home and needs intermittent skilled nursing care, physical therapy and/or speech therapy or continues to need occupational therapy. The patient is under my care, and I have authorized the services on this plan of care and will periodically review the plan.
27. Attending Physician's Signature and Date Signed	28. Anyone who misrepresents, falsifies, or conceals essential information required for payment of Federal funds may be subject to fine, imprisonment, or civil penalty under applicable Federal laws.

continues

Exhibit 7–3 continued

ADDENDUM TO:	☒ PLAN OF TREATMENT	☐ MEDICAL UPDATE		Page: 2

1. Patient's HI Claim No.	2. SOC Date 010101	3. Certification Period From: 100196 To: 100296	4. Medical Record No. 0054064	5. Provider No. 317067

6. Patient's Name TEST, NON CHAT	7. Provider Name COMM MED CTR HOME HEALTH

8. Item No.	
21	ORDERS FOR DISCIPLINE/TREATMENTS CONTINUED: within 2 weeks of discharge. I in OTC meds high in sodium. Community resource referrals to include O.P. dietary counseling, lighthouse, care coordination, smoking cessation, Center for Health Promotion.
22	( ) MSW: Eval plus 1 PRN long range planning, community resource referrals. GOALS/REHABILITATION DISCH/PLANS CONTINUED: Maintains daily weight/symptom log, recognizes individual S/S or CHF and verbalizes how/when to contact MD. States consequences of noncompliance to treatment.

9. Signature of Physician	10. Date

11. Optional Name/Signature of Nurse/Therapist	12. Date

Source: This form has been reprinted from Health Care Financing Administration. The content is from Community Medical Center.

Exhibit 7–4 Departmental Representation for Pathway Development

Title	Requirement
Physician	Mandatory
Nursing	Mandatory
Cardiopulmonary Services	Mandatory/Guest
Radiology	Mandatory/Guest
Laboratory	Mandatory
Rehabilitation Services	Mandatory/Guest
Nutrition	Mandatory
Case Management	Mandatory
Center for Family & Senior Services	Guest
Center for Women's Health	Guest
Clergy	Guest
Clinic	Guest
Clinical Resources	Guest
Decision Support	Guest
Diabetes Education	Guest
Education Services	Guest
Patient Education & Documentation	Guest
Finance	Guest
Home Health/Hospice	Guest
Infection Control	Mandatory/Guest
Legal	Guest
Lighthouse Network—Care Coordination	Guest
Medical Records	Guest
Neurodiagnostics	Guest
Operative Services	Guest

Guest: Invited for at least one meeting or as a Task Force Member.

A survey allows for participation of all pathway "users" and alerts the team to issues and barriers not previously considered, while promoting greater acceptance of the finished product. Do not make the mistake of ignoring these solicited responses. If at all possible, and when feasible, consider them now for a more successful implementation later.

The sixth step, education, is just as important as pathway content and cannot be overemphasized. The team should utilize the educational services team member for guidance in this step, when considering curriculum and teaching methods. All groups that may be affected by the pathway should be identified, specifically the knowledge or skills that are needed, the way the information will be presented, and how training and feedback will occur. Before submission for organizational approval, the pathway should be reviewed to determine if the final draft represents a process that includes "best practice" with consideration of the appropriate utilization of resources. As soon as the patient care pathway meets the final approval process, the trial phase can begin.

After 3 to 6 months of trial, sufficient numbers of patients should have completed the pathway, to allow for collection of data. The original team, who can recommend further improvements, can then analyze data according to the patient outcome measures and project goals. This is an opportunity to revise the work before introduction on a wider scale. The early data generally provide a sense of focus in introduction to physician groups and often inspire confidence that the pathway has been well constructed and is dynamic in nature.

The final (and often most neglected) step is ongoing monitoring. The patient care pathway process and data collected should be collated and monitored at least, but not limited to, quarterly. Findings should be reported to the appropriate departments or team for follow-up and revision.

Pathway Advantages

The acute care pathway (Exhibit 7–A–1 in Appendix 7–A) has several advantages to patient care delivery. It ensures continuity of care, is multidisciplinary in approach, and is based on "best current practice" (as identified through literature review), and knowledge of the population served. Since the acute care pathway is a comprehensive plan of care, some institutions may want to consider the use of this document in lieu of the "nursing care plan," and/or other individual disciplines' patient care plans. Pathways are also extremely useful tools for physicians. From the resident to the seasoned attending physician, pathways serve as an excellent standard or foundation for care. Pathways can also help eliminate the yo-yo effect that occurs when one physician's orders are changed by a partner due to individual preferences only to be reversed on the following day by a third partner.

Another specific advantage, if available in the institution, is the automation of pathways. Availability of the pathways on our Medical Information System (MIS) continues to contribute to physician acceptance and utilization. Automation allows easy order entry and transcription, decreases errors, and improves response times as the orders move directly to the clinical staff. Once the order is given to initiate the patient care pathway, individual ordering screens that are identical to the written pathway appear. A simple cursor (light pen) selection of either the individual or "all of the above orders" expedites the process. Having an automated template to follow decreases variation and ultimately increases quality. Some of our physicians have embraced the concept so positively that they have developed automated standing order sets, which have streamlined the pathway ordering process even further (Exhibit 7–A–2 through 7–A–19).

Another potential benefit with mixed reviews in the literature is the possibility that pathways decrease the risk of

Exhibit 7–5 Patient Care Pathway Development Process Steps

Steps:	Responsible Person:
1. Develop Mission Statement	1. Team Members

2. Data Collection
  • baseline length of stay
  • baseline resource consumption
  • baseline care delivery
  • baseline care delivery practices/chart review
  • literature review
	2. Team Leader assigns

3. Process Flow	3. Team Leader and Member

4. Develop Pathway Categories
  • Treatments
  • Medications
  • Laboratory
  • Other tests
  • Respiratory
  • Dietary
  • Consults
  • Key nursing activities
  • Rehabilitation services
  • Patient/Significant other education
  • Discharge planning
  • Expected patient outcomes
	4. Team Leader and Members as assigned

5. Survey Customers
  • Patient/Family/Significant other
  • Physicians
  • Nursing
  • Rehabilitation
  • Laboratory
  • Radiology
  • Respiratory/Cardiac
	5. Team Leader assigns

6. Education Plan	6. Team Leader, Members, and Education Department
7. Approval Process	7. Team Leader

8. Pathway Trial
  • duration
  • location(s)
	8. Team Leader and Members

9. Collect Data
  • *pathway acceptance form
  • patient/significant other education form
  • *discharge outcome evaluation form
    *To Clinical Coordinator/Nurse Manager for review
    ↓
    To Case Management for entry into MAXSYS
    *Follow these steps
	9.
	  • upon admission
	  • RN caring for patient
	  • RN caring for patient

10. Evaluation data • draw conclusions, recommendations, actions, follow-up	10. Team Leader and Members

11. Ongoing monitoring
  • monitor process and data, at least but not limited to, quarterly
  • report findings to appropriate departments	

Source: Copyright © Community Medical Center.

lawsuits. "By adopting critical pathways and closely adhering to nationally accepted guidelines, you can substantially reduce your risk of facing a malpractice lawsuit and increase your chances of winning if you are sued. . . ."[8(p.78)] However, before we rush to point this out to our physician colleagues, consider that this article also warns that deviation from the pathway must be carefully documented to "avoid any appearance that you breached the standard of care."[8(p.78)] Our physicians are well aware of this point and it is often the focus of much concern regarding pathways.

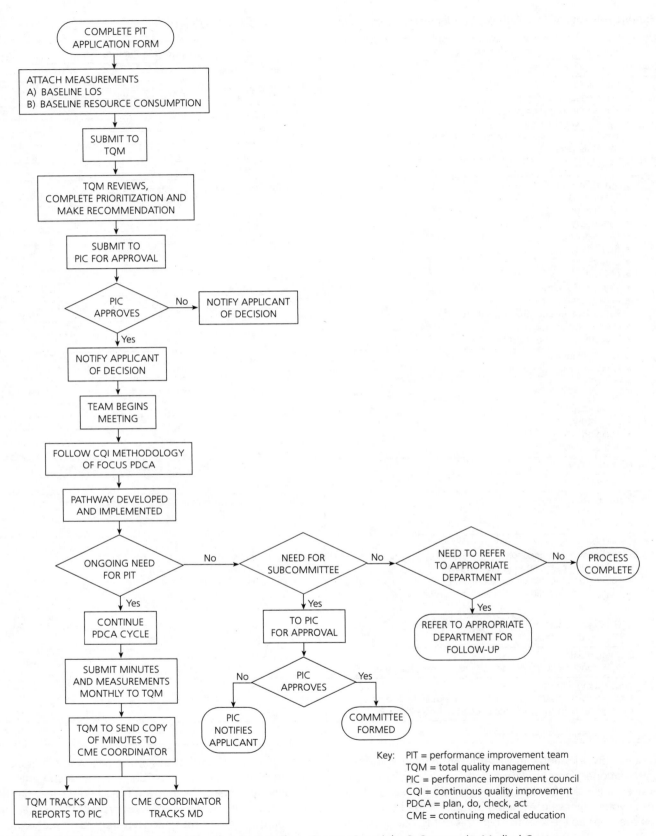

Figure 7–3 Patient Care Pathway Development Process Flow. *Source:* Copyright © Community Medical Center.

The pathway usually includes orders for routine measures, such as analgesics, sleeping aids, and laxatives, that in the past have caused the patient and physician needless discomfort while a midnight phone call was made by the nursing staff. The pathway also provides an excellent teaching tool, as patient fears are eased by knowing what to expect for each day of hospitalization, and provides a guide for making discharge plans. Routine uncomplicated care is able to progress independently of the time the physician makes rounds and level of care transitions are smoother as demonstrated by our CHF home and acute care pathways. Finally, the creation and evolution of what we now call our clinical service teams has genuinely improved the knowledge of interdisciplinary roles and creates a synergy that teams are able to identify. One example of a significant change in attitude came when one of our cardiologists (who had never utilized home health services in the past) stated "Home care is the key to this thing" in a meeting involving our CHF pathway. When the team is engaged in the work, the barriers between disciplines are less visible and the focus is on clinical standards of care.

Pathway Implementation and Acceptance

Along with most of our colleagues across the country, we too experienced much resistance from physicians and other practitioners during the initiation of pathways. Utilization varied considerably, and continues to do so, with highest pathway use in areas of obstetrics and total joint replacement, and the lowest with more complex diagnoses, such as abdominal aortic aneurysm. As previously noted, considerable benefits were evident as these multidisciplinary groups, originally charged with formulating the pathways, evolved into highly functioning teams. These teams, now working to measure outcomes and revise pathways, have become the driving forces for change, quality, and innovation within their service areas. As in any major project implementation, the support of senior management is essential, and in our case support from the vice-president for patient care services and executive director was key in the success of these early efforts. There was also an attempt to create teams around groups of physicians who were more willing to attempt to change while individual physician concerns were examined and addressed on an ongoing basis. This process continues even now.

Patient Education and Pathways

Although CHF is not one of the pathways that we further developed into a patient education tool, plans to do so are being considered. Pathways used in this way allow the patient to actively participate in the plan of care, make adaptations as appropriate, and they are useful communication tools.

CASE MANAGEMENT

Case management is a multidisciplinary continuous process in which patients are assisted through various levels of care to optimize health potential. Newell defines case management according to the Individual Case Management Association (ICMA) and the Case Management Society of America:

> Case Management is a collaborative process which assesses, plans, implements, coordinates, monitors and evaluates options and service to meet an individual's health needs through communication and available resources to promote quality, cost effective outcomes.[9(p.3)]

Case management at CMC is an integrated, comprehensive model. Case managers, who are registered professional nurses, are assigned by physician and therefore follow patients from admission to discharge regardless of their bed status (ie, critical care vs medical surgical). Therefore, unless the patient is admitted and discharged within 48 hours, a case manager "manages" the case. In some cases, case managers are even assisting in short stay or observation status patients.

Case managers wear many hats. They perform a variety of comprehensive functions that include assessment, documentation, discharge planning, utilization review, DRG assignment, and quality monitoring. In performance of these functions, their communication is essential to assisting the patient along the continuum. At the very least, communication with the patient regarding discharge planning needs, as well as enhanced communication to the managed care companies, provides the information needed to ensure payment and responsibility in moving the patient to an alternate level of care.

Social workers are an integral part of the case management department. They are assigned to teams of case managers and social worker. The social worker handles the complex discharge disposition cases as well as guardianship and new extended care placement processing, and living wills as well as complex family psychosocial interactions.

Physician advisors are a resource to all case managers and social workers. Referrals to the physician advisor can occur at any time. These referrals can be as simple as assistance with a physician on a discharge order or as complex as a quality occurrence that will also be referred to the specific department quality review. Physician advisors also assist with any utilization review issue. The vice-president for medical quality conducts weekly LOS meetings with case managers and social workers to assist and guide them. He or she also mentors the physician advisors. Pathways are utilized by physician order. There are several physicians who automatically order pathway usage on their patient.

Other physicians request they be asked whether or not the patient should be placed on pathway. The registered nurse on the unit manages pathway implementation. He or she requests the order from the physician and follows the daily guidelines for usage. Variances are collated by the pathway coordinator and reported to the utilization review committee. The registered nurse on the unit, also as part of the admission assessment, incorporates discharge planning, which assists case managers in their internal assessment of the patient.

Cases are measured by acuity daily. The acuity tool measures productivity of each case manager and social worker and assists in decision making with reassignment of case load. The productivity tool measurement averages between 95% and 105% biweekly. The tool has been used successfully for 1 year and has been revalidated once during the year.

Case management has successfully met challenges with an almost 70% Medicare population by assisting the patient to an alternate level of care quickly without compromising the quality of care.

OUTCOMES

Outcome measurement at our facility rests with a number of associates who are responsible for various aspects of the work. The structure is shown in Table 7–1.

Summation of Outcomes Measured

Clinical Outcomes

Patient care pathways have been in place in our hospital for over 4 years. Since that time, measurement of clinical outcomes has evolved from the measurement and analysis of many daily variances to the development of a few main specific and measurable diagnosis-specific outcomes (Exhibit 7–6). These outcomes were, and continue to be, developed by the multidisciplinary clinical service teams with a direct interest and involvement in the specialty.

Table 7–1 Outcomes—Overall Measurement and Responsibility

Outcome Measurement	Responsibility	Measured/Reported	Reviewed/Reported
Financial	VP, Finance	Monthly	Executive Director Department Heads
Length of Stay	Director, Decision Support	Quarterly and as requested	Utilization Medical Executive Committee Board of Trustees
Pathway	Manager, Pathways/Outcomes	Quarterly	Clinical Service Teams Utilization Review PIC
Disease Management	Disease Management Coordinator	Quarterly	Clinical Service Teams Outcomes Steering Continuum of Care Medical Executive Committee Performance Improvement Council Board of Trustees
Patient Satisfaction	Director, Patient Satisfaction	Monthly	Executive Director Department Heads Process Improvement Council Medical Sections Board of Directors System Vice-President of Patient Satisfaction System Performance Improvement Council
Physician Satisfaction	Director, Patient Satisfaction	Annually	
Employee Satisfaction	System VP, Human Resources	Annually	System Vice-President Human Resources Executive Directors
Emergency Department	Director, Emergency Department	Quarterly	Executive Director Hospital Services Quality Performance Improvement Council Board of Trustees System Performance Improvement Council
Medical	Coordinator, Quality Resource Services	Monthly/Quarterly	Performance Improvement Council Medical Executive Committee Board of Trustees

Exhibit 7–6 Patient Care Pathway Outcome Measurements

Community Medical Center
Toms River, New Jersey

Patient Care Pathway Outcome Measurements

Congestive Heart Failure (127)

Admission Date: _____

Discharge Date: _____

Addressograph

	YES	NO	KEY
Readmission	☐	☐	☐
In-hospital mortality	☐	☐	☐
Patient discharged on ACE inhibitor	☐	☐	☐
Time on telemetry			
Discharged by Day 4	☐	☐	☐

If questions above are answered **NO**, give reason in **KEY** box:

Patient:

P1	Refused
P2	Not appropriate for patient
P3	Complications—remains in pathway
P4	Complications—pulled out of pathway
P5	Surgery—pulled out of pathway
P6	Confused
P7	Does not follow directions or treatment regimen
P8	Language barrier
P9	Unable to be discharged
P10	Other: (**SPECIFY**)

Caregiver:

C1	MD did not order
C2	RN did not follow
C3	LPN did not follow
C4	Delayed discharge order
C5	Lab delay/problem
C6	Radiology delay/problem
C7	Consult not called
C8	Consult not done within 24 hours
C9	Other: (**SPECIFY**)

System:

S1	Transferred to another unit
S2	Waiting for test/study
S3	Service not available on WE
S4	Service not available on nights
S5	Patients in ER > 24 hours
S6	Equipment not available
S7	Patient treatment started late, delayed DC
S8	No nursing home bed available
S9	Other: (**SPECIFY**)

Family/Placement:

F1	No family available
F2	Family slow to seek placement
F3	Family refuses placement
F4	Guardianship
F5	Other: (**SPECIFY**)

Source: Copyright © Community Medical Center.

For example in the area of CHF, we expect that for most patients, telemetry will be discontinued in 2 days, the patient prepared for discharge by day 4, on an ACE inhibitor unless contraindicated. Patients are not expected to require hospitalization for this diagnosis for at least the next 30 days.

Similarly, we are currently refining all of our pathways and outcomes to include broad, specific, and measurable outcomes that reflect "best practice" based on current research. Prioritization of this work is done by admission volume and physician pathway utilization.

Financial Outcomes

"Outcomes management (OM) is supported by a premise which suggests that utilization of best provider practice drives health care cost reduction. Practitioners utilizing an OM framework are not directly concerned with reduction. Instead, continuous improvement of interdisciplinary provider practice is pursued through research methods which target the outcomes of care on a designated population."[11] Clinical service teams request and review our average cost per admission pre- and post-pathway development. These data are obtained from the director of decision support, utilizing our clinical cost accounting system, and include an ancillary cost breakdown. Although the team might investigate obvious high-cost items and research the use of the most cost-effective medications, there is clearly an emphasis on quality and "best practice" in the development of pathways and accompanying outcomes. We have found though, that in all but a few cases, costs *do* go down as a byproduct of the process. Our costs per pathway are reported on a quarterly basis, along with utilization patterns, average length of stay (ALOS) and comparisons to available benchmarks, ie, state data, geometric average Medicare data, etc (Table 7–2).

Patient Satisfaction Outcomes

Community Medical Center, an affiliate of the SBHCS, measures patient satisfaction utilizing Press Ganey Associates Inc. One hundred percent of all patients who are hospitalized, or seek emergency services receive a patient satisfaction questionnaire within 1 week of discharge. The results are mailed directly to Press Ganey, which provides monthly reports with benchmarking comparisons to similarly sized facilities. These data, in addition to feedback from patient letters and focus groups, provide the basis for improvement efforts within our hospital and health system.

Table 7–2 Community Medical Center Patient Care Pathway Quarterly Report

Pathway	Quarter	ALOS	Medicare ALOS	M & R Criteria	Total # Eligible	Total % on Pathway
1. AAA	1st Quarter '97	(9.5 non)	8.7	4.0	15	0%
2. AMI	1st Quarter '97	3.75 (4.5 non)	4.9	4.0	43	7%
3. Unstable Angina	1st Quarter '97	2.0 (2.1 non)	3.1	2.0	37	5%
4. Adult Asthma	1st Quarter '97	2.0 (4.0 non)	3.8	2.0	20	5%
5. Pedi Asthma	1st Quarter '97	3.3 (1.9 non)	3.0 NJ State ALOS	2.0	37	27%
6. C-Section	1st Quarter '97	3.5 (3.6 non)	3.4	2.0	64	63%
7. Chemo	1st Quarter '97	1.8 (2.3 non)			213	4%
8. Chest Pain	1st Quarter '97	1.2 (1.9 non)	2.3	1.0	27	11%
9. CHF	1st Quarter '97	5.4 (5.3 non)	5.2	2.0	273	15%
10. Lap Chole	1st Quarter '97	(3.7 non)			41	0%
11. COPD	1st Quarter '97	4.0 (5.7 non)	5.3	3.0	86	5%
12. CVA	1st Quarter '97	6.4 (6.6 non)	6.0	6.0	151	38%
13. DKA	1st Quarter '97	3.0 (4.1 non)	4.7	2.0	24	13%
14. Fractured Hip	1st Quarter '97	5.4 (6.6 non)	8.2	3.0	71	59%
15. GI Bleed	1st Quarter '97	6.1 (4.6 non)	4.7	1.0	94	5%
16. Hysterectomy	1st Quarter '97	2.3 (2.6 non)	3.5		112	25%
17. Normal Newborn	1st Quarter '97	2.1 (3.8 non)	3.1		390	99%
18. Pneumonia ECF	1st Quarter '97	6.3 (5.8 non)	6.2	2.0	18	44%

Source: Copyright © Community Medical Center.

Staff Satisfaction Outcomes

Employee satisfaction is measured annually on a systemwide basis. Over 20,000 associates received the first survey in May 1997 inquiring about satisfaction with organizational communication, our individual work environment, and understanding of policies and procedures. Results are tabulated in aggregate with process improvement activities and accolades planned for both our system and individual centers.

Physician Satisfaction Outcomes

Surveys were mailed to all attending physicians within the system in the spring of 1997. Results are tabulated individually and in aggregate for our nine system hospitals. Plans are currently in place to work with interested physician groups to gain a deeper understanding of satisfaction results and satisfaction improvement methods.

CHF Outcomes: The Early Results

Length of Stay and Resource Consumption

An initial step in pathway construction and implementation is the examination of resource consumption by service line. As previously mentioned, our admission and readmission volume was among the highest in the nation, so there was considerable interest in this pathway. Resources in each category per ALOS were measured be-

fore and after implementation of the CHF pathway. In our case the ALOS was reduced by approximately 2 days (Figure 7–4), while overall resource consumption was reduced by $600 per admission (Figure 7–1).

To date our quarterly reports consist of pathway and nonpathway average LOS and comparisons to state, Medicare, and Milliman and Robertson guidelines (Table 7–2). In hindsight, there are two areas of measurement we now wish we had pursued—outcomes measurements with *nonpathway* patients, and more in-depth examination of pathway daily variances. Physicians who chose not to utilize the pathways for personal reasons were exposed to them in the introduction process along with the "best practice" research upon which they were based. Since our data collection efforts focused on pathway patients, we have no data to support our hypothesis that these pathways have influenced overall practice, akin to the "Hawthorne Effect," which could partially account for the overall changes in ALOS. This also poses a potential quality issue, for if practice patterns *vary* between pathway and nonpathway patients, we should be examining and encouraging those patterns that produce the best patient outcomes. Another area of measurement that we now need to "catch up" on is the analysis of daily variances for individual pathways. As we continuously evaluate the efficacy and effectiveness of our actions, in the process of updating our pathways, we could now utilize our own data in addition to current literature.

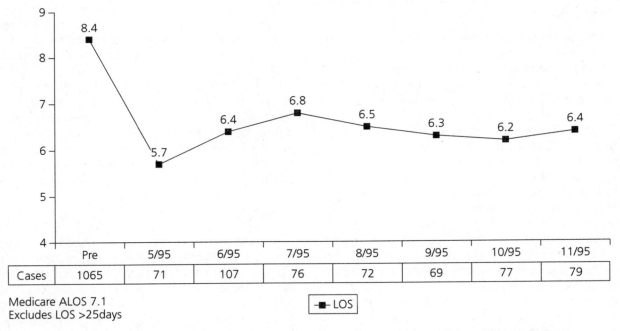

	Pre	5/95	6/95	7/95	8/95	9/95	10/95	11/95
Cases	1065	71	107	76	72	69	77	79

Medicare ALOS 7.1
Excludes LOS >25days

Figure 7–4 Pathway LOS Analysis CHF DRG 127 Excludes Expirations. *Source:* Copyright © Community Medical Center.

Outcomes Efforts in CHF—Current and Future Initiatives

Outcomes management would draw on four rapidly maturing techniques. First, it would place greater reliance on standards and guidelines that physicians can use in selecting appropriate interventions. Second, it would routinely and systematically measure the functioning and well being of patients, along with disease specific clinical outcomes, at appropriate time intervals. Third, it would pool clinical and outcome data on a massive scale. Fourth, it would analyze and disseminate results from the segment of the database most appropriate to the concerns of each decision maker. This should also allow the entire outcomes management system to be modified continuously and improved with advances in medical science, changes in people's expectations, and alterations in the availability of resources.[10(p.1552)]

With this charge from Dr. Paul Ellwood, considered the father of Outcomes Management, we began to create our own program in January 1997. Initially, we identified the outcomes data currently being collected, the software program utilized, the data source, the owner of the data, and the group(s) that were reviewing the data. There are numerous sources and systems to investigate simply to get a current picture of our outcomes efforts to date.

In analysis of the types of outcomes data currently being reported, and the outcomes data needs of the organization, we outlined our own plan. Highlights of the plan are as follows:

1. Create an outcomes steering committee consisting of the data managers within the organization, to include home health, geriatric services, clinics, and extended care facilities. This group would prioritize studies, critique the study design, and analyze the data for validity.
2. Create a distribution process for the reporting and flow of outcomes data to all stakeholders, with mechanisms for process improvement across the continuum of care.
3. Create a template for the organization of data that is based on disease and incorporates all possible outcomes currently being collected within the organization (Exhibit 7–7). (We struggled considerably with format until we reviewed the work of Levknecht et al in the March 1997 issue of *Journal on Quality Improvement*.[12])
4. Revise pathway outcomes, beginning with CHF, and automate the collection of data, wherever possible through the hospital's MIS.

5. Prioritize the diseases that will be addressed by the program. In our case we are starting with CHF and will look at the top five diagnoses of highest volume over the next 2 years. The outcomes steering committee is currently prioritizing the next two diseases from among our "top 5."
6. Begin studies of functional status within our patient population. We have chosen Velocity Infomatics to capture and analyze this data and will be using the Health Status Questionnaire as the basic functional assessment tool with the addition of disease-specific tools as needed.

The comprehensive disease report will be "owned" and maintained across the continuum of care, and at this level decisions will be made to address opportunities for improvement that present through data analysis. For example, an ambitious clinical service team that decides to lob 3 days from the LOS may earn raves from the finance department until it is discovered that the disease-specific patient satisfaction scores have plummeted, the mortality rate has doubled, the rate of complaint has risen, and the readmission rate has increased by 20%. However, representatives from all service areas can easily address subtler changes in outcomes through utilization of one comprehensive report. For example, a trend in patient satisfaction showing that CHF patients were not prepared concerning what to expect from home care can easily be addressed by the case management department together with the patient education committee and the Home Care Coordinator.

This data can also be used to justify care decisions to managed care companies. It is hard to find a practicing physician or nurse who is not concerned about the effect that managed care is having on quality of care across the continuum. Through the use of pathways and outcomes we can quantify and report the results of the care we feel is best for our patients. Although not yet tested, it would seem impossible for a managed care company to argue against reimbursement for our CHAT Home Care Pathway in a public forum when the results are reviewed. It is this type of research—and not our feelings or hunches—that will bring our patients the kind of health care they need.

CONCLUSION

As more and more demands are placed on health care providers for alternate levels of care, programs like our CHF model represent the beginning of our efforts to utilize resources more effectively—and prove that those efforts make a difference in the lives of those we serve. Patient care pathways are a cornerstone in the structure that will make up the future of health care. As more hos-

Exhibit 7–7 Disease Management Report

Community Medical Center
An Affiliate of The Saint Barnabas Health Care System
Outcomes/Pathways
Disease Management Report   Congestive Heart Failure   Fourth Quarter, 1997

CHF Patient Case Mix
N = 350
Average Age =   Male/Female =
% smokers =
Average # of Comorbidities =
Secondary diagnoses: HTN =   IDDM =   NIDDM =
                   CAD =   ESRD =   Chr Renal =
Old NY Heart Association Classification:
            I =   II =
            III =   IV =   (N =   )
Average LV Ejection Fx =   Mortality rate at 1 year =
% on ACE at DC =   Readmission rate =
% not on ACE at DC =   Readmission rate =
% on Dobutrex =   Read =
% on Primacor =   Read =

Clinical Outcomes
Acute
Discharged on ACE Inhibitors   97%
Telemetry D/C in 2 days   90%
Mortality   1%
Rehospitalization   w/i 30d   1%
                w/i 6 mo   1%
Home Health
CHAT Readmission Rate: (N =   )   3 months =   6 months =
Non CHAT Readmission Rate

Cost and Utilization

	Pathway	Non-Pathway
ALOS		
Avg # of visits (HHP)		
Cost per admission		
Avoidable days		
Denials		
System benchmark		
Quadra med benchmark		
M & R guidelines		

Improvement Activities This Quarter
Pilot of CHF pathway on MIS
Lighthouse presentations on CHF
Dr. Michael Hess presentation
20% Improvement on use of ACE inhibitors
Pathway outcome criteria defined
Inpatient outcome data sysem revised with IS&T

Patient Satisfaction
(Press Ganey top score = 5, pop = CHF Health Status and HHP)
                         CHF pop   Overall CMC
Likelihood of recommending hospital
Explanation of tests and treatments
Nurses attention to special needs
Time physician spent with you
Adequacy of advice for home care
Adequacy of home care arrangements

Functional Health Status (HSQ)
(Top Score 100)

	Initial	*3 Month*	*p*
	(n = 50)	(n = 100)	
Physical function			
Role			
Pain			
Mental			
Emotion			
Social function			
Energy			
GHP			
Depression screen			

Living with Heart Failure Questionnaire
(Scale 0–100, with 100 most severe)

3 month	6 month
(N =)	(N =)

KEY: MIS = medical information system; IS&T = information support and technology; GHP = general health perception; ALOS = average length of stay; M&R = Milliman and Robertson; HTN = hypertension; IDDM = insulin dependent diabetes mellitus; NIDDM = non–insulin dependent diabetes mellitus; CAD = coronary artery disease; ESRD = end stage renal disease; Chr Renal = chronic renal

Source: Copyright © Community Medical Center.

---

pitals move toward a managed care environment and consolidated or integrated systems, the need for a standardized set of "best practices" that is developed by a multidisciplinary team will become critical to the success of these organizations.

Future plans for managing patients with CHF are to implement revised measurable patient-centered outcomes and utilize them to guide our care and practice decisions across all of the levels of our health system. We have also formed a CHF process improvement team focusing on management of this disease in our long-term care facilities. The results of this team's work will complete the care continuum for our path-based practice. Certainly we expect to see a continued shift to the home care setting in the provision of patient care. However, we are more aware of the need to plan this care with all members of the continuum of services and to institute an ongoing measure of our progress and ensure appropriate utilization of services. System changes such as the need for outpatient CHF centers, home health cardiac nursing specialists, community-based case management, telemanagement and telemonitoring, and the need for increased utilization of cardiac rehab are already on the table for discussion. These management strategies can be implemented at relatively low cost, while offering enormous payback potential. Incremental investments in outpatient care reduce patient need for expensive acute care services, producing a dramatically lower total cost of CHF care.[4]

There is no powerful leader or group that can command reforms in health systems. . . . Normally in devising health policies, we are pragmatically inclined to think small and to act smaller—independently. We are faced with a health and professional crisis that compels us to think big and act bigger—together.[10(p.1555)]

REFERENCES

1. Rapaport M. How much does health care cost? *Consumer Digest.* 1997;May/June:46–49.

2. US Department of Health and Human Services, National Institutes of Health, National Heart, Lung and Blood Institute, (IT), *Data Fact Sheet,* 1996.

3. Ignatauicius DA, Hausman KA. *Clinical Pathways for Collaborative Practice.* Philadelphia: WB Saunders Co; 1995:19–25.

4. Cardiology Roundtable. *Beyond Four Walls: Cost Effective Management of Chronic Congestive Heart Failure.* Washington, DC: The Advisory Board Company; 1994.

5. Roark MK. Critical pathways. *Health Care Resource Manage.* 1997;Jan/Feb:12–15.

6. Scholtes PR. *The Team Handbook: How to Use Teams to Improve Quality.* Madison, WI: Joiner Associates Inc; 1991:5–31.

7. Mozena JP, Emenck CF, Black SC. *Clinical Guideline Development, An Algorithm Approach.* Gaithersburg, MD: Aspen Publishers Inc; 1996:11.

8. American Health Consultants. Clinical pathways can help you prevent/win malpractice lawsuits. *Hosp Case Manage.* 1997;5:77–79.

9. Newell M. *Using Nursing Case Management To Improve Health Outcomes.* Gaithersburg, MD: Aspen Publishers Inc; 1996:3.

10. Ellwood, P. Shattuck lecture—outcomes management a technology of patient experience. *JAMA.* 1988:549–1556.

11. Wojner AW. *Outcomes Management: Application for Clinical Practice.* St. Louis, MO: Mosby–Year Book. In press.

12. Levknecht L, Marionice B, Schriefer John, Schreifer Janice. Combining case management, pathways, and report cards for secondary cardiac prevention. *J Qual Improve.* 1997;23(3):162–174.

■ Appendix 7–A ■

Clinical Pathways and Sample Pathway Screens

Exhibit 7–A–1 Community Medical Center Patient Care Pathway (CHF)

(4 Day LOS) DATE		ER—Admission	Day 2	Day 3	Discharge Day 4
TREATMENTS	***	Telemetry for 24 hours	Discontinue telemetry/		Consider discharge
		I&O q 4 hours and chart	_____	– – – – – – – – – –	_____/
	***	HL*	Discontinue HL	– – – – – – – – – –	_____/
	***	V/S and temp q 4 hours then reevaluate	q 8 hours	V/S q 12 hours	_____/
		Daily weight by 7 AM	_____	Weigh before breakfast	_____/
		Old chart to floor/			
CONSULTS		Cardiology, if indicated*			
		Pulmonary, if indicated*			
LABS	***	U/A with micro	Chem 7	CBC with diff	
	***	Chem 20	Magnesium	Chem 7	
	***	Cardiac enzymes every 8 hours times 2	_____	– – – – – – – – –	_____/
	***	CBC with diff	– – – – – – – – – – – –/		
	***	Coag profile			
	***	Magnesium			
	***	Thyroid profile in AM		– – – – – – – – –	_____/
	***	Dig level, if indicated			

continues

Exhibit 7–A–1 continued

(4 Day LOS) DATE	ER—Admission	Day 2	Day 3	Discharge Day 4
OTHER TESTS ***	EKG	Echo/Doppler		
***	Chest x-ray	*Consider cardiac cath or NEST		
RESPIRATORY CARE ***	Continuous nasal oxygen*	Consider changing to lower dose O2	Consider discontinuing oxygen	/
***	Pulse oximetry 30 mins. after O2	Pulse oximetry 30 mins. after O2, change	—————————	——————/
***	Consider ABGs if O2 sat < 92			
MEDICATIONS ***	Consider: IV Dig*	Consider changing IV medications to PO if indicated*	—————————	——————/
***	IV diuretic* or oral			
***	ACE Inhibitor*	—————————	Adjusting ACE Inhibitor*	Discharge on ACE Inhibitors
***	Nitropaste*	Change Nitropaste to oral or patch*	—————————	——————/
***	K+ supplement*			
***	Heparin SC*	Sleeper prn* (Ambien)	—————————	——————/
***	Pericolace*	Stool softener*	—————————	——————/
***	Norvasc			
KEY NURSING ACTIVITIES (Impaired Gas Exchange)	Auscultate breath sounds q 4 hours	q shift	—————————	——————/
	Monitor respiratory rate, rhythm, depth and effort q 4 hours	q shift	—————————	——————/
	Elevate HOB 45 degrees	——————————/		
	Assess mental status q 4 hours	q shift	—————————	——————/
(Fluid Volume Deficit)	Assess extremities q 12 hours for: edema temperature cyanosis	—————————	—————————	——————/
	Restrict fluid intake to 1500 cc unless otherwise ordered	—————————	—————————	——————/
	Assess for abdominal distension q 12 hours	q shift	—————————/	

continues

Exhibit 7–A–1 continued

(4 Day LOS) DATE	ER—Admission	Day 2	Day 3	Discharge Day 4
(Decrease Cardiac Output) (Anxiety)	Bed rest with commode or BR and BRP w/assist Increase activity, as indicated Provide opportunity for verbalization of fears, questions and concerns	_ _ _ _ _ _ _ _ _ _ _ _ _ _ _ _	_ _ _ _ _ _ _ _ _ _ _ _ _ _	_____/
DIETARY	Na restricted and low cholesterol diet, as tolerated Instruction on diet modification	_ _ _ _ _ _ _ _ _ _ _ _ _ _ _ _	_ _ _ _ _ _ _ _ _ _ _ _ _ _ _ _ _ _ _ _ _ _ _ _ _ _ _ _	_____/ _____/
PATIENT/ SIGNIFICANT OTHER EDUCATION	Orient to unit routine Explain treatment/ procedures	_ _ _ _ _ _ _ _ _ _ _ _ _ _ _ _	Reorient x 4 as needed Give booklet "Recovering from Congestive Heart Failure" disease process	_____/ Follow-up care and appointment w/MD Review meds, S/S to report and diet
DISCHARGE PLANNING	Case Management consult Consider CHAT, Non-CHAT program		Coordinate other disciplines needed for discharge	Discharge with computerized instructions customized for patient/ family Referral for smoking cessation
EXPECTED PATIENT OUTCOMES		V/S stable Verbal and nonverbal behavior demonstrates decreased anxiety Output exceeds intake	Verbalizes or demonstrates ability to rest comfortably	Outcome Criteria for discharge: V/S WNL for patient Complete measurement tool

*** ER Order
* MD order required
/ Stop
-- Continue daily
__ Hold until day specified
Definitions: CHAT = congestive heart failure assessment team; I&O = intake and output; HL = heparin lock; V/S = vital signs; temp = temperature; q = every; U/A = urianalysis; chem = chemistry; CBC w/diff = complete blood count with differential; coag = coagulation; K+ = potassium; Dig = digoxin; EKG = electrocardiogram; NEST = nonexercise stress test; O2 = oxygen; ABG = arterial blood gas; IV = intravenous; po = by mouth; ACE = angiotension converting enzyme; HOB = head of bed; BR = bed rest; BRP = bathroom privileges; Na = sodium; WNL = within normal limits

Source: Copyright © Community Medical Center.

Exhibit 7–A–2 Medical Information System—Patient Care Pathway Sample Screens—CHF

```
TRNG -0493          SAINT BARNABAS HEALTHCARE SYSTEM—DEVELOPMENT
MATRIX # 2868    HOSP# 3      08/12/97      01:41 PM


              1         2         3         4
    1234567890123456789012345678901234567890
    ------------------------------------------
                                            01
    CMC        PATIENT CARE PATHWAYS        02
                                            03
    THE DATE OF ACTIVATION OF THE PATIENT   04
       CARE PATHWAY MUST BE PROVIDED        05
                                            06
                                            07
                                            08
    IF PATIENT ALREADY ON PATHWAY, SELECT   09
       *INDEX CHF PILOT ONLY FOR            10
              DAYS 2, 3 & 4                 11
                                            12
    ---------------------------------------13
                                            14
         *CURRENT DATE: MM/DD/YY            15
                                            16
         LAST DATE: ___/___/___            17
                                            18
                                            19
                                            20

    ------------------------------------------
```

Exhibit 7–A–3 Medical Information System—Patient Care Pathway Sample Screens—CHF

```
TRNG -0494          SAINT BARNABAS HEALTHCARE SYSTEM—DEVELOPMENT
MATRIX # 6716    HOSP# 3      08/12/97      01:41 PM
DYNAMIC MATRIX

              1         2         3         4
    1234567890123456789012345678901234567890
    ------------------------------------------
    CMC        PATIENT CARE PATHWAYS        01
                                  08/12/97  02
    DRG    LOS    PROCESS NAME               03
    *110A  8DAYS  AAA-RUPTURED, POST-OP      04
    *110B  8DAYS  AAA-ELECTIVE, PRE-OP       05
    *110C  8DAYS  AAA-ELECTIVE, POST-OP      06
    *098A  3DAYS  ASTHMA-ADULT               07
    *098B  3DAYS  ASTHMA-PEDI                08
    *371A  48HRS  CESAREAN SECTION, PRE-OP   09
    *371B  48HRS  CESAREAN SECTION, POST-OP  10
    *410A  3DAYS  CHEMOTHERAPY-3DAY          11
    *410B  4DAYS  CHEMOTHERAPY-4DAY          12
    *143.  23HRS  CHEST PAIN                 13
    *127   4DAYS  CHF                        14
    *088.  5DAYS  COPD                       15
    *014.  7DAYS  CVA                        16
    *236.  8DAYS  FRACTURED HIP              17
    *175A  4DAYS  GI BLEED-LOWER             18
    *175B  3DAYS  GI BLEED-UPPER             19
                              *CONTINUE      20
    ------------------------------------------
```

Exhibit 7–A–4 Medical Information System—Patient Care Pathway Sample Screens—CHF

```
TRNG -0495          SAINT BARNABAS HEALTHCARE SYSTEM—DEVELOPMENT
MATRIX # 10970   HOSP# 3       08/12/97      01:41 PM


            1         2         3         4
12345678901234567890123456789012345678990
------------------------------------------
CMC               CHF PROTOCOL                01
                                              02
         IF NOT ORDERED IN THE ED, SELECT     03
CBC W/DIFF STAT.....................** 04
EKG WITH RHYTHM STRIP-INDICATIONS: CAD  05
 414.9, SCHEDULING: STAT, INSTRUCTIONS, 06
 LEAVE ON CHART....................** 07
X-RAY: 40401023 CHEST, INDICATIONS: CHF 08
 428.0, SCHEDULE STAT, PRECAUTIONS:     09
 DO NOT REMOVE ELECTRODES...........** 10
PULSE OXIMETRY, 30MIN AFTER O2          11
 INITIATED........................** 12
ARTERIAL BLOOD GAS-ADMITTED PATIENT     13
 IF O2 SAT IS <92.................** 14
                                        15
NASAL CANNULA, ___LPM CONTINUOUS        16
                                        17
                                        18
                                        19
                           *CONTINUE 20
------------------------------------------
```

Source: Copyright © Saint Barnabas Health Care System.

Exhibit 7–A–5 Medical Information System—Patient Care Pathway Sample Screens—CHF

```
TRNG -0496          SAINT BARNABAS HEALTHCARE SYSTEM—DEVELOPMENT
MATRIX # 10969   HOSP# 3       08/12/97      01:41 PM


            1         2         3         4
12345678901234567890123456789012345678990
------------------------------------------
CMC               CHF PROTOCOL                01
                                              02
CHEM 20, PREP L2, TODAY.............** 03
MAGNESIUM TODAY....................** 04
THYROID PROFILE TODAY..............** 05
COAG PROFILE TODAY.................** 06
CHEM 7 DAILY X2 START TOMORROW......** 07
U/A AUTO MICRO, RAND URINE, TODAY...** 08
             ALL OF THE ABOVE......** 09
                                        10
 IF NO CARDIAC PROFILE DONE IN ED SELECT 11
    CARDIAC PROFILE STAT &THEN Q8H X2   12
                  OR                    13
  IF CARDIAC PROFILE DONE IN ED SELECT  14
       CARDIAC PROFILE SPACED AT        15
    8HOUR INTERVALS FROM INITIAL DRAW   16
    *CARDIAC PROFILE TODAY AT           17
    *CARDIAC PROFILE TOMORROW AT        18
                                        19
*DIGOXIN LEVEL              *CONTINUE 20
------------------------------------------
```

Source: Copyright © Saint Barnabas Health Care System.

Exhibit 7–A–6 Medical Information System—Patient Care Pathway Sample Screens—CHF

```
TRNG -0497        SAINT BARNABAS HEALTHCARE SYSTEM—DEVELOPMENT
MATRIX # 10973   HOSP# 3        08/12/97       01:41 PM


                    1         2         3         4
         12345678901234567890123456789012345678 90
         ------------------------------------------
         CMC              CHF PROTOCOL              01
                                                    02
         HEPARIN LOCK INSERT ACCORDING TO          03
          HOSPITAL POLICY CON'T TILL DC'D          04
         SALINE 0.9% FLUSH 2CC IV PUSH DAILY AT    05
          8AM.....................................** 06
         TELEMETRY X24HRS.......................** 07
         VITAL SIGNS: T-P-R Q4H................** 08
         VITAL SIGNS: BP-LYING Q4H............** 09
         VITAL SIGNS: BP STANDING Q4H.........** 10
         RECORD I&O: Q4H & CHART..............** 11
         WEIGH PATIENT ON ADMISSION & DAILY AT     12
          7AM BEFORE BREAKFAST.................** 13
         ACTIVITY: BED REST & BRP W/ASSIST....** 14
         DIET: 2GM NA: LOW CHOLESTEROL........** 15
         ACETEMINOPHEN TAB 325MG GIVE #2, PO       16
          Q4H PRN.............................** 17
                                                    18
                     ALL OF THE ABOVE....** 19
                                    *CONTINUE 20
         ------------------------------------------
```

Source: Copyright © Saint Barnabas Health Care System.

Exhibit 7–A–7 Medical Information System—Patient Care Pathway Sample Screens—CHF

```
TRNG -0498        SAINT BARNABAS HEALTHCARE SYSTEM—DEVELOPMENT
MATRIX # 10974   HOSP# 3        08/12/97       01:41 PM


                    1         2         3         4
         12345678901234567890123456789012345678 90
         ------------------------------------------
         CMC         CHF PROTOCOL                  01
                                                    02
         INITIATE CASE MANAGEMENT PROTOCOL FOR     03
          CHF                                       04
                                                    05
         DISCHARGE PLANNING SERVICES: HOME         06
          SERVICES: HOME HEALTH NURSE SERVICES:    07
         CMC HHP                                    08
         01                                         09
          INITIATE HOME SERVICES REFERRAL FOR      10
           CHAT                                     11
          INITIATE HOME SERVICES REFERRAL FOR      12
           NON-CHAT                                 13
                                                    14
                                                    15
                                                    16
                                                    17
                                                    18
                                                    19
                                    *CONTINUE 20
         ------------------------------------------
```

Source: Copyright © Saint Barnabas Health Care System.

Exhibit 7–A–8 Medical Information System—Patient Care Pathway Sample Screens—CHF

```
TRNG -0499        SAINT BARNABAS HEALTHCARE SYSTEM—DEVELOPMENT
MATRIX # 2039     HOSP# 3      08/12/97      01:41 PM


                1         2         3         4
       12345678901234567890123456789012345678 90
       --------------------------------------------
CMC                    CHF PROTOCOL                 01
                                                    02
             SELECT ONE OF THE FOLLOWING            03
       (2D ECHO) ECHOCARDIOGRAM TO ECHO LAB         04
         WITHOUT NURSE/WITHOUT MONITOR              05
         INDICATION: LV FXN V71.8                   06
         SCHEDULE ON __/__/__                       07
                          OR                        08
       ECHO/DOPPLER TO ECHO LAB WITHOUT NURSE/      09
         WITHOUT MONITOR INDICATION: MITRAL         10
         REGURGITATION 746.6                        11
         SCHEDULE ON __/__/__                        12
                                                    13
                                                    14
                                                    15
                                                    16
                                                    17
                                                    18
                                                    19
                               *MEDICATIONS 20
       --------------------------------------------
```

Exhibit 7–A–9 Medical Information System—Patient Care Pathway Sample Screens—CHF

```
TRNG -0500        SAINT BARNABAS HEALTHCARE SYSTEM—DEVELOPMENT
MATRIX # 2098     HOSP# 3      08/12/97      01:42 PM


                1         2         3         4
       12345678901234567890123456789012345678 90
       --------------------------------------------
CMC                CHF PROTOCOL                 01
                                 *CUR.ORD/DC 02
         DIGOXIN INJ 0.125MG/ML CHECK APICAL      03
                                                  04
         DIGOXIN INJ 0.25MG/1ML CHECK APICAL      05
                                                  06
         FUROSEMIDE INJ 20MG/2ML                  07
                                                  08
         FUROSEMIDE INJ 40MG/4ML                  09
                                                  10
         FUROSEMIDE INJ 80MG/8ML                  11
                                                  12
         POTASSIUM CL SLOW RELEASE 10MEQ (ORAL    13
           DOSE SHOULD NOT BE CRUSHED) GIVE #1    14
                   SELECT ROUTE/SCHEDULE          15
         IV PUSH      DAILY START TODAY           16
         PO           DAILY START TOMORROW        17
                      STAT      BID     Q8H       18
                      &THEN     TID     Q12H      19
                                  *CONTINUE 20
       --------------------------------------------
```

Exhibit 7–A–10 Medical Information System—Patient Care Pathway Sample Screens—CHF

```
TRNG -0501          SAINT BARNABAS HEALTHCARE SYSTEM—DEVELOPMENT
MATRIX # 2129    HOSP# 3         08/12/97      01:42 PM


            1         2         3         4
1234567890123456789012345678901234567890
----------------------------------------
CMC                 CHF PROTOCOL                01
                                                02
 NITROGLYCERIN OINT 2% _____ INCH(S)            03
  CHECK BP APPLY TO CHEST WALL                  04
                                                05
 HEPARIN INJ 5000UNITS/ML                       06
                                                07
 HEPARIN INJ 7500UNITS                          08
                                                09
 DOCUSATE SOD. W/CASANTHRANOL GIVE #1           10
                                                11
 AMLODIPINE 5 MG TAB CHECK BP GIVE #1           12
                                                13
 ZOLPIDEM TAB 5 MG GIVE #1                      14
                SELECT ROUTE/SCHEDULE           15
   PO    DAILY START TODAY                       16
   SC    DAILY START TOMORROW                    17
         STAT    Q6H    QHS        BID           18
         &THEN   Q8H    QHS PRN    TID           19
                 Q12H        *ACE INHIBIT        20
----------------------------------------
```

Source: Copyright © Saint Barnabas Health Care System.

Exhibit 7–A–11 Medical Information System—Patient Care Pathway Sample Screens—CHF

```
TRNG -0502          SAINT BARNABAS HEALTHCARE SYSTEM—DEVELOPMENT
MATRIX # 2312    HOSP# 3         08/12/97      01:42 PM


            1         2         3         4
1234567890123456789012345678901234567890
----------------------------------------
CMC                 CHF PROTOCOL                01
                                                02
ACE INHIBITORS                                  03
 LISINOPRIL 5 MG TAB CHECK BP GIVE              04
  #1/2, PO STAT & THEN DAILY                    05
                                                06
 ENALAPRIL MALEATE TAB 2.5 MG CHECK BP          07
  GIVE #1, PO STAT & THEN DAILY                 08
                                                09
 CAPOTEN CAPTOPRIL TAB 12.5 MG CHECK BP         10
  GIVE #1/2, PO STAT & THEN DAILY               11
                                                12
 COZAAR 25 MG CHECK BP GIVE #1, PO STAT         13
  &THEN DAILY                                   14
                                                15
 COZAAR 50 MG CHECK BP GIVE #1, PO STAT         16
  &THEN DAILY                                   17
                                                18
                                                19
                    *CARDIOLOGY CONSULT         20
----------------------------------------
```

Source: Copyright © Saint Barnabas Health Care System.

Exhibit 7–A–12 Medical Information System—Patient Care Pathway Sample Screens—CHF

```
TRNG -0503          SAINT BARNABAS HEALTHCARE SYSTEM—DEVELOPMENT
MATRIX # 4700    HOSP# 3      08/12/97      01:42 PM


             1         2         3         4
12345678901234567890123456789012345678 90
-------------------------------------------
CMC                   INDEX                    01
                                               02
CHF:    *DAY 2                                  03
        *DAY 3                                  04
        *DAY 4    (VIEW ONLY)                   05
                                               06
                                               07
                                               08
                                               09
                                               10
                                               11
                                               12
                                               13
                                               14
                                               15
                                               16
                                               17
                                               18
                                               19
                                               20

-------------------------------------------
```

Source: Copyright © Saint Barnabas Health Care System.

Exhibit 7–A–13 Medical Information System—Patient Care Pathway Sample Screens—CHF

```
TRNG -0504          SAINT BARNABAS HEALTHCARE SYSTEM—DEVELOPMENT
MATRIX # 2376    HOSP# 3      08/12/97      01:42 PM


             1         2         3         4
12345678901234567890123456789012345678 90
-------------------------------------------
CMC               CHF PROTOCOL                 01
                                               02
         THE FOLLOWING ARE RECOMMENDED         03
                  GUIDELINES                   04
           FOR THE TREATMENT OF CHF            05
                                               06
DISCONTINUE TELEMETRY                          07
DISCONTINUE HEPARIN LOCK                        08
CHANGE IV DIGOXIN & DIURETIC TO PO             09
INCREASE ACE INHIBITOR DOSE DEPENDING          10
 ON BLOOD PRESSURE AND POTASSIUM LEVEL         11
EVALUATE O2 AND ACTIVITY ORDERS                12
                                               13
                                               14
TO DISCONTINUE CURRENT ORDERS. SELECT          15
            *CUR. ORD/DC                        16
TO CONTINUE DAY 2 ORDERS, SELECT               17
            *DAY 2                              18
                                               19
                                               20

-------------------------------------------
```

Source: Copyright © Saint Barnabas Health Care System.

Exhibit 7–A–14 Medical Information System—Patient Care Pathway Sample Screens—CHF

```
TRNG -0505           SAINT BARNABAS HEALTHCARE SYSTEM—DEVELOPMENT
MATRIX # 2387    HOSP# 3      08/12/97      01:42 PM


              1         2         3         4
12345678901234567890123456789012345678 90
-------------------------------------------
CMC         CURRENT ORDERS BY CATEGORY         01
                                               02
                                               03
IVs.......................................** 04
MEDICATIONS...............................** 05
NURSING/CONSULT/MISC....................** 06
RESPIRATORY.............................** 07
                                               08
                                               09
                                               10
                                               11
                                               12
                                               13
                                               14
                                               15
ALL CURRENT ORDER....................** 16
      EXCEPT LAB/RADIOLOGY                      17
                                               18
                                               19
                                               20

-------------------------------------------
```

Source: Copyright © Saint Barnabas Health Care System.

Exhibit 7–A–15 Medical Information System—Patient Care Pathway Sample Screens—CHF

```
TRNG -0506           SAINT BARNABAS HEALTHCARE SYSTEM—DEVELOPMENT
MATRIX # 2402    HOSP# 3      08/12/97      01:42 PM


              1         2         3         4
12345678901234567890123456789012345678 90
-------------------------------------------
CMC             CHF PROTOCOL                   01
                                               02
  NASAL CANNULA, ___LPM CONTINUOUS            03
                                               04
  PULSE OXIMETRY, 30 MIN AFTER O2             05
    INITIATED                                  06
                                               07
  CBC W/DIFF TOMORROW                          08
                                               09
  ACTIVITY, AS TOLERATED                       10
                                               11
          IF NOT ORDERED, SELECT:              12
RESPIRATORY CARE CONSULT TO INSTRUCT           13
   PATIENT ABOUT SMOKING CESSATION             14
   EDUCATION PROGRAM                           15
                                               16
                                               17
                        *DIGOXIN/LASIX  18
                        *ACE INHIBITORS  19
                                               20

-------------------------------------------
```

Source: Copyright © Saint Barnabas Health Care System.

Exhibit 7–A–16 Medical Information System—Patient Care Pathway Sample Screens—CHF

```
TRNG -0507          SAINT BARNABAS HEALTHCARE SYSTEM—DEVELOPMENT
MATRIX # 2401    HOSP# 3       08/12/97      01:42 PM


            1         2         3         4
   12345678901234567890123456789012345678 90
   ---------------------------------------
   CMC               CHF PROTOCOL              01
                                              02
    DIGOXIN TAB 0.125MG CHECK APICAL          03
      GIVE #1, PO                             04
                                              05
    DIGOXIN TAB 0.25MG CHECK APICAL           06
      GIVE #1, PO                             07
                                              08
    FUROSEMIDE TAB 20MG GIVE #1, PO           09
                                              10
    FUROSEMIDE TAB 40MG GIVE #1, PO           11
                                              12
    FUROSEMIDE TAB 80MG GIVE #1, PO           13
                                              14
                 SELECT SCHEDULE              15
       DAILY START TODAY        BID           16
       DAILY START TOMORROW    Q8H            17
       STAT                    Q12H           18
       &THEN                                  19
                          *ACE INHIBITORS 20
   ---------------------------------------
```

Source: Copyright © Saint Barnabas Health Care System.

Exhibit 7–A–17 Medical Information System—Patient Care Pathway Sample Screens—CHF

```
TRNG -0508          SAINT BARNABAS HEALTHCARE SYSTEM—DEVELOPMENT
MATRIX # 3788    HOSP# 3       08/12/97      01:42 PM


            1         2         3         4
   12345678901234567890123456789012345678 90
   ---------------------------------------
   CMC               CHF PROTOCOL              01
                                              02
   INCREASE DOSE OF ACE INHIBITORS IF NO      03
   S/S, SYSTOLIC BP * 100 AND K+ WNL          04
                                              05
    ENALAPRIL MALEATE TAB 5MG CHECK BP        06
      GIVE #1, PO BID                         07
                                              08
    CAPOTEN CAPTOPRIL TAB 12.5MG CHECK BP     09
      GIVE #1, PO BID                         10
                                              11
    LISINOPRIL 5MG TAB CHECK BP GIVE #1,      12
      PO DAILY START TODAY                    13
                     OR                       14
    LISINOPRIL 5MG TAB CHECK BP GIVE #1,      15
      PO DAILY START TOMORROW                 16
                                              17
                                              18
                                              19
                                              20
   ---------------------------------------
```

Source: Copyright © Saint Barnabas Health Care System.

Exhibit 7–A–18 Medical Information System—Patient Care Pathway Sample Screens—CHF

```
TRNG -0509          SAINT BARNABAS HEALTHCARE SYSTEM—DEVELOPMENT
MATRIX # 4510    HOSP# 3        08/12/97      01:42 PM


              1         2         3         4
      1234567890123456789012345678901234567890
      ------------------------------------------
CMC                CHF PROTOCOL                  01
                              *CUR.ORD/DC        02
      INCREASE DOSE OF ACE INHIBITORS IF NO      03
      S/S, SYSTOLIC BP * 100 AND K+ WNL          04
                                                 05
      ENALAPRIL MALEATE TAB 10MG CHECK BP        06
       GIVE #1, PO BID                           07
                                                 08
      LISINOPRIL 10MG TAB CHECK BP GIVE #1,      09
       PO DAILY START TODAY                      10
                        OR                       11
      LISINOPRIL 10MG TAB CHECK BP GIVE #1,      12
       PO DAILY START TOMORROW                   13
                                                 14
      CAPOTEN CAPTOPRIL TAB 25MG CHECK BP        15
       GIVE #1, PO BID                           16
                                                 17
                                                 18
                                                 19
                                                 20

      ------------------------------------------
```

Source: Copyright © Saint Barnabas Health Care System.

Exhibit 7–A–19 Medical Information System—Patient Care Pathway Sample Screens—CHF

```
TRNG -0510          SAINT BARNABAS HEALTHCARE SYSTEM—DEVELOPMENT
MATRIX # 4697    HOSP# 3        08/12/97      01:42 PM


              1         2         3         4
      1234567890123456789012345678901234567890
      ------------------------------------------
CMC                CHF PROTOCOL                  01
                                                 02
                   VIEW ONLY                     03
                                                 04
      DISCHARGE PATIENT TODAY                    05
      DISCHARGE ON ONE OF THE FOLLOWING MEDS:    06
          VASOTEC 20MG PO BID                    07
          LISINOPRIL 10MG PO DAILY              08
          COZAAR 50MG PO DAILY                   09
          CAPOTEN 25MG PO BID                    10
                                                 11
                                                 12
                                                 13
                                                 14
                                                 15
                                                 16
                                                 17
                                                 18
                                                 19
                                                 20

      ------------------------------------------
```

Source: Copyright © Saint Barnabas Health Care System.

■ 8 ■

Stroke Outcome Improvement

Deborah Summers, Patricia A. Soper, and Marilyn Rymer

STROKE OUTCOME IMPROVEMENT

Saint Luke's Hospital is a 642-bed urban, tertiary care, teaching hospital in metropolitan Kansas City, MO. In April 1992, Saint Luke's developed a Collaborative Care program supported by tools such as clinical paths as a means to improve outcomes and ensure best clinical practice processes. This chapter describes the development of the Collaborative Care program for stroke patients, highlights the Cerebrovascular Accident (CVA) clinical path and patient path, and describes the outcomes from these efforts.

In 1992, as managed care began to penetrate the Kansas City community it was clear that for Saint Luke's Hospital to retain its quality reputation and remain competitive in the marketplace, clinical outcomes needed to be maintained or enhanced while costs had to be more carefully controlled. In response to this challenge, the Collaborative Care program set out to adapt the principles of internal case management through implementation of a collaborative, interdisciplinary care delivery model designed for specific patient populations. Collaborative practice teams were formed to focus on the care of certain patient groups, defined generally by diagnosis-related groups (DRGs). Patients with stroke (DRG 14) were selected among the first classifications to be included since they met our initial criteria of high volume, high risk, high cost/loss, and clinician readiness.

Nationally, stroke is the leading cause of adult disability and the third leading cause of death in the population over age 60. Direct stroke care from the initial hospitalization and rehabilitation through long-term care costs $17 billion annually.[1,2] A study by Wentworth and Atkinson demonstrated that by using a stroke path and associated physician standing orders, the mean length of stay declined from 7.0 days to 4.6 days providing a sav-

ings of approximately $450,000 per year over a period of 4 years.[3] Similarly, a study by Jorgensen et al found that the use of a stroke care unit and associated guidelines resulted in reductions in length of stay (30%), mortality, and readmissions.[4]

The diagnosis of stroke is typically associated with high risk for morbidity and mortality related to the complexity of the disease, secondary complications, and the potential for difficult or fragmented communications among multiple caregivers. In 1992, treatment options were limited to medical support and rehabilitation. At Saint Luke's Hospital the stroke patient population represents a high volume. In 1992, DRG 14 ranked 12th in volume (inpatient discharges) and 8th in 1996. Saint Luke's Hospital treatment costs were higher in 1992 than benchmarked hospitals and DRG 14 was ranked 7th by (total) loss to profit. In 1992 the Saint Luke's average length of stay (ALOS) for these patients was 8.0 days, lower than the HCFA mean of 9.8, but higher than benchmarked hospitals. In addition, patient, physician, and staff satisfaction scores were lower than targeted. In 1995, the Health Care Financing Administration (HCFA) reimbursement for acute stroke hospitalization was $4826.[5] At the same time, newly approved thrombolytic therapy alone cost approximately $2300 per patient.

The neuroscience department at Saint Luke's continues to be respected in the region as a center of excellence, drawing referrals from a wide geographic radius. When the Collaborative Care program was initiated, a neuroscience care unit was in place, however, stroke patients were cared for throughout the hospital leading to inconsistencies of medical and nursing management. The neuroscience clinical staff and physicians demonstrated a readiness to embark on a new course to improve outcomes and processes. A stroke clinical path had been developed in the past, but was unsuccessfully implemented. One neurologist agreed to be a champion of the effort and the nurses displayed an interest and willingness to commit the time and energy necessary to design an improved care delivery system.

PATHWAY AND PROTOCOLS

A stroke collaborative practice team formed and included a neurologist; neuroscience staff nurses; physical, speech, and occupational therapists; a nutritionist; a social worker; a chaplain; and the neuroscience unit nurse manager. Others participated with the team: the medical/surgical clinical nurse educator, the collaborative care coordinator, a utilization review specialist, a medical records specialist, and a financial specialist. Members of the team were selected by the neurologist, the nurse manager and the neuroscience clinical nursing practice committee. Team members agreed to a 2-year commitment with the goal of initiating the Collaborative Care program in neuroscience and ensuring its continued success.

Goals of the Collaborative Care program were as follows:

- Maintain/improve patient outcomes.
- Increase continuity of care and reduce practice variation while allowing for individual patient needs.
- Increase patient and family understanding of, participation in, and satisfaction with care.
- Identify opportunities and provide mechanisms for quality improvement.
- Promote appropriate, cost-effective resource utilization.
- Operationalize the value-added concept.

Roles were carefully defined and periodically revised. The collaborative care director facilitated for the team extensive education and rationale regarding the current and projected health care financial environment, case management concepts, data collection and review tools, clinical path models, and effective team skills. The nurse manager provided necessary resources (ie, time available for staff to participate, support of the program and the team). The physician and medical surgical clinical nurse educator provided clinical expertise, scientific literature, support, and assistance with team process including consensus-building and conflict resolution. The clinical staff provided their knowledge of this patient population and the clinical care processes involved. The medical records, utilization review, and finance members provided data necessary to accurately define our stroke population, ie, trends in patient volume, case-mix demographics, resource utilization, ALOS, clinical outcomes, cost and reimbursement figures, and marketplace demands.

An all-day meeting was held off site to facilitate this process and to define goals and timelines. At this meeting the team reviewed and discussed the stroke case mix and financial data and formulated assumptions about our patient's needs. Using chart audits and empirical data, a stroke clinical path was drafted that reflected current practice. At subsequent meetings the involved disciplines reviewed the literature for examples of "best practice," examined clinical paths obtained from other hospitals, and referred to a path previously written but not implemented. Collaboratively they revised the path draft to reflect improved outcomes, starting with discharge goals and then adding daily patient goals and clinical interventions to achieve those goals within the desired timeline. The expectation was that the path would accommodate the needs of 80% of our stroke patient population. The process of review, revision, and validation continued, and each draft was sent to the neurologists and interdisciplinary staff for comment. When the team reached consensus, the final draft was prepared as our interdisciplinary plan of care for stroke patients.

The clinical path format changed over time as staff suggested improvements (Appendix 8–A, Exhibit 8–A–1). Page 1 of Exhibit 8–A–1 defines the general plan of care. This begins with specific patient information such as physician, date of admission, etc. The targeted ALOS was defined by the team and written on the first path as 7 days (1 day less than actual ALOS). On later revisions the targeted ALOS was omitted from the path because physicians believed that hindered earlier discharge in some cases. The common nursing diagnoses for stroke patients are listed next. For each diagnosis the general outcomes expected for these patients and the interdisciplinary staff interventions required to meet those outcomes are defined.

Pages 2 and 3 of Exhibit 8–A–1 outline the general plan of care in detail. The vertical columns refer to designated time frames. In the case of the stroke path this is days; however, with some paths the time elements may be minutes, stages, weeks, etc. The last column defines patient outcomes required for discharge. The horizontal rows refer to care categories (eg, ASSESSMENT) and these are standardized throughout the hospital with minor variations to accommodate specific patient populations. Within this grid daily patient outcomes are defined to ensure that the discharge outcomes are achieved within the defined time frame. Also included are the physician and staff interventions required to assist the patient in achieving each goal. For example, to resolve the mobility diagnosis (page 1) and to achieve the discharge outcome of "assists with bed to chair transfer" (page 3, MOBILITY/ADLs row) the patient must begin "rehab at bedside" on day 2 (page 2, second column, MOBILITY/ADLs row).

In developing the clinical path we found it most effective to first define the general care plan for the stroke patient, then determine expected discharge goals (page 3, last column) and finally fill in the grid each day working toward discharge. Over time the goals and interventions in the grid change as the clinicians and patients determine more effective and efficient processes to achieve improved outcomes. Each path revision has the revision date noted on the first page, lower right-hand corner. A copy of the original path and each revision is maintained for reference.

The clinical path is used as a guide rather than as a standard of care and is expected to be individualized for each patient. As such, interventions on the path that require a physician order are not instituted until the order is written. To standardize and streamline that process, the neurologists instituted standing orders to reflect the path interventions.

DOCUMENTATION

Utilizing the path for interdisciplinary documentation was an initial goal of the team but was not achieved until the path had been in use for a number of months. Because this was a significant change in charting procedures for all disciplines, negotiation and consensus from the practice committees of each discipline was required. In each grid there is either a box or a line preceding a statement. A line precedes patient outcomes and requires the initials of the provider verifying this outcome or a notation defined in the symbol key at the bottom of the page. Signatures are authenticated on the bottom of the same page. A box precedes provider interventions and requires a notation as indicated in the symbol key. When an item is not done, a variance note is required in the progress notes and on the variance report form attached to the path. Blank lines following a statement in the grid allows for individualization of the path. For example, in the TESTS/LABS row on day 1 there is a blank line following admission oxygen saturation (ADM O_2 sat) so the actual oxygen saturation value can be entered. To standardize procedures and to facilitate documentation, various protocols were developed by the team. All protocols are available on the unit for staff reference but the content is not repeated in the path. A transfer record concludes the path to ensure complete yet efficient documentation (page 4).

VARIANCES

The variation report, attached to the clinical path, is not a permanent part of the medical record so is removed when the patient is discharged. We use colored paper for this sheet to more easily identify it as a quality monitoring tool. All disciplines utilize this tool to record variances (positive and negative) from expected outcomes or interventions. These variation reports are collected by the team, and the data are collated and analyzed quarterly to identify trends or high-risk situations that may provide an opportunity for improvement. The team then develops and institutes a quality improvement plan related to the identified opportunity and monitors improvement progress.

PATIENT PATHWAY

Originally the team shared the clinical path with patients and families but soon found that it was too detailed and difficult to understand. With assistance from patients and families the team developed a patient path designed as an educational tool for patients. It outlines the plan of care through hospitalization and discharge and indicates what the patient can expect from the staff and facility and what patient responsibilities are expected. This tool has been revised periodically based on process changes and as a result of patients' feedback regarding their learning needs and their desire for involvement, accountability, and satisfaction with the care experience (Exhibit 8–A–2).

CARE/CASE MANAGEMENT

Clinical nurses at the bedside assume responsibility for the care management of each stroke patient. All patients are managed through the use of the clinical path. The physician standing orders are required for all stroke patients on the neuroscience unit. Progress on the path is evaluated each shift, additional interventions are initiated if necessary to reduce variation, and path progress is discussed in shift report. The patient path is provided to, and reviewed with, the patient and family as soon as possible after admission and again at least on a daily basis during hospitalization.

For ease of use by the interdisciplinary care team, the clinical path is located in the patient's chart at the bedside. Physicians examine the path during daily rounds. Nurses, the social worker, the nutritionist, and the physical, speech, and occupational therapists utilize the path as a care plan, shift report, and documentation tool. The utilization review specialist analyzes the path daily as part of concurrent chart review and enters pertinent information into a database via laptop computer. The clinical nurse specialist reviews the path when intervening for a specific patient. Nurses who float to the neuroscience unit rely on the path to define the individualized plan of care. The path is utilized as a learning tool for nurses, medical students, and physician residents orienting to the neuroscience unit.

The stroke collaborative practice team progressed rapidly with paths and protocols in place. New diagnostic and treatment modalities became available and the team agreed to become involved with national research initiatives. In 1993 we articulated our vision of a stroke center at Saint Luke's Hospital dedicated to providing a continuum of outstanding and aggressive care and services regionally for stroke patients and families. Extensive research and administrative support produced a business plan for the stroke program, which was approved by the board of directors. The previous neuroscience clinical educator completed graduate studies, transitioned to the roles of clinical nurse specialist, and stroke center coordinator. Comprehensive staff education occurred while remodeling and equipment installation of the medical intensive care unit and contiguous neuroscience unit produced a state-of-the-art stroke center, which opened in January 1994. Internal and external marketing resulted in overall increase in volume of stroke patients in the hospital and concentration of these patients on the stroke center, enhancing the continuity of care and the competency of staff.

CARE TEAM

In the past, a multidisciplinary clinical team on the neuroscience unit met weekly to discuss each patient's progress toward discharge. As length of stay declined this process became less effective in timely assessment and intervention. In 1995 a neuroscience care coordination team (care team) was formed to replace the weekly interdisciplinary discharge planning rounds. This care team consists of the neuroscience clinical nurse specialist as leader, the neuroscience social worker, and the neuroscience utilization review specialist. They communicate daily with each other and with the physicians and clinical staff to review the progress and needs of each stroke patient and to assist the direct care providers to meet specific patient needs. The care team formally rounds at least three times per week and is particularly valuable in cases where patients are not progressing as well as expected on the path.

OUTCOMES MEASUREMENT

Until recently, stroke has been America's most neglected disease, despite the fact that every minute someone suffers a stroke in this country. New significant therapeutic modalities and interventional technologies to reverse damage from a stroke has recently changed the treatment of acute stroke patients. Improvement in the patient's functional outcome with minimization or reversal of the stroke is expected and should lead to a reduction in cost of stroke care and patient's length of stay. The long-term impact on patient's functional outcome and the stroke care cost remains unknown related to the December 1995 approval of tissue plasminogen activator (tPA) for acute ischemic stroke.[6]

Consumers and payers are holding health care providers accountable for measuring and reporting clinical, financial, and psychosocial outcome data. Health care teams must provide data demonstrating improved functional patient outcomes as well as increased efficiency, timeliness, appropriate resource utilization, and maximized cost containment. The clinical path can successfully guide the interdisciplinary management for stroke patients.

Outcome measurement should occur throughout the continuum of stroke care from community education related to risk factors and stroke symptoms through acute care, rehabilitation, and home health. A comprehensive outcome measurement program can direct the need for services of stroke patients during hospitalization and throughout the rehabilitation phase and provide the framework for optimizing the care of future stroke patients.

At Saint Luke's Hospital the vision of the stroke collaborative practice team is to provide a continuum of interdisciplinary expert care to patients in a cost-effective manner and with measurable outcomes. Outcome measurement is a concurrent and retrospective process occurring throughout the continuum of care, from community education to a patient's admission to the hospital and continuing through the rehabilitation and home care phase. Our outcome measurement program consists of four categories of data including: discharge outcomes, clinical outcomes, patient satisfaction outcomes, and fiscal outcomes. The program is evolving and has grown from an initial set of key measures to an extensive outcome measurement data profile gathered for each stroke patient.

Discharge Outcomes

The team identified discharge outcome criteria focusing on patient safety and continuity of care through and beyond hospitalization. Hospital criteria focus on the patient and family's ability to successfully transition to the next level of care whether it be home, rehabilitation, or skilled nursing facility. Discharge planning must be individualized to assist each patient to achieve his or her highest potential, considering neurological deficits, related to activity, nutrition, living environment, emotional, and spiritual components. Discharge outcomes are listed on each clinical path and the staff document the progress and achievement of these outcomes at each level of care. Quarterly, the team reviews aggregate path data and evaluates the frequency that discharge criteria are met, the length of time required to meet the discharge outcomes, and the identified reasons that outcome achievement was delayed or did not occur.

Discharge expected outcomes address the patient's and family's ability to verbalize (1) signs and symptoms of a recurrent stroke and appropriate actions to take, (2) means to modify lifestyle to reduce further stroke risk, (3) follow-up labs or physician appointments, (4) purpose of discharge medications, schedule, and possible food–drug interactions, (5) maintenance of adequate nutrition and the discharge diet, (6) reestablished bowel and bladder program, (7) risks for injury, safety measures, and use of adaptive equipment, and (8) support systems and available community resources.

Outcome Measurement Instruments

The stroke team researched available measurement options and chose well-validated, standardized instruments recommended in *Post-Stroke Rehabilitation Clinical Practice Guideline*.[7] The purpose of these instruments is twofold: (1) to ensure reliable and standardized assessment and documentation of each stroke patient's condition and progress over time, and (2) to determine patient's needs, expected outcomes, and appropriate plan of care including services and level of care.

The nine instruments chosen are the

1. National Institutes of Health Stroke Scale, which measures stroke deficits[8]
2. Orpington Prognostic Scale, which measures stroke severity[9]
3. Barthel Index, which measures disability and activities of daily living[10]
4. Functional Independence Measure (FIM), which measures disability and activities of daily living[11]
5. Instrumental Activities of Daily Living (IADL), which measures disability and activities of daily living[12]
6. Fugl-Meyer, which measures motor function[13]
7. Duke Mobility Scale, which measures motor function[14]
8. Geriatric Depression Scale[15]
9. Medical Outcomes Study (MOS 36) Short Form Health Survey, which measures health status and a quality of life measure[16]

Collection of patient data via the stroke instruments occurs at points of admission and discharge for each level of care, ie, intensive care unit, acute unit, rehabilitation unit, and home care. All members of the interdisciplinary team use the instruments to document patient assessment and progress toward defined outcomes. This requires comprehensive staff education in use of the instrument and validation of competency for all care team members.

The team developed a comprehensive stroke database to define our stroke population and to capture the treatments and outcomes of all Saint Luke's Hospital and Home Health of Saint Luke's stroke patients. The data items are listed in Table 8–1. Data collection began in June 1996, and once there is stabilization of the data collection process, formal reporting of patient's clinical and functional outcomes will occur. Aggregated results of this data set is intended to be used for future research that leads to improvements in the care and outcomes for all stroke patients. Design of this outcome measurement project, while time consuming, will provide future rewards especially in evaluating the value of new stroke therapies and may allow better prediction of resource consumption at the differing levels of care.

Clinical Outcomes

Based on professional practice standards found in the current stroke literature, the team collaboratively defined the patient clinical indicators required to evaluate the quality of clinical practice outlined in the standing orders and clinical path. This was particularly important since stroke patients were being admitted to the newly opened stroke center. The clinical review specialist conducts a concurrent chart review on 100% of the stroke patients to capture occurrences related to these clinical indicators (Exhibit 8–1). Data are reviewed quarterly by the team for similar review and recommendations. Reports are forwarded to the neurology department quality committee

---

Table 8–1 Stroke Comprehensive Data Set Collection

Patient Demographics	
Function Level Prior to Stroke	Medical/Family/Surgical History
Smoking/Alcohol History	Coexisting Conditions
Medications at Time of Stroke	Admission & 24-Hour Post-Admission Vitals
Diagnosis	Infarct Location
Diagnostic Studies and Results	Stroke Mechanism
Treatment Regimen	New/Current Neurological Deficits
Complications: ie, pneumonia, urinary tract infection	Functional and Activities of Daily Living Scores FIM, Barthel, IADL, Mobility Measurements
Stroke Severity Scores NIH Stroke Scale, Orpington Prognostic Scale	Quality of Life and Depression Scores

Source: Copyright © Saint Luke's Hospital of Kansas City.

and the hospital performance improvement committee to detect trends and improvement opportunities.

Initial results of the clinical indicator review demonstrated that some patients were not being fed within the first 48 hours. The length of stay for stroke patients can increase by 24 days if they are without adequate nutrition within the first 3 days. Unfavorable outcomes may occur, such as fever of unknown origin that leads to a costly septic workup and activity intolerance that delays the patient from transferring to the rehabilitative level of care.[17,18] The team researched this issue, developed practice changes and revised the clinical path and physicians' standing orders. Further description of this change is found later in this chapter.

The clinical outcome measurement program has progressed over time. During the first quarter of 1996 the team began to track additional routine clinical indicators including patients readmitted to the hospital within 30 days of discharge, discharge disposition, and utilization of home health. Readmission data initially appeared to be 10%. Retrospective chart review found that all but 3% of the patients were readmitted for either a scheduled carotid endarterectomy or other vascular related diseases such as angina or peripheral vascular disease. Retrospective chart review showed that four patients may have benefited from a home health program and one patient needed formal rehabilitation instead of in the home.

Further evaluation of stroke patients' discharge dispositions showed that only 3% of the 50% percent discharged from the acute unit received home health. The national average of stroke patients going home from acute units with home health is in the range of 20% to 25%. These evaluations prompted the development of the Home Health of Saint Luke's Hospital Neuroscience Program. The hospital home health coordinator is consulted on the first day of admission. In addition, the hospital home health coordinator now rounds with the care team daily to assist in identifying possible home health options and patient eligibility. These practice changes were incorporated into the clinical path and standing physician orders. Subsequent data show an increase in the appropriate use of home health and demonstrate no readmissions related to discharge or home health needs (Figure 8–1).

Patient Satisfaction Outcomes

The stroke center and the revised stroke clinical path were initiated at the same time (January 1994). The clinical nurse specialist and the unit clinical nurses interviewed the first 50 patients admitted to the stroke center to evaluate patient satisfaction. The satisfaction survey was a 5-point Likert scale assessing the patient's perception of care provided in the stroke center, admission process, hospital physicians, nursing staff, home health staff, and facility operations. The survey showed a 97% overall patient satisfaction with the stroke center experience, and 96% of the patients indicated they would recommend the stroke center to their families and friends. Since stroke patients were previously located throughout the institution, there was no stroke patient–specific satisfaction data prior to 1995.

Dissatisfactors identified through the survey were two. Patients did not like their food, and they had financial concerns with questions about their insurance benefits during acute care and rehabilitation. Improvements made as a result of the patient satisfaction survey were: (1) a dietitian or dietitian technologist educated the stroke patients on the low fat and low salt diet and discussed with the patient's family ideas to improve the taste of this diet, and (2) financial information was placed in the patient's education folder and the patient path was revised to explain that the social service worker would discuss any financial concerns on day 2 of the hospitalization. While one-on-one surveys were time consuming, they provided valuable information to improve the patient care.

Currently the stroke patients are surveyed with all the neuroscience patients and survey results are reported quarterly. The System Marketing and Research Department conducts the survey using a survey form mailed to patients soon after discharge. The survey measures (5-point Likert scale) the patient's perception of hospital admission process, hospital departments, hospital personnel including physicians and nursing staff, facility operations, and the

---

Exhibit 8–1 Stroke Patient Clinical Indicators

1. Elevated systolic BP > 180 and /diastolic BP > 110 for greater than 12 hours
2. Undiagnosed/untreated fever 100.5 greater than 12 hours
3. Unaddressed new arrhythmia greater than 2 hours
4. Blood sugar greater than 200 on two successive peripheral blood glucose
5. Unaddressed oxygen saturation < 90
6. New neurological changes
7. Not out of bed for longer than 48 hours
8. No enteral nutrition for longer than 48 hours
9. Falls
10. Pneumonia on chest x-ray
11. Deep vein thrombosis
12. Urinary tract infection
13. Skin breakdown

Source: Copyright © Saint Luke's Hospital of Kansas City.

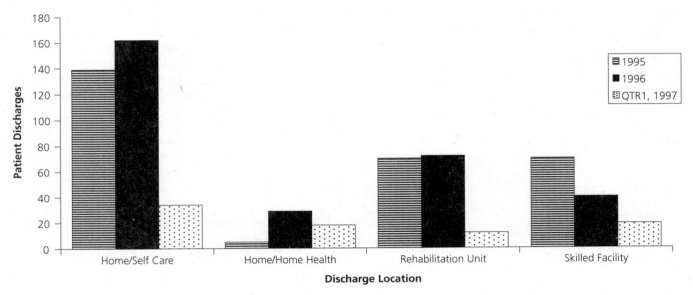

Figure 8–1 Discharge Disposition. *Source:* Copyright © Saint Luke's Hospital of Kansas City.

hospital discharge process. The questions relate to patient perception with the following: overall hospital experience, reliability, responsiveness, competence, access, and empathy. Questions regarding patient care focus on the sensitivity, responsiveness, and availability of the care providers and their success in providing adequate explanations and skilled care. Patient satisfaction with the neuroscience physicians and nurses is consistently rated 90% to 94%.

Financial Outcomes

Since implementation of our clinical path there has been a steady decline in the inpatient cost for stroke care despite an increase in stroke admissions (Figures 8–2, 8–3, and 8–4). Cost reduction is attributed to shortened length of acute hospital stay, awareness of and attention to resource utilization, improved efficiency and timeliness of testing, and the development of a comprehensive home health program.

The average length of stay for our stroke patients in 1993 ranged from 6.73 to 8.03 days with an average of 7.5 days and an average cost of $7451. By 1996, the average length of stay decreased to 6.5 days (Figure 8–2) and the average cost reduced to $6667 (Figure 8–3). This 11% decrease in cost is a conservative estimate since the costs are non–inflation adjusted. All departmental costs except pharmacy have been reduced since utilization of the clinical path (Figure 8–4). A rise in our pharmacy costs is expected with the increasing use of thrombolytics, which range from $2000 to $2500 per patient.

The diagnosis of stroke is associated with high costs related to LOS and to expensive radiology imaging studies. In June of 1996, the team concentrated efforts to evaluate radiology resource utilization. A retrospective chart review evaluated whether diagnostic testing such as cerebral arteriograms, computed tomography (CT) scans, and magnetic resonance imaging (MRI) scans were indicated for determining diagnosis and if such testing made a difference in the patient's treatment plan. In all patients' charts reviewed, the team found that arteriograms were appropriately indicated. They determined that the CT scan was the most effective initial examination and MRI was useful only if posterior lesions were expected. The team discovered that MRIs were often done on day 2 or 3 and the MRI results did not provide additional information that changed the treatment regimen beyond the initial CT scan results. Therefore, the MRI was deleted from the clinical path and is ordered individually when a specific clinical need is identified.

Laboratory utilization was evaluated in a similar fashion. Retrospective chart review found evidence that some of the laboratory studies ordered did not necessarily contribute added value to the treatment of the stroke patient. A baseline stroke panel consisting of electrolytes, creatinine, blood urea nitrogen, glucose and coagulation profile was developed and is now done initially for each patient. The expensive extended chemistry panels are ordered only if indicated for a patient. The average lab cost has decreased from $514 per patient in 1993 to $364 in the first quarter of 1997 (non–inflation adjusted).

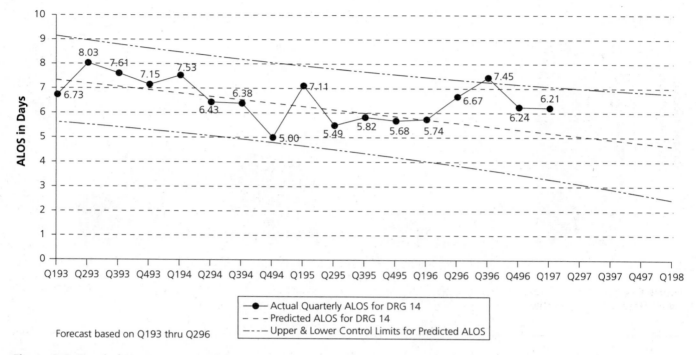

Figure 8–2 Trend of Average Length of Stay for DRG 14 from Q193 through Q197 with Forecast through Q198. *Source:* Copyright © Saint Luke's Hospital of Kansas City.

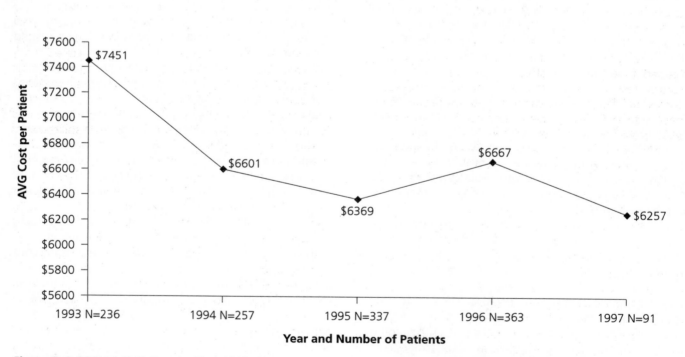

Figure 8–3 DRG 14 AVG Cost per Year 1993–QTR 1 1997. *Source:* Copyright © Saint Luke's Hospital of Kansas City.

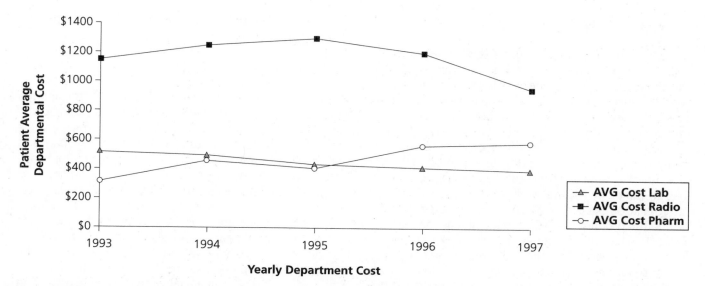

Figure 8–4 Saint Luke's Stroke Center of Kansas City, DRG 14 Average Cost Lab, Radiology, Pharmacy. *Source:* Copyright © Saint Luke's Hospital of Kansas City.

The team continues to review the literature and to incorporate recommendations into the clinical path and standing orders. Recent research validates that the transesophageal echocardiogram (TEE) is the test of choice to detect cardioemboli as the stroke mechanism. Prior to these research findings, common practice was to first order a transthoracic echocardiogram (TTE) for each patient and then perform a TEE. The stroke team and cardiologists reviewed the research and changed practice. Currently a TTE is performed if the patient's history indicates a need for further cardiac testing. Such attention to appropriate resource consumption has decreased the average radiology cost per patient from $1147 (non–inflation adjusted) in 1993 to $900 in the first quarter of 1997.

Physician development and use of the clinical path and associated standing orders standardizes the patient care and guides each physician in diagnostic testing and treatment. The sharing of clinical and cost data with our physicians has led to enhanced partnerships, more effective and efficient clinical processes, and improved patient outcomes.

As the care coordination team reviews each patient's progress by daily rounds with the clinical nurse and physician, they discuss discharge plans and assist in evaluating whether the patient is a candidate for formal rehabilitation, skilled facility, or home care. The utilization review specialist monitors for potentially avoidable days and works with the team to minimize such occurrences. The most frequent potential avoidable days are related to diagnostic testing and to social issues. Some diagnostic tests are not readily available on the weekend, ie, carotid duplex scan and TEE. Initially the team was concerned and speculated that patients admitted on Friday or the weekend had a longer length of stay. When aggregate data were reviewed they revealed that these patients had no longer length of stay than patients admitted any other day of the week. Furthermore, the data indicated that the treatment plan would not have been altered if the weekend testing had been done. The team learned that practice changes cannot be made on speculations alone but must be driven by valid data.

The team determined that patients had an increased length of stay while waiting for available rehabilitation or skilled beds in our hospital and at outside facilities. Furthermore, patient transfer is often delayed while families make the difficult decisions regarding the next level of care. These social issues may delay stay, but must be addressed with compassion; families need guidance and information. The social worker attempts to return patients to their communities when appropriate services and support systems exist. The team holds interdisciplinary patient and family care conferences to assist the family in making these decisions. Hospital-based insurance reviewers are included in the conference to inform the family of their care benefits. Early anticipation of discharge needs and involvement of all patient, family, and caregivers has led to a decrease in the potentially avoidable days since 1994.

Variance Analysis

Many of the practice changes discussed earlier are a result of variance data analysis. Another successful practice change came about when the team identified significant variance (delay) in meeting patients' nutritional needs by hospital day 2. During a subsequent quality improvement session, the stroke collaborative practice team questioned the logic of our process for feeding tube insertion. They identified 6 steps that occurred before the patient received a feeding tube: (1) patient traveled to radiology for the swallow screening; (2) patient returned to stroke center; (3) speech pathologist read the results and brought the report to the unit as soon as possible; (4) clinical nurse paged physician or called the office; (5) physician responded, received the report, gave the nurse further orders; (6) patient returned to radiology for tube placement. The process of waiting for each team member to complete his or her step was very timely, often involving more than one day. In addition, the patient was transported twice, which was inefficient use of the staff time and dissatisfying to the patient and family. An interdisciplinary team (radiology, speech pathologist, physician, and clinical nurse) met to discuss how this process could be improved. The consensus was that steps could be eliminated if the feeding tube could be placed in radiology after a swallow study demonstrated the need. The physicians agreed to explain this to the patient and family, obtain their permission prior to the study, and leave orders on the chart for the radiologist and speech pathologist dependent upon test results. Now a decision is made prior to the procedure, the patient is transported only once, manpower is reduced, and time spent by radiology, speech pathology, and neurologists is greatly reduced. This was one of the most satisfying practice changes for all involved, especially the patient and family.

Initially, all variances found by the staff were recorded and analyzed. Many of these were related to system processes. Improving these processes was the responsibility of the team in collaboration with other caregivers. Once many of the care processes were identified and streamlined, the team concentrated on patient variances and development of the outcome management program, but documentation of every variance was confusing and time consuming for the staff. Data overload made it difficult to identify variances that actually made a difference in patient outcomes and clinical processes. Currently the team is identifying four to six key variances that will be tracked to determine if patient outcomes can be improved. At this point the variances chosen include: activity intolerance, decreased level of consciousness, urinary incontinence, bowel incontinence, delayed nutrition, bed availability, and patient or family reluctant or undecided regarding discharge plans.

CONCLUSION

The development of teams dedicated to improving the outcomes for specific patient groups at Saint Luke's Hospital has been a challenging and rewarding experience for all involved. In the past 5 years many of the goals originally defined for the collaborative care program have been realized. The need for and the enthusiasm to continue this quest is even more obvious and powerful today. The formal program has concluded and the processes developed are now the way we do business, the way we deliver patient care. Undoubtedly the advantages to this collaborative care approach are many: the clinical paths, patient paths, variance reduction, improved clinical and financial outcomes, processes and systems, enhanced patient and family understanding of, involvement in, and accountability for the plan of care, and higher staff competency. But the principal value lies in the ability to focus on the needs of our stroke patients and to design for them the highest level of care and services to improve their health outcomes.

The future for us presents more challenges. We believe that outcome measurement must occur throughout the continuum of stroke care to identify opportunities to continually evaluate and improve stroke care. In addition, the effectiveness of community education must be measured in relation to client risk factor modification and stroke prevention. We recently began a project of stroke risk assessment and education in the community. Our nurses are targeting health care providers and the public by presenting forums in the hospital, in health centers, in rural areas, and in businesses. A clinical path is being developed for home care and nursing home skilled care units as an extension of the hospital path.

A significant barrier to improving outcomes lies in the labor-intensive method of documenting and collecting data from the sources mentioned earlier. The Saint Luke's–Shawnee Mission Health System has embarked on a comprehensive project of computerization that is only in the beginning stages of implementation. We intend to have the capability of computerized documentation, and data collection, aggregation, and analysis. While the clinical paths have saved hours of charting each day, computerized charting will significantly streamline the processes.

Finally, our goal is to partner with our consumers, payers, and purchasers to develop systems of outcomes measurement that provides even greater information on which to base clinical practice decisions. While the changes with managed care provide challenges to us, they also offer opportunities to improve care, outcomes, and costs. Through collaborative practice teams, these opportunities are not only possible, but certain and positive.

REFERENCES

1. Matchar DB, Duncan PW. Cost of stroke. *Stroke Clin Update.* 1994;5:9–12.

2. Gorelick PB. Acute ischemic stroke and transient ischemic attack: a costly business and a strategy to reduce costs. *J Stroke Cerebrovasc Dis.* 1995;5:1–5.

3. Wentworth DA, Atkinson DA. Implementation of an acute stroke program decreases hospitalization costs and length of stay. *Stroke.* 1996;6:1040–1043.

4. Jorgensen HS, Nakayama H, Raaschoy HO, Larsen K, Hubbe P, Olson JS. Effect of a stroke unit: reduction in mortality, discharge rate to nursing home, length of hospital stay and cost. *Stroke.* 1995;26:1178–1182.

5. *St. Anthony DRG Guide Book, 1996.* Reston, VA: Saint Anthony Publishing; 1995.

6. The National Institute of Neurological Disorders and Stroke rt-PA Stroke Study Group. Tissue plasminogen activator for acute ischemic stroke. *N Engl J Med.* 1995;333:1581–1587.

7. Gresham GE, Duncan PW, Stason WB, et al. *Post-Stroke Rehabilitation. Clinical Practice Guideline, No. 16.* Rockville, MD: US Dept of Health and Human Services, Public Health Service; May 1995. Agency for Health Care Policy and Research publication 95-0662.

8. Brott T, Adams HP, Olinger CP, et al. Measurements of acute cerebral infarction: a clinical examination scale. *Stroke.* 1989;17:731–737.

9. Kalra L, Date P, Crome P. Evaluation of a clinical score for prognostic stratification of elderly stroke patients. *Age Ageing.* 1994;23:492–498.

10. Mahone FL, Barthel DW. Functional evaluation: the Barthel Index. *Md Med J.* 1965; 14:61–65.

11. Guide for the uniform data set for medical rehabilitation (Adult FIM), version 4.0. Buffalo, NY 14214: State University of New York at Buffalo; 1993.

12. Lawton MP. Assessing the competence of older people. In: Kent D, Kastenbaum R, Sherwood S, eds. *Research Planning and Action for the Elderly.* New York: Behavioral Publication; 1972.

13. Fugl-Meyer AR, Jaasko L, Leyman I, Olsoon S, Steglind S. The post stroke hemiplegic patient, I: a method for evaluation of physical performance. *Scand J Rehabil Med.* 1975;7:13–31.

14. Studenskie S, Duncan P, Chandler J, et al. Predicting falls: the role of mobility and non-physical factors. *J Am Geriatric Soc.* 1994;42:297–302.

15. Yesavage JA, Brink TL, Rose TL, et al. Development and validation of a geriatric depression screening scale: a preliminary report. *J Psychiatr Res.* 1982–83;17:37–49.

16. Ware JE, Sherbaurne CD. The MOS 36 item short-form health survey (SF-36), I: conceptual framework and item selection. *Med Care.* 1992;30:473–483.

17. Nyswonger GD, Helmchen RN. Early external nutrition and length of stay in stroke patients. *J Neurosci Nurs.* 1992;24:220–223.

18. Finestone HM, Greene-Finestone LS, Wilson ES, Teasell RW. Prolonged improvement rate in malnourished stroke rehabilitation patients. *Arch Phys Med Rehabil.* 1996;77:340–345.

■ Appendix 8–A ■

Clinical and Patient Pathways with Acronym Key

The following is a key to the acronyms used in the clinical pathway and patient pathway that follow:

ABG—arterial blood gas
ADM—admission
A/S/C—assist/self/complete
BID—twice a day
BM—bowel movement
BP—blood pressure
BRP—bathroom privileges
BSC—bedside commode
C&S—culture & sensitivity
CNS—clinical nurse specialist
CT—computed tomography
Diast—diastolic
DVT—deep vein thrombosis
EKG—electrocardiogram
I & O—intake & output
INR—International Normalized Ratio
NIH—National Institutes of Health

OT—occupational therapy
PBG—peripheral blood glucose
PEG—percutaneous endoscopic gastrostomy
PRN—if needed
PT—physical therapy
Pt/SO—patient/significant other
PTT—partial prothrombinplastin time
q—every
QID—4 times a day
sat—saturation
SNF—Skilled Nursing Facility
Syst—systolic
TED—thromboembolic device
TEE—transesophageal echo
W1—0700–1900 shift
W2—1900–0700 shift
WNL—within normal limits

Source: Copyright © Saint Luke's Hospital of Kansas City.

Exhibit 8–A–1 Clinical Pathway for Stroke Patient (DRG 14)

SAINT LUKE'S HOSPITAL OF KANSAS CITY
Clinical Path—Stroke Patient (DRG 14)

Admission date: _____ Medical Diagnosis: _____

Consultants: _____ Allergies: _____ Allergies to iodine/dye   Yes   No

Nursing Diagnosis: **IMPAIRED MOBILITY RELATED TO NEUROMUSCULAR IMPAIRMENT**

Initiated Date _____   Modified: _____

                          Resolved: _____

Expected Outcome:
1. Patient will attain the highest degree of strength, endurance, and mobility possible within the confines of the disease process.

Nursing Assessment/ Interventions (Practice Standards)
1. Assess the patient's functional capacity daily.
2. Assess for complications of immobility PRN.
3. Support and instruct patient as needed to perform activity as allowed by condition.
4. Instruct and/or assist pt/SO with optimal level of mobility.

Nursing Diagnosis: **ALTERED NUTRITION RELATED TO SWALLOWING DIFFICULTY & INCREASED METABOLIC NEEDS**

Initiated Date _____   Modified: _____

                          Resolved: _____

Expected Outcome:
1. Patient will maintain caloric intake equal to meet metabolic requirements.
2. Maintains weight within 3–5 lbs. of admission weight unless other parameters established by physician.
3. Patient/significant other verbalizes or demonstrates positioning and swallowing techniques to decrease risk of aspiration.

Nursing Assessment/ Interventions (Practice Standards)
1. Assess hydration (skin turgor, moist mucous membrane) every shift.
2. Assess nutritional intake PRN.
3. Assess weight every other day.
4. Dietician consult as appropriate.
5. Instruct/supervise patient in safe swallowing techniques according to swallowgram report.
6. Instruct pt/SO on nutritional plan.

Nursing Diagnosis: **IMPAIRED ADJUSTMENTED RELATED TO DISBABILITY**

Initiated Date _____   Modified: _____

                          Resolved: _____

Expected Outcome:
1. Patient expresses understanding of the disease process and adapts to a new health status.
2. Patient participates in health care regimen and plans care activities.
3. Patient identifies inability to cope and demonstrates new coping strategies.

Nursing Assessment/ Interventions (Practice Standards)
1. Encourage patient to express feelings. Provide a safe, nonthreatening environment.
2. Begin teaching patient and caregivers the skills needed to safely manage care.
3. Encourage patient to plan and participate in care activities such as time of treatments, personal hygiene, and rest periods.
4. Encourage compliance and adjust to optimum wellness.
5. Obtain a consultation with mental health specialist if patient develops depression or other psychiatric problems.

PATIENT IMPRINT

SAINT LUKE'S HOSPITAL OF KANSAS CITY
COLLABORATIVE CARE PROGRAM

The suggested plan represents the initial desired course of treatment and goals of recovery. These are representative or average guidelines only and should be reviewed periodically by the attending physician and other involved disciplines. Deviations are generally expected and revisions to the plan should be made as warranted.

Page 1   FORM# NSP.15.003 (9/97)

continues

Exhibit 8–A–1 continued

CARE CATEGORY	DATE: _____ DAY: ___1___ W1 W2	DATE: _____ DAY: ___2___ W1 W2	DATE: _____ DAY: ___3___ W1 W2
ASSESSMENT	❏ ❏ Neuro assessment q 2h x 4, then q 4h and PRN (flowsheet) ❏ ❏ Vital signs q 2h x 4, then q 4h and PRN ❏ ADM NIH stroke score _____ ❏ Pre-stroke Barthel score _____ ❏ ADM Barthel score _____ ___ ___ BP Syst < 180; Diast < 110	❏ ❏ Neuro assessment q 4h and PRN (flowsheet) ❏ ❏ Vital signs q 4h and PRN ___ ___ BP Syst < 180; Diast < 110	❏ ❏ Neuro assessment q 4h and PRN (flowsheet) ❏ ❏ Vital signs q 4h and PRN
CONSULT	❏ Social Service entered ❏ Nutrition screening entered ❏ PT/OT entered ❏ Speech Screening entered ❏ Rehab medicine entered	❏ Social Service referral received ❏ Rehab medicine consultation initiated	❏ Psych CNS evaluation ❏ Social Service assessment completed ❏ Discuss placement options with Social Services ❏ PT screening completed ❏ OT screening completed ❏ Speech screening completed
TESTS/LABS	❏ Stroke panel ❏ CT Scan ❏ ❏ PBG _____ QID if diabetic ❏ ADM O₂ sat/PRN _____ % ❏ ❏ ABG if O₂ sat < 90 ❏ EKG ❏ Other _____	❏ Swallowgram entered ❏ Arteriogram ❏ Other _____ ❏ TEE ❏ Carotid Duplex ❏ ❏ PBG QID if diabetic ❏ ❏ O₂ sat PRN ❏ ❏ Bladder scan	❏ Swallowgram ❏ ❏ PBG QID if diabetic ❏ ❏ Bladder scan ❏ Other _____ ❏ Discuss long-term feeding option with family/physician
MEDICATIONS	❏ ❏ Anti-Coagulant _____ ___ ___ PTT therapeutic/Weight-based Heparin Protocol	❏ ❏ Anti-Coagulant _____ ___ ___ PTT therapeutic/Weight-based Heparin Protocol	❏ ❏ Anti-Coagulant _____ ___ ___ PTT therapeutic/Weight-based Heparin Protocol
TREATMENTS	❏ ❏ O₂ _____ ❏ ❏ I & O ❏ ❏ Telemetry ❏ ❏ Saline lock/IV ❏ ❏ DVT precautions TED/Cuff (circle) ❏ ❏ Suction PRN	❏ ❏ O₂ _____ ❏ ❏ I & O ❏ ❏ Telemetry ❏ ❏ Saline lock/IV ❏ ❏ DVT precautions TED/Cuff (circle) ❏ ❏ Suction PRN	❏ ❏ O₂ _____ ❏ ❏ I & O ❏ ❏ Telemetry ❏ ❏ Saline lock/IV ❏ ❏ DVT precautions TED/Cuff (circle) ❏ ❏ Suction PRN
MOBILITY/ADLs	❏ ❏ Bedrest ❏ ❏ Turn q 2° ❏ ❏ Restraint/bed alarm ❏ ❏ Siderails x _____	❏ ❏ ADLs A / S / C ❏ ❏ Bedrest ❏ ❏ Restraint/bed alarm ❏ ❏ Rehab at bedside ❏ ❏ Siderails x _____ ___ ___ Tolerates chair x 15 min BID	❏ ❏ ADLs A / S / C ❏ ❏ Restraint/bed alarm ❏ Needs assistive device Yes/No ❏ ❏ Siderails x _____ ___ ___ Tolerates chair x 30 min BID
NUTRITION	❏ ❏ Diet / A / S / C _____ _____ Weight _____ ___ ___ Demonstrates risk for swallowing problems/aspiration Yes/No If Yes, notify physician for nutrition orders/DH placement	❏ ❏ Diet / A / S / C _____ ❏ ❏ Nutrition screen completed —patient needs: supplemental Feeding _____ Dobhoff _____	❏ ❏ Diet / A / S / C _____ ❏ ❏ Supplemental feeding _____ Weight _____ ___ ___ Demonstrates safe swallowing techniques
ELIMINATION	❏ ❏ Assess for BM schedule ❏ ❏ Assess bladder function BRP/BSC _____ Foley _____ Incontinent _____	❏ ❏ Consider laxative if no BM recorded ___ ___ Demonstrates normal bowel function ___ ___ Demonstrates normal bladder function	❏ ❏ If incontinent: post void straight residual cath—to lab for C&S ❏ ❏ BM recorded ___ ___ Demonstrates normal bowel function ___ ___ Demonstrates normal bladder function
TEACHING AND DISCHARGE PLANNING	___ Pt/SO receives and understands purpose of stroke education book ___ Pt/SO receives path and agrees on plan of care ___ Pt/SO understands safety measures/mobility techniques	___ ___ Pt/SO demonstrates safety measures/mobility techniques ___ ___ Demonstrates using alternative communication techniques	❏ ❏ Post-hospital care needs discussed with Social Services ___ ___ Demonstrates safety measures/mobility techniques
PATIENT IMPRINT	SIGNATURE KEY: _____ _____ _____ _____		❏ ❏ = Interventions ___ ___ = Expected Outcomes _____ _____

SYMBOL KEY: "Initials" on a line means done and findings as expected
"✔" in a ❏ box means an intervention or item was completed
"O" in a ❏ box or on a line indicates the item was not pertinent to that shift
"*" in a ❏ box or on a line indicates an item was not done (not as expected)

Exhibit 8–A–1 continued

CARE CATEGORY	DATE: _____ DAY: 4 W1 W2	DATE: _____ DAY: 5 W1 W2	DATE: _____ DAY: 6 W1 W2	DISCHARGE EXPECTED OUTCOMES Date/Initial
ASSESSMENT	☐ ☐ Neuro assessment q 8h/ PRN (flowsheet) ☐ ☐ Vital signs q 8h/PRN	☐ ☐ Neuro assessment q 8h/ PRN (flowsheet) ☐ ☐ Vital signs q 8h/PRN	☐ ☐ Neuro assessment q 8h/ PRN (flowsheet) ☐ ☐ Vital signs q 8h/PRN ☐ NIH stroke score _____	___ Verb signs/symptoms to report to physician ___ Verb signs/symptoms of stroke ___ Verb own risk factors ___ Discharge Barthel score
CONSULT	☐ Consult utilization review to obtain SNF/Rehab evaluation	☐ Re-evaluate if SNF/Rehab criteria not met Day 4	☐ Re-evaluate if SNF/Rehab criteria not met Day 5	
TESTS/LABS	☐ ☐ PBG qid if diabetic ☐ Protime/INR ☐ ☐ Bladder scan	☐ ☐ PBG qid if diabetic ☐ Protime/INR	☐ ☐ PBG qid if diabetic ☐ Protime	___ Verb understanding need of follow-up labs if ordered
MEDICATIONS	☐ ☐ Anticoagulant _____	☐ ☐ Anticoagulant _____	☐ ☐ Anticoagulant _____	___ Pt/SO verbalizes understanding of discharge medication/schedule and food drug interaction
TREATMENTS	☐ ☐ I & O ☐ ☐ Saline lock/IV ☐ ☐ DVT precautions TED/ Cuff (circle) ☐ ☐ Suction PRN ☐ ☐ Consider PEG ☐ ☐ Telemetry _____	☐ ☐ I & O ☐ ☐ Saline lock/IV ☐ ☐ DVT precautions TED/ Cuff (circle) ☐ ☐ Suction PRN ☐ ☐ Telemetry _____	☐ ☐ I & O ☐ ☐ Saline lock/IV ☐ ☐ DVT precautions TED/ Cuff (circle) ☐ ☐ Suction PRN ☐ ☐ Telemetry _____	___ Verb follow-up physician appointment
MOBILITY/ADLs	☐ ☐ ADLs A / S / C ☐ ☐ Restraint/bed alarm ☐ ☐ Siderails x ___ ___ Tolerates chair x 30 min BID	☐ ☐ ADLs A / S / C ☐ ☐ Restraint/bed alarm ☐ ☐ Siderails x ___ ___ Tolerates up ad lib ___ ___ Tolerates chair x 45 min BID	☐ ☐ ADLs A / S / C ☐ ☐ Restraint/bed alarm ☐ ☐ Siderails x _____ ___ ___ Tolerates up ad lib ___ ___ Tolerates chair x 1h BID	___ Assists with bed to chair transfer ___ Verb & demonstrates ability to perform care at home and/or support or community resources
NUTRITION	☐ ☐ Diet / A / S / C _____ ☐ ☐ Supplemental feeding ___ Demonstrates risk for swallowing techniques	☐ ☐ Diet / A / S / C _____ ☐ ☐ Supplemental feeding ___ Weight _____	☐ ☐ Diet / A / S / C _____ ☐ ☐ Supplemental feeding	___ Patient demonstrates understanding of discharge diet ___ Maintains admission weight unless other parameters established by physician
ELIMINATION	☐ ☐ BM recorded ___ Demonstrates normal bowel function ___ Demonstrates normal bladder function	☐ ☐ BM recorded ☐ ☐ Consider urology consult if incontinent ___ Demonstrates normal bowel & bladder function	___ ___ Demonstrates normal bowel & bladder function	___ Re-established usual elimination pattern ___ Verb need to discuss with physician if incontinence occurring
TEACHING AND DISCHARGE PLANNING	☐ ☐ Rehab: assess discharge equipment & service needs ☐ ☐ Coumadin video watched ☐ ☐ Coumadin teaching brochure & tape given to patient	___ ___ Demonstrates using safety measures/mobility techniques/adaptive equipment ___ ___ Patient verbalizes understanding of activity plan, nutrition plan, and swallowing plan ___ ___ Patient verbalizes understanding anticoagulation therapy precaution ___ ___ Verb own risk factors to future strokes, i.e. smoking, ↑ cholesterol, A-fib, ↑ blood pressure	☐ Discharge to SNF/Rehab/ Home (circle) ☐ ☐ Reinforce D/C teaching ☐ ☐ Suggest stroke support group ☐ ☐ Discharge therapy/ equipment needs met by Rehab ___ ___ Verb ways to modify life style to ↓ risk of future stroke	___ Demonstrates understanding of risks for injury, safety measures, & use of adaptive equipment ___ Identifies support systems ___ Pt/SO demonstrates the ability to practice new coping behaviors for adapting to physical or mental changes ___ Verb ways to modify lifestyles

SIGNATURE KEY:	☐ ☐ = Interventions ___ ___ = Expected Outcomes	Time: _____ ☐ All nsg dg resolved ☐ Valuables with Pt Mode: ☐ Wheelchair ☐ Stretcher Accompanied by: _____ Discharge Nurse Initials: _____

PATIENT IMPRINT   **SYMBOL KEY:** "Initials" on a line means done and findings as expected
"✔" in a ☐ box means an intervention or item was completed
"O" in a ☐ box or on a line indicates the item was not pertinent to that shift
"*" in a ☐ box or on a line indicates an item was not done (not as expected)

Exhibit 8–A–1 continued

SAINT LUKE'S HOSPITAL
TRANSFER NOTE

STROKE PATIENT (DRG 14)

Transfers Outcomes   Date: _____   Time: _____

Transfer to: ❑ E8   ❑ E9   ❑ Other _____   ❑ Belongings with patient

_____ Nursing diagnosis reviewed and discussed with transfer unit   _____ Advance Directive status discussed
_____ Vitals WNL for patient   _____ Established nutrition plan
_____ Neuro status unchanged   ❑ oral
_____ IV fluid/rate _____   ❑ dobhoff
         Amount left in bag _____   ❑ PEG
_____ Fall potential discussed   _____ Transfers with assist
_____ Last BM date _____   ❑ one man
_____ Voiding adequately   ❑ two man
_____ Patient/family verbalizes understanding of transfer   ❑ total
                                                                          _____ Braden scale _____

Transfer Nurse Signature: _____

PATIENT IMPRINT   **SAINT LUKE'S HOSPITAL OF KANSAS CITY**
                                                             COLLABORATIVE CARE PROGRAM

Exhibit 8–A–2 Patient Path for Stroke (Average Hospital Stay Range: 6 days)

SAINT LUKE'S HOSPITAL OF KANSAS CITY

Patient: _____

Saint Luke's Hospital's staff and doctors are dedicated to giving you the best possible care. This path is to help you and your family become more involved in your care. Since each person is an individual, your care may differ from this general guideline.

HEALTH TEAM MEMBERS

Registered Nurses (RN):

Are responsible for your nursing care. They plan your care with all health team members as well as with you and your family.

Patient Service Associate (PSA):

Will assist you with meal selection and serving, maintain the cleanliness of your room, and respond to your "hospitality" needs.

Nutrition Services:

Works with your doctor to meet your nutritional needs. They may check your eating habits and teaching you about your diet.

Patient Care Technician (PCT):

Will help you with personalized care and daily needs. They may perform procedures such as blood pressure, temperature, drawing of blood, and EKGs.

Physical, Occupational, and Speech Therapists:

Will be responsible for checking your mobility and coordination, ability to perform self care tasks, and communication skills. The therapist will plan your therapy and set goals to assist in discharge.

Social Worker:

Will provide counseling, information, education, and referrals to you as you adjust to the impact of your illness or treatment. They also assist you in making plans for post-hospital care.

Patient Representative:

Is your personal conneciton to the hospital system. If you have a question about your hospital stay, please call (816) 932-2328.

Hospital Chaplain:

Is available 24 hours a day to provide spiritual guidance and emotional support for you and your family.

Exhibit 8–A–2 continued

Stroke Center Check List
Phone: (816) 932-2261

Patient and family involvement during recovery following the stroke:

    ☑ As completed:

❑   1. Review stroke path to understand treatment plan.

❑   2. Review Stroke Center Education Booklet to understand role of each team member involved in stroke recovery.

❑   3. Discuss with Social Worker continued care options, financial concerns, and resources.

❑   4. Discuss patient's health care wishes and verbalize to health care team (Advance Directive/Living Will).

❑   5. Obtain appropriate education pamphlets located at entrance and exit of Stroke Center.

❑   6. Discuss with nurse how to participate in patient/self care.

❑   7. Family attends at least one physical therapy session to discuss physical rehabilitation treatment plan.

❑   8. Family attends at least one occupational therapy sessions to learn techniques on daily tasks such as bathing, dressing, and grooming.

❑   9. Family attends at least one speech therapy session to learn techniques to improve communication.

❑   10. Obtain nutrition education materials from Nutrition Services.

❑   11. Patient/family learn own risk factors that cause stroke.

❑   12. Patient/family learn signs/symptoms of stroke.

❑   13. Patient/family discuss ways to modify lifestyle and lower risk of stroke.

For more information about stroke or stroke support groups, call:
National Stroke Association 1-800-STROKES
American Heart Association 1-800-AHA-USA1
Local American Heart Association 1-913-648-6727

FORM# NS.15.103 (9/97)

continues

Exhibit 8–A–2 continued

CARE CATEGORY	HOSPITAL STAY GOALS	DAY 1	DAY 2	DAY 3
ASSESSMENT	• Stabilize and monitor symptoms	• Neurological and physical exams • Heart monitor • Blood pressure, heart rate, temperature	• Neuro and physical exams • Heart monitor • Vital signs	• Neuro and physical exams • Vital signs
TESTS/LABS		• Brain scans • Lab work, blood sugar • Oxygen concentration	• Tests to visualize flow of blood in neck, brain, and heart	• May have a swallowing test
FOOD AND DRINK	• Maintain nutritional status	• May or may not eat solid food	• May or may not eat solid food • If unable to eat may be fed by a tube placed in the stomach • If concerns about swallowing and speech, talk to Speech Therapist	• If concerns about diet ask to talk to dietitian
ACTIVITY	• Move toward self care	• Bedrest • Turned from side to side every 2 hours	• Physical Therapist will check mobility at bedside	• Therapy in the Rehabilitation Department (4th floor) • Chair 2 times a day
STROKE TEAM MEMBERS	• Begin rehabilitation process		• Dietitian; Physical, Occupational, and Speech Therapists; Social Worker	• Rehabilitation physician will plan therapy
MEDICATION	• To reduce risk of further complications	• May be put on blood thinner either by pill or intravenously	• Blood thinner continued	• Blood thinner continued
DISCHARGE PLANNING AND LEARNING	• Discuss physical needs, home environment, and equipment necessary to return home	• Stroke Education Booklet • Nurse will discuss plan of care during hospitalization, and safety measures • Inform nurse of normal bowel and bladder habits	• Nurses and therapist will educate on stroke disease	

The suggested plan represents the initial desired course of treatment and goals of recovery. These are representative of average guidelines only. They should be reviewed periodically by the attending physician and other involved care providers. Deviations are generally expected and revisions to the plan should be made as warranted.

FORM# NS.15.103 (9/97)

continues

Exhibit 8–A–2 continued

CARE CATEGORY	DAY 4	DAY 5	DAY 6	PATIENT/FAMILY DISCHARGE GOALS
ASSESSMENT	• Neuro and physical exams • Heart monitor	• Neuro and physical exams • Heart monitor	• Neuro and physical exams • Vital signs	• Understands signs and symptoms and own risk factors of stroke
TESTS/LABS	• Blood thinner test			• Understands what follow-up lab and next doctor appointment required
FOOD AND DRINK				• Understands discharge nutrition plan and swallowing instructions
ACTIVITY	• Chair 2 times a day • Therapy in Rehab Department			• Understands mobility techniques and safety measures to take at home • Understands activity and exercise options
STROKE TEAM MEMBERS	• Discuss continued care options with Social Services		• Social Services/Home Health will discuss discharge/rehabilitation needs	• Understands further home rehabilitation needs
MEDICATION				• Understands purpose, medication schedule, and any adverse effects before going home
DISCHARGE PLANNING AND LEARNING	• Blood thinner medication education • Ask questions of nurse concerning medications • Discuss admission to Skilled Nursing or Rehabilitation Unit	• Education on risk factors that cause stroke and signs/symptoms of stroke	• Discuss ways to lower risk of future strokes • Stroke Support Group brochure	• Understands to talk to physician if depression continues or worsens when discharged • Understands to discuss with physician if unable to control bladder
YOUR QUESTIONS AND COMMENTS				

FORM# NS.15.103 (9/97)

continues

Exhibit 8–A–2 continued

LEARNING FOR DAILY LIVING
Stroke Discharge Instruction Sheet

Activity	**Nutrition**
◆ Rehabilitation activity as recommended by your therapists	◆ Eat well-balanced meals, follow the Food Guide Pyramid
◆ Resume sexual activity as before	◆ Low fat, low cholesterol
◆ If being treated for hypertension monitor blood pressure weekly for four weeks, then monthly	◆ Follow swallowing plan/guidelines recommended by Speech Therapist
	◆ Weight yourself weekly to make sure eating adequate nutrition
	◆ Limit alcoholic drinks to two or less a day
	◆ No salt added if being treated for cardiac disease/hypertension

Rehabilitation Home/Outpatient Needs (*Social Worker/Home Health will complete*)

Call your Social Worker _____ at 932-2169 if you have problems or questions about your therapy program.

Physical Therapy: ❑ Home Health ❑ Outpatient ❑ None

_____

Occupational Therapy: ❑ Home Health ❑ Outpatient ❑ None

_____
Saint Luke's Inpatient Rehab Department: 932-2020
Saint Luke's Outpatient Rehab Department: 932-3344

Speech Therapy: ❑ Home Health ❑ Outpatient ❑ None

_____
Saint Luke's Speech Therapy Department: 932-3262

Home Needs: ❑ Home Health _____

❑ Services _____
Saint Luke's Home Health Department: 932-6940

Special Instructions

_____ Follow-up Lab work _____

FORM# NS.15.103 (9/97)
continues

Exhibit 8–A–2 continued

When to call for medical advice:

CALL 911 immediately if you experience any of the stroke warning signs:

- ◆ Numbness or weakness of face, arm or leg
- ◆ Difficulty speaking or understanding
- ◆ Dizziness when accompanied by one or more warning signs
- ◆ Sudden blurred or decreased vision
- ◆ Loss of balance

Call doctor for medical advice if:

- ◆ Systolic blood pressure consistently greater than 160 or if diastolic consistently greater than 90

Risk factors to modify by medical treatment and life style changes

_____ High blood pressure   _____ Smoking

_____ Heart disease   _____ Obesity

_____ Elevated blood cholesterol   _____ Decreased alcohol intake

P = Pamphlet/Handout   D = Discussion   V = Video   NA = Not applicable

---

Medication

- ◆ Take medications as directed, never stop taking medication unless advised by physician
- ◆ If *Coumadin* follow the instructions provided in the Coumadin pamphlet

---

Medication, Dose, Side Effects

_____

_____

_____

_____

Appointments

- ◆ Call Dr. _____'s office at _____ to make an appointment in _____ weeks.

Source: Copyright © Saint Luke's Hospital of Kansas City.   FORM# NS.15.103 (9/97)

An Acute Myocardial Clinical Pathway: Design, Outcomes, and Lessons Learned

Jean Barry-Walker, Becky Hassebrock, Kelley McLaughlin, and Krista Smeins

The use of clinical pathways is an effective tool for quality improvement with reductions in length of stay (LOS), case costs, and enhancement of various clinical quality outcomes reported in the literature.[1,2] This chapter describes the University of Iowa Hospitals and Clinics's (UIHC) approach to the design and implementation of a pathway for patients admitted with a diagnosis of acute myocardial infarction (AMI). A description of the developmental process and the three-part AMI CareMap® (AMIC) is included. The methods for tracking variances plus outcome data comprise the last two sections of this chapter.

The AMIC project, which was undertaken in late 1992, was one of the first clinical pathway initiatives at the UIHC. This diagnosis was selected based on DRG 122 LOS and charge data that indicated this DRG had significant financial losses and an extended LOS. A multidisciplinary team involving physician champions, nursing, physical therapy, the cardiac rehabilitation service, social service, dietary, pharmacy, and pathology was instrumental in designing a system of clinical quality management of the AMI population. This system included both a caregiver and a lay version of the AMIC, a research-based intervention to promote patient participation in decision making, preprinted AMIC orders, an "at home" pathway, and a variance tracking and monitoring process.

THE AMIC DEVELOPMENTAL PROCESS

At the beginning of this project, facilitating structures for the development and implementation of CareMaps® were limited but did include a three-day orientation program for staff nurse case managers. Additionally, a financial management report had been designed, which allowed tracking of the LOS and costs (based on a cost-to-charge ratio) by patient population for the following:

- separate cost categories for general and special care services: costs associated with the daily room rate for the hospital stay on a general and/or specialty unit
- separate cost categories for pharmacy, laboratory, and radiology
- operating room: costs for operating procedures, anesthesia, and postanesthesia care
- supplies: costs related to medical supplies including such things as heart valves, implants, braces, intravenous administration sets, blood transfusion sets
- other: costs associated with ancillary services such as electrocardiograms, treadmill laboratory, cardiac rehabilitation services, medical gases

This information can be provided both at the individual and group level of analysis. A "drill down" to more specific financial information is also possible. Outliers that are defined as having costs greater than two standard deviations above the mean are removed from the final analysis and listed at the end of the report.

Two intermediate-level, monitored cardiology patient care units were selected to participate in this project as a significant number of AMI patients are directly admitted to these units. An audit of 30 patient records was conducted to describe current practice for AMI patients cared for on the two intermediate units. Data from this audit indicated a wide range of patient acuity and marked variation in practice. It also showed that many of the AMI transferred from their local rural hospitals to the AMC were 1 to 2 days post-AMI at the time of the transfer; this fact would present a challenge to determine the appropriate point on the AMIC to start these transferred patients.

Three staff nurse case managers were selected and attended, along with their respective nurse managers, the 3-day staff nurse case manager orientation program. Following these sessions, they assumed primary accountability for the development and implementation of the AMIC. Administrative support was provided by the nurse managers, the cardiovascular advanced practice nurse, and two nursing administrators. In addition, physician champions were identified. The cardiology service had over 30 medical staff and these champions were essential in introducing this new concept to the medical team and overcoming resistance when the use of clinical pathways was declared to be "cookbook medicine" by various members of the interdisciplinary team.

A cardiovascular multidisciplinary CareMap® team was convened with representatives from all relevant disciplines. A philosophy of "full inclusion" was a cornerstone of this project with over 20 individuals participating on the team. This group is chaired on a rotating basis by the staff nurse case managers who are responsible for agenda development, scheduling and conducting of the team meetings, and ensuring that all necessary between-meeting work is completed. These meetings were implemented in early 1993 and continue to be convened on a semiannual basis, with much work done by pathway-specific subgroups that meet on a more frequent basis.

The staff nurse case managers were charged with drafting a copy of an AMIC that demonstrated current practice. Following this, each discipline reviewed best practice research and national clinical standards, when available. Benchmarking to identify best practice in other institutions was also performed. Each discipline provided specific content for their portions of the map with the entire cardiovascular team reviewing and reaching consensus on the final draft of the AMIC. This developmental process was quite prolonged as the organization had a steep learning curve regarding the systems needing to be designed and implemented in order to create and implement a CareMap®, to track and monitor variances, and to measure outcomes.

The AMIC presented in this chapter was implemented as a pilot on the two monitored intermediate units in early 1993; it has gone through three revisions since then. Originally, it was thought that exclusion criteria that differentiated between uncomplicated and complicated AMIs should be used to determine specific patients who could appropriately be placed on the AMIC. These have since been dropped and all patients who are ruled in with an AMI are placed on this pathway. After many lengthy dialogues, the cardiovascular team determined that the benefits that the AMIC provided outweighed any benefits gained from differentiating patients a priori based on severity of illness. These benefits include reduction of variation in practice, improvement in first-year resident education, an increase in patient education, an improvement in communication and collaboration

among all cardiovascular team members regarding the patient's plan of care, and a marked enhancement of shift-to-shift care coordination on the part of the nursing staff who use the AMIC for end-of-the-shift report on the two intermediate units. Concerns related to analysis and interpretation of outcome data regarding patients with varying degrees of severity of illness were resolved by separating patient data into four different groups, each of which have a different profile of acuity.

In 1992 the AMIC project was a grass-roots initiative and there was limited institutional understanding and acceptance of the value of CareMaps® as a clinical resource management strategy. Many of the necessary system supports for pathway development, implementation, and monitoring are now in place. A full-time case management coordinator oversees the institutional program of clinical pathway development. There are now resources to enter this variance information into a database with reports being generated for review and analysis by the cardiovascular team. Also, the time it takes to design and implement a CareMap® has been reduced from approximately 1 year to a few months.

VERSIONS OF THE AMIC

Two different versions of the AMIC have been created. There is a three-part caregiver AMIC, which includes sections for problem statements and goals, the 6-day critical path, and a patient education documentation section. The caregiver AMIC also has accompanying preprinted orders. The second version is the lay version of the AMIC. In addition, a homegoing pathway has been designed for patients and families.

Caregiver AMIC

The actual caregiver AMIC is four pages long and is accompanied by preprinted physician orders (see Appendix 9–A, Exhibit 9–A–1). The first page is divided into two sections with the first section providing space for the names of the staff physician and the staff nurse case manager assigned to the patient. There is space for secondary diagnoses, dates of initial review of the pathway for appropriateness for the specific patient, and the date that the AMIC was initially reviewed with the patient. Also, there are check boxes to indicate if and when the pathway was discontinued, if the patient was transferred to another pathway or placed on concurrent pathways, and finally, if the patient was returned to the AMIC. The second section incorporates the multidisciplinary plan of care (MPC). The MPC notes both patient problems and goals/outcomes of care for three phases of the patient's LOS: the day the AMI is ruled in; the cardiac rehabilitation phase, which is implemented from the day cardiac rehabilitation is started through to discharge; and, finally, the day of discharge phase, which identifies desired outcomes for each of the four listed problems.

The second and third page of the caregiver AMIC show a 6-day critical path with standardized path categories of consults, tests, treatments/interventions, medications, diet, activity, teaching, and discharge planning. At the top of page 2, cues with check boxes provide space for documentation regarding when and where the patient was originally ruled in as an AMI. Knowing when the patient was ruled in as an AMI assists in determining where to place the patient on the critical path. A teaching code can be found at the top of page 3 for the topics listed below in the teaching category of the critical path. Page 4 (not shown) allows for documentation of additional educational and audiovisual materials used and provides space for anecdotal documentation related to patient education. By incorporating the information into the teaching category of the critical path and on page 4 of the AMIC, two other documentation forms and redundant documentation were eliminated. Finally, there is a set of approved preprinted, standing orders that facilitates the ordering of tests, medications, etc, which are consistent with the AMIC critical path. These orders can be changed if the patient condition warrants a different course of therapy.

Lay Version of the AMIC

After the caregiver AMIC had been in use for a year, a subgroup made up of the staff nurse case managers and a nursing administrator began the design process for a lay version of the AMIC. Prior to this, the nursing administrator had done a thorough review of the research literature related to patient desire for participative decision making and information sharing and the impact of these two processes on patient satisfaction.[3–7] The research findings indicated the following:

- Patients have a preference for information sharing between themselves and their caregivers.
- Patients have a preference for some involvement in decision making regarding their plan and goals of care.
- These preferences are influenced by factors including disease severity, health status, age, socioeconomic status, level of education, ability to communicate, and willingness to participate.
- Patients wishing to be involved in treatment decisions were significantly more hopeful.
- Involvement in decision making and information sharing has been positively associated with patient satisfaction.

- Specific interventions to promote involvement in decision making and information sharing include assisting patients to identify and communicate specific concerns and questions related to treatment options and to identify mutual goals for the plan of care.

A formal research utilization project was initiated with research findings reviewed with members of the cardiovascular team and a research-based intervention designed to accompany the lay version of the AMIC (see Exhibit 9–1). The model seen in Figure 9–1 summarizes the nursing diagnoses, interventions, and outcomes that were measured before and after the lay map and research-based intervention were implemented. These outcomes will be described in another section of this chapter.

The lay map, which was designed for reading comprehension at the eighth grade level, is composed of an introduction letter to the patient (not shown), a lay version of the goal statements, a 7-day critical path (day 7 correlates with day 6 on caregiver AMIC; this heading is now being reformatted into just day 6) and a final page (not shown) for patient questions and concerns (see Appendix 9–A, Exhibit 9–A–2). The consult/appointment section of the critical path was kept quite general in terms of when

Exhibit 9–1 The Research-Based Intervention To Promote Patients' and Families' Participation in Care

Definition: Ensuring active involvement of patient and family in health care provision, decision making, and evaluation of outcomes.

Defining Activities:
1. Determine patient's and family's desire for, ability to assume, and comfort with participative role in decision making, provision of care, and evaluation of outcomes.
2. Promote patient's and family's attempts at independent decision making through
   - assisting patient and family to identify areas of concern
   - assisting patient and family to formulate questions for health care team
3. Facilitate collaborative decision making and serve as patient advocate within the health care system through
   - assisting patient and family to participate with the health care team in the mutual identification and selection of goals of care
   - assisting patient and family in the formulation and implementation of strategies to attain goals
4. Review plan of care with patient and family at least twice daily.
5. Identify social/environmental factors and personal health habits that have a potential impact on recovery.
6. Identify and encourage use of community resources.
7. Identify patient and family learning needs; plan and implement appropriate patient/family education activities.

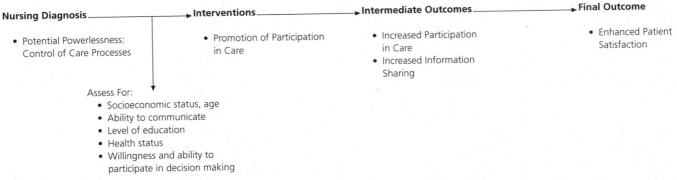

Figure 9–1 The Promotion of Patients' and Families' Participation in Care Model (Model Illustration by D.E. Walker)

they would occur in order to allow as much scheduling flexibility for the different disciplines as possible. Also, the test category of the critical path was kept quite general in terms of what type of test might be performed and on what day of the hospital stay these tests might occur. The medical staff requested this due to concerns of creating false expectations for, and conflict with, patients and families. Also, the medication category was written in general terms to decrease the frequency with which the map would have to be revised and to allow flexibility in ordering different brands of drugs within any specific drug category (eg, medications for sleep). Blank lines accompany the drug nitroglycerin, which allows patients to write in what their specific warning signs are that indicate a need for this drug.

As previously noted, the last page of the pathway has space for patients to note questions they may have for various members of the interdisciplinary team. Staff nurses review the lay map twice daily and assist the patient to identify any questions or concerns they may have; these are then noted on the last page. Research has shown that when patients are coached in terms of identification of questions and concerns, information sharing increases and so does patient satisfaction.[8] To ensure that this important twice-daily review is performed, cues with check boxes are included in the teaching category of the caregiver AMIC. This is not tracked as a variance at this time.

The "At Home" Path

In late 1996, the staff nurse case managers and the unit-level quality assurance coordinators developed a homegoing map (see Exhibit 9–2). This map was designed to provide the patient and family with essential "need-to-know" information to assist with the transition from the acute inpatient unit into the community setting. There are five categories of information provided for the patient and family: (1) activity levels that provide guid-ance for the continuation of cardiac rehabilitation and information on how to assess if the exercise is too strenu-ous; (2) health facts that focus on cardiac risk factors and how to check pulse rates; (3) medication advice on how to evaluate the need for nitroglycerin and how to administer this drug if the patient experiences chest pain; (4) self-care guidelines for bathing and diet; and (5) a section on reasons to call for more information and when to call the doctor. Other valuable information is contained in this pathway including the phone number of the intermediate unit; the date on which the patient's nurse will do a postdischarge call; the patient's discharge high and low density lipoprotrein blood levels; and the patient's return clinic appointment date and time.

By using this user-friendly map, the cardiovascular team hopes to increase at-home understanding and compliance with the discharge regime and to further improve patient satisfaction. Many patients have expressed appreciation of the at-home path and the postdischarge follow-up telephone call. The nursing staff report that the patients have many questions and concerns following discharge and frequently have requested a second follow-up call.

APPROACH TO VARIANCE TRACKING AND MONITORING

The process and procedure for tracking and monitoring variances has evolved over a 3-year period with variances being tracked both on a concurrent and retrospective basis.

Concurrent Review: Caregiver AMIC

All members of the cardiovascular team assigned to care for a specific AMI patient are expected to chart their variances from care on a daily basis. With the exception of the medical staff who are still poor at documenting on

Exhibit 9–2 At-Home Path to Recovery from a Heart Attack: Things To Do Each Day

Activity	Health	Medications	Self-Care	Reasons To Call for More Information
♥ Do exercises as prescribed by CHAMPS (Cardiac Rehab). ♥ Utilize local cardiac rehab facilities if available. ♥ Resume sexual activity when ready. ♥ Know when you are overdoing it: – chest pain – unusual shortness of breath – palpitations – irregular or slow heartbeat – weakness – faintness, dizziness, lightheadedness – leg pain or cramps ♥ Take rest periods 1/2 hour after meals as needed.	♥ Check your pulse for 1 minute at the same time each day (normal: 50 to 100 beats/minute). Your pulse is _____ ♥ Risk factors that you can control: – use of tobacco – high blood fats (cholesterol) – high blood pressure – overweight target weight _____ – lack of regular exercise – diabetes – poor stress management ♥ Return appointment: _____ _____	♥ Take your medications as prescribed. Do not stop without discussing with your doctor. ♥ **If your heart symptoms occur:** sit or lie down • put one (1) nitroglycerin tablet under your tongue • wait three (3) minutes **If chest pain has not subsided:** • take one (1) more nitroglycerin tablet • wait three (3) minutes **If you still have chest pain:** • take one (1) more nitroglycerin tablet • wait three (3) minutes **If chest pain continues, call 911!!!**	♥ Shower/bathe as instructed. ♥ Practice reading food labels for fat intake, cholesterol, and sodium levels. ♥ Eat healthy! Try new recipes. ♥ Follow no-added-salt, low cholesterol diet. ♥ Your cholesterol level is: Good (HDL) _____ Bad (LDL) _____	The 4JCW Nursing Station phone number at UIHC is (319) 356-3591. ♥ If you have any questions or concerns that are not emergent write them down. Your nurse will call you in 2 to 3 days _____ Call your doctor if: ♥ Your heart rate (pulse) is less than 50 beats/minute or greater than 100 at rest consistently. ♥ You experience light-headedness or dizziness. ♥ You experience increases in swelling of your feet or hands or abdomen. ♥ You experience an increase in your heart symptoms: – You experience an increase in shortness of breath. – You get angina after less exertion than usual. – Your angina is more severe or lasts longer. – Keep a written record of your heart discomfort episodes. ♥ You experience an increase in weight of 5 pounds or more in a week. – Weigh yourself weekly and keep a written record.

Source: Reprinted with permission from *American Journal of Nursing,* Vol. 96, No. 10, © 1996 Lippincott-Raven Publishers and North Carolina Baptist Hospitals.

the variance tracking form, the team has been successful in meeting this documentation goal. In nursing shift report, the patient's progress on the critical path is discussed along with variations from the path. The nurse reviews the critical path with the patient and family, and with various members of the multidisciplinary team, including the medical staff, paying close attention to what needs to occur that day, the day following, and by the time of discharge. Variations from the critical path are discussed, clarified and resolved when possible. Some of the medical staff were initially uncomfortable with nursing staff reviewing the critical path and questioning the rationale for variations from the path. The use of interdisciplinary clinical pathways required the introduction and eventual mastery of new behaviors for all members of the care team. Over the last 3 years, trust among the various disciplines has grown, collaborative practice has been enhanced, and the role of the nursing staff with ongoing monitoring of the critical path has been accepted. The medical staff is now much more willing to meet with the charge nurse to

assist in answering questions regarding variations in practice, writing order changes as needed, and planning for the next day's plan of care.

Retrospective Review

Concurrent review helps to ensure that each patient placed on the pathway is receiving care consistent with that plan of care. Retrospective review of aggregated data is a continuous quality improvement method that allows the care team to critically analyze the program of care for any given population. Aggregate data regarding the presence and degree of variation from a "best practice" pathway are reviewed and interpreted by the care team and improvements in the overall quality of care are designed and implemented. Cost and LOS are tracked consistently since the initiation of the first version of the AMIC in 1995. Also, all variances from the AMIC critical path were tracked initially by the staff nurse case managers. This was quickly stopped as it was very labor intensive and much of the collected data did not provide meaningful information for quality management.

Over a period of 18 months, the AMIC was redesigned multiple times and key markers of quality for the AMIC were identified (see Exhibit 9–A–3 in Appendix). These key markers are based on recommendations from the American College of Cardiology and the American Heart Association. This was a time-consuming and, at times, frustrating part of the clinical pathway development process for the staff nurse case managers. They had to do quite a bit of behind-the-scenes conflict management as the medical staff attempted to reach consensus on what the key markers should be. Many one-on-one meetings were needed and flexibility and patience were required. While difficult to go through, this was one of the most creative times during the AMIC developmental process. What made this so challenging was that prior to the introduction of the clinical pathway, there had been limited need among the disciplines to reach consensus on overall programs of care. Each discipline functioned as a "vertical silo," making individual decisions regarding patient care and each had limited accountability as a member of an integrated front-line care team. This vertical silo mental model is a well-known organizational characteristic of academic medical centers and is thought to be a primary source of practice variation. In time, not only was physician consensus reached regarding key markers, but also, much resistance to the use of clinical pathways was worked through, and the various cardiovascular team members learned to work together as an integrated team, to manage and resolve conflict, and to use the skill of consensus building. The staff nurse case managers grew tremendously in their leadership, communication, and project management skills.

Data Collection and Analysis

Key marker variances are documented on a preprinted form, which is a permanent part of the patient record. Since the majority of the key markers occur at the time of discharge, the nursing staff complete the tracking record at this time. The original tracking record remains in the chart and a Xerox copy is given to the staff nurse case managers. Two days a month the staff nurse case managers are given administrative time, part of which is used to code the data from the tracking record. "Type" refers to either critical path category (eg, tests) or the problem/goal section, "time frame" refers to which day on the critical path the event should have occurred, "event" refers to the specific variance (eg, telemetry discontinued), and "source" refers to the reason for the variance (eg, patient condition). The data are submitted to the Office of Clinical Outcomes and Resource Management

(CORM) for data entry and analysis with feedback to the cardiovascular team twice a year.

In mid-1996, the APR DRG system, which is a retrospective severity-of-illness system, has been used to stratify the patients into four levels of severity (minor, moderate, major, and extreme). The majority of the patients placed on the AMIC fall within the minor to moderate range (approximately 62%) with 34% being categorized as major and 4% as extreme. Cost and LOS data can also be categorized into four major clinical groups that the cardiovascular team determined were clinically relevant to the AMI patient population. These categories are

- AMI with no revascularization
- AMI with coronary bypass
- AMI with angioplasty
- AMI with cardiac catheterization

The ability to both severity-adjust and differentiate financial and LOS information based on the above allows all AMI patients to be placed on the AMIC and outcome data to be aggregated and analyzed in to meaningful, clinically relevant groups.

COST AND LOS OUTCOMES OF THE CAREGIVER AMIC

Due to the multiple revisions of the AMIC, the changes made in what variances were tracked and the addition of the APR DRG system in 1996, it is not possible to give a full 3 years' worth of outcome measurements. The LOS and cost data will be reported in two different manners. First, aggregate 1993 pre-AMIC data will be compared to aggregate 1994 and 1996 data. Additionally, 1994 disaggregated data for the clinical group AMI with no revascularization will be compared to 1996 data for the same clinical group. The AMIC policy precludes the use of actual cost and LOS numbers so only percent change is reported for these time periods.

Figure 9–2 shows line graphs comparing inpatient LOS and cost data for a 4-month period in 1993, a 3-month period in FY 1994–1995, and for the first three quarters of FY 1996–1997. This graph shows significant decreases in both case costs and LOS when comparing 1993 pre-AMIC data to post-AMIC data. The increase in both costs and LOS for the third quarter of FY 1996–1997 was due to the LOS increasing by 3 days and an increase in costs in the categories of special care and "other."

Figure 9–3 shows a comparison of the FY 1994–1995 vs FY 1996–1997 data for the clinical group: AMI with no revascularization. Figure 9–4 shows a comparison of the FY 1994–1995 vs FY 1996–1997 LOS and cost data for the clinical group: AMI with angioplasty. The cardiovascular team determined that the number of AMI patients

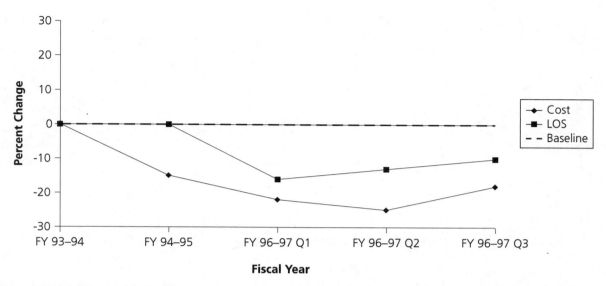

Figure 9–2 Percent Change in LOS and Costs for Aggregated AMI Population with Data from FY 1993–1994 Serving as the Baseline for Comparison

who also go on to have coronary bypass surgery is low and that LOS and cost information for this subpopulation are more relevant to the CareMap® specific for coronary bypass surgery. This subgroup has been dropped as a population to be studied by the AMIC team. There is no FY 1994–1995 data available for the clinical group: AMI with cardiac catheterization so no longitudinal comparison can be made. With the costs and LOS for FY 1994–1995 serving as the baseline, all three periods in FY 1996–1997 for

both clinical groups show decreases in LOS and costs. The data for AMI with no revascularization are most encouraging, showing only slight variability in the percent of decrease in costs. The decrease in the percent of change between period three and the FY 1994–1995 baseline for the AMI with angioplasty is due to marked increases in the expenses associated with both the radiology and supplies cost categories. This is believed due to two outliers with costs in excess of the 90th percentile.

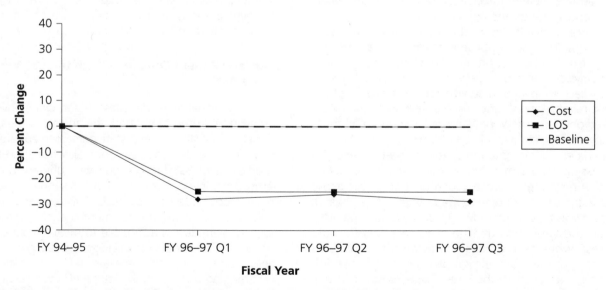

Figure 9–3 Percent Change in the First Three Quarters of FY 1996–1997 in LOS and Cost for Clinical Group: AMI with No Revascularization with Data from FY 1994–1995 Serving as the Baseline

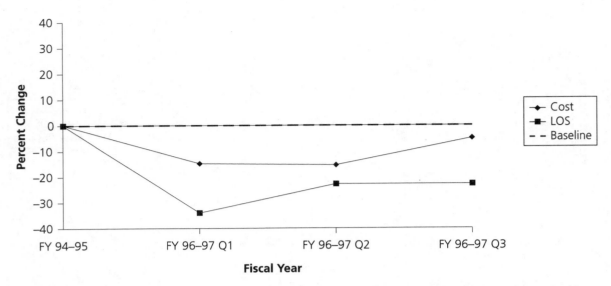

Figure 9–4 Percent Change in the First Three Quarters of FY 1996–1997 LOS and Cost for the Clinical Group: AMI with Angioplasty with Data from FY 1994–1995 Serving as the Baseline

Clinical Outcomes

Data have been collected for the last 3 years in terms of variances from the critical path. At the initiation of the original CareMap®, the audit showed significant problems with the cardiac rehabilitation component of care: (1) poor exercise progression, especially on day 2 of the critical path; (2) low percentage of referrals of inpatients to the cardiac rehabilitation program for predischarge evaluation and referral to the posthospitalization program, and (3) inpatient exercise progression being interrupted on weekends and holidays when physical therapy staff were unavailable. Additionally, problems were noted with the overall documentation of the inpatient phase of the exercise program. An ad hoc group made up of physical therapy (inpatient cardiac rehabilitation), the cardiac rehabilitation service (phase II or outpatient cardiac rehabilitation), and the unit-level case managers focused on meshing the patient exercise goals of physical therapy and the cardiac rehabilitation program. They designed a new documentation system so that the exercise progression status of patients could easily be determined. A new referral process for inpatients to the phase II program was also implemented. Finally, the phase II program was able to provide 10 hours of nursing coverage on the weekends and holidays so patients could progress with their exercise programs. Figures 9–5 through 9–7 show via line graphs the progressive improvement achieved for the three problems noted above.

With the latest iteration of the caregiver CareMap®, only one round of data is available to measure the per-

cent compliance with some key markers for the fourth quarter of FY 1996–1997. This descriptive information is available in Exhibit 9–3. The cardiovascular team was pleased with the percent compliance on many of the key markers and will monitor closely, looking for opportunities to further improve the rate of compliance. For the time being, the team is focusing on the variation found with the key marker: Cardiac rehabilitation BID. This variance is believed due to a number of vacant registered nurse positions on this service; aggressive recruitment, hiring, and orientation is occurring. The team expects improvement in compliance with this key marker within the next few months.

Staff and Patient Outcomes with the Lay Version of the AMIC

A longitudinal pretest/posttest design with convenience sampling was used to evaluate the impact of the lay map on patient satisfaction and staff nurse perceptions of the model of care delivery. Two subscales from the Blegen-Goode Patient Satisfaction tool were utilized: the three-item Communication subscale and the four-item Partnership in Decision-Making subscale. This patient satisfaction tool was selected because its unit of analysis is the interdisciplinary team, it had a subscale to measure patient satisfaction with partnership in decision making, and it had established psychometric properties. It has a 5-point Likert scale ranging from strongly disagree (1) to strongly agree (5). This tool was administered for the 5 months preceding the implementation of the lay version of the AMIC; a

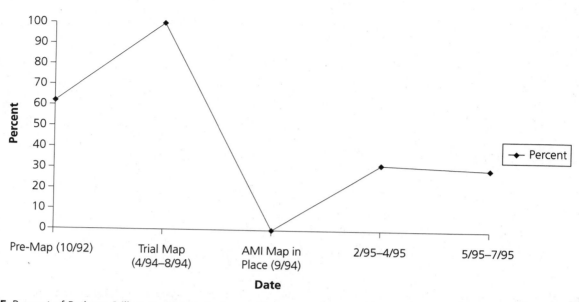

Figure 9–5 Percent of Patients Still on Bedrest or Bedrest with Bedside Commode on Day 2 after Rule-in as an AMI

sample of 32 usable surveys was collected. It was also administered postimplementation until a sample of 32 usable surveys was collected.

A survey tool was designed by the research team to measure staff nurse perceptions of the care delivery model being used on the two study units. Sixteen items were divided into five separate subscales, which included Care Planning, Patient Knowledge, Patient Advocacy, Interdisciplinary Team Relations, and Accountability. A 5-point Likert scale was used ranging from strongly dis-

agree (1) to strongly agree (5). No psychometric properties were established for this tool beyond content validity. This tool was mailed to all staff nurses preimplementation of the lay version of the AMIC with a total of 28 surveys returned. The survey was again mailed to all staff nurses on the two study units 4 months postimplementation of the lay version of the AMIC with 27 surveys returned. An additional data collection period 8 months postimplementation was originally planned. However, one of the two units was actively engaged in redesign of

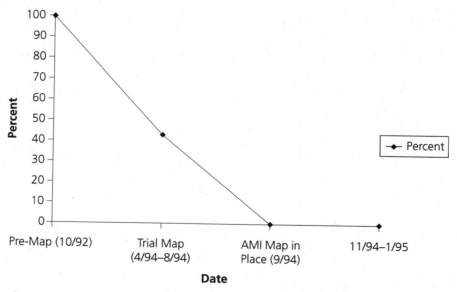

Figure 9–6 Percent Occurrence of Patients NOT Being Referred to Phase II Cardiac Rehabilitation for Postdischarge Evaluation

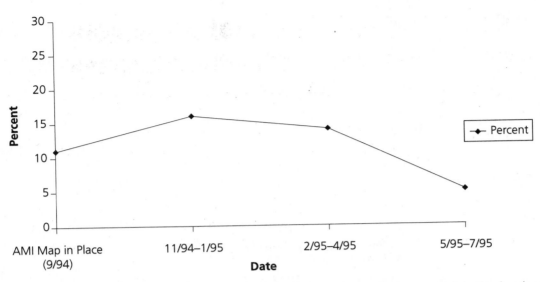

Figure 9–7 Percent Occurrence of Patients' Cardiac Rehabilitation Exercise Program Being Interrupted on Weekends and Holidays

their care delivery model. Also, during this time, a major change in unit and divisional leadership occurred and there was reaggregation of some of the cardiology patient populations, which affected both of the study units. Because of this, the research team decided not to do the second postimplementation measurement.

DATA ANALYSIS

Independent t tests were used for analysis of the data from both the patient satisfaction survey and staff nurse perceptions of the care delivery model survey. Exhibit 9–4 provides the results of data analysis for the patient satisfaction survey. No statistically significant changes were noted when comparing pretest results with posttest results. Pre-

test scores were quite high for the two subscales, which indicated a high level of patient satisfaction preimplementation. While the comparison of pretest vs posttest results was not statistically significant, the posttest scores for the partnership in decision making did decrease. As noted above, the postimplementation period was marked by multiple changes and a high level of job stress for the registered nurses. This may have affected the consistency with which the staff nurses were able to implement the key steps in the research-based intervention.

Exhibit 9–5 provides the results of data analysis for the comparison of the pretest and posttest scores from the staff nurse survey. Again, the outcome of the independent t tests was not statistically significant. Given the degree of change and job stress experienced by the staff

Exhibit 9–3 Percent Compliance with Key Markers for the Latest Iteration of the AMI CareMap®. See Exhibit 9–A–3 for more specific explanation of key markers.

Type of Variance	Specific Key Marker of Quality	% Compliance	Source of Variance
Tests	Fasting lipid profile	92%	None given
Tests	≤ 2 CBCs	38%	Patient condition
Tests	≤ 3 O_2	46%	MD Choice
Treatments/Interventions	d/c telemetry	73%	Patient condition; MD choice
Treatments/Interventions	Cardiac Rehab BID	69%	Coverage on weekends/holidays; Patient condition
Medications	Anticoagulant RX at time of discharge	100%	
Medications	ACE Inhibitor if LV EF > 40%	88%	None given
Medications	Beta blocker at discharge	92%	None given
Teaching	Pt education scores	84%	Patient condition
Discharge Planning	Risk Stratification	85%	Patient condition

Exhibit 9–4 Comparison of Pretest and Posttest Results for Patient Satisfaction Data. Independent *t* tests show no statistical significance at p level of .05.

Data Analysis: Patient Satisfaction Survey		
Patient Satisfaction Subscale	Pretest Results (n = 32)	Posttest Results (n = 32)
• Communication	Mean: 4.72	Mean: 4.80
• Partnership in Decision Making	Mean: 4.78	Mean: 4.70

nurses during the postimplementation period, the research team was relieved that the postimplementation scores did not drop precipitously. However, given the relatively low scores for the care planning, interdisciplinary team relations and accountability subscales, it is felt that there is room for improvement in the design of the unit-level care delivery models.

CONCLUSION

The last 3 years have been a time of intense and fruitful learning for the cardiovascular CareMap® team. During this time, many valuable lessons have been learned by this team:

- Engage all affected disciplines in the dialogue and planning of the CareMap®. This will create a sense of ownership from the very beginning. The cardiovascular team meetings still have about 20 people attending.
- Before beginning any work on the CareMap®, identify key physician champions who are willing to assume a leadership role for the duration of the planning and implementation of the CareMap®.
- Eliminate as much redundant charting as possible as this is a major source of dissatisfaction when implementing a pathway; begin plans to computerize as

soon as possible.
- Use preprinted orders as part of the overall clinical pathway package as the use of these greatly increases compliance with the critical path.
- Identify key markers of quality as soon as possible. This gives meaning and focus to the variance tracking and analysis process and greatly increases the possibility of continuous quality improvement activities. Also, having meaningful data that heighten the likelihood of continuous improvement goes a long way in convincing the "nonbelievers" about the effectiveness of clinical pathways.
- Give lots and lots of positive feedback, recognizing that not everyone will be able to change immediately. Capitalize on the team members who are supportive of the change to CareMaps® and don't give up on the resisters!

Much remains to be done in terms of the UIHC's approach to the design of clinical practice guidelines. Plans are in place to computerize the CareMaps® so that variance tracking can occur behind the scenes and the problem of redundant charting is eliminated. Also, the use of decision-making algorithms must be explored as a method to improve the clinical management of specific aspects of the critical path. To further manage quality, cost, and LOS, the UIHC is refining its overall approach to

Exhibit 9–5 Comparison of Pretest and Posttest Results for RN Survey Data. Independent *t* tests show no statistical significance at p level of .05.

Data Analysis: RN Survey		
RN Survey Subscales	Pretest Results (n = 28)	Posttest Results (n = 27)
• Care Planning	Mean: 3.66	Mean: 3.67
• Patient Advocacy	Mean: 4.22	Mean: 4.30
• Interdisciplinary Team Relations	Mean: 3.79	Mean: 3.77
• Accountability	Mean: 3.73	Mean: 3.80

care management of patient populations in terms of defining process, roles, and accountabilities of all members of the unit-level multidisciplinary teams. The introduction of a full system of clinical practice guidelines is a very complex method for organizational transformation, challenging highly ingrained paradigms of care. It is challenging and exhilarating work with multiple opportunities for both professional and personal growth.

REFERENCES

1. Ireson CI. Critical Pathways: effectiveness in achieving patient outcomes. *J Nurs Adm*. 1997;27(6):16–23.
2. Gustfson DH, Risberg L, Gering D, et al. *Case Studies from the Quality Support System: AHCPR Research Report No. 97-0022*. Rockville, MD: US Dept of Health and Human Services; 1997.
3. Bull MJ. Patients' and professionals' perceptions of quality in discharge planning. *J Nurs Care Qual*. 1994;8(2):47–61.
4. Carmel S. Satisfaction with hospitalization: a comparative analysis of three types of service. *Soc Sci Med*. 1985;2(1):1243–1249.
5. Cassileth BR, Zupkis RS, Sutton-Smith MV. Information and participation preferences among cancer patients. *Ann Int Med*. 1980;92:832–836.
6. Faden RR, Becker C, Lewis C, Freeman J, Faden AI. Disclosure of information to patients in medical care. *Med Care*. 1981;XIX(7):718–733.
7. Marquis SM, Davies AR, Ware JE. Patient satisfaction and change in medical care provider: a longitudinal study. *Med Care*. 1983;21(8):821–829.
8. Graham O, Schubert W. A model for developing and pretesting a multi-media teaching program to enhance self-care behavior of diabetes patients. *Pat Ed Counseling*. 1985;7(1):53–64.

■ Appendix 9–A ■

CareMaps®, Care Goals, and Tracking Record

Exhibit 9–A–1 Multidisciplinary CareMap® Acute Myocardial Infarction

	DATE
	HOSP. NO.
	NAME
Map ID No. 05006	BIRTH DATE
Page 1	ADDRESS
	SS#
• File most recent sheet of this number ON BOTTOM	If not imprinted, please print date, hosp. no., name and location

Unit(s) _____

Case Manager(s) _____

_____

_____

Physician(s) _____

_____

Date/Time/Name of MD/RN—Initial Review _____

Date/Time and Name of RN Who Reviewed with Patient _____

❑ Discontinue Map (date) _____

❑ Transfer to Other Map (date) _____

❑ Concurrent Map _____

❑ Return to This Map (date) _____

Secondary Diagnoses _____

_____

Age _____

Discharge Date _____

Other Map Name _____

Other Map Name _____

Side tabs: A–20 / B Clin. Notes / C Laboratory / D X-Ray Exam / E Consultation / F Spec. Exam / G Therapy / H Pathology / I Diagnosis

MULTIDISCIPLINARY PLAN OF CARE (Critical path is on reverse side)

	INTERMEDIATE GOALS		OUTCOMES
	Day of Rule-In	Cardiac Rehabilitation	Day of Discharge
Date			
Patient Problem	Met — Unmet	*Put initials in Met or Unmet box.* Met — Unmet	Met — Unmet
1. Potential for decreased cardiac output and/or cardiac dysrhythmia	Free of chest discomfort or shortness of breath and hemodynamically stable without cardiac dysrhythmia	Tolerates activity progression	Discharged from Phase I cardiac rehabilitation hemodynamically stable without cardiac dysrhythmia
2. Potential for chest discomfort	Verbalizes chest discomfort appropriately to RN	Reports chest discomfort to care provider and obtains relief	Free of chest discomfort
3. Knowledge	Verbalizes: –Reason for hospitalization –Activity restriction –When to notify RN –Rationale for any procedures	Briefly verbalizes understanding of applicable MI topics and tests	Verbalizes relevant information from discharge teaching forms
4. Potential for anxiety	Recognizes and verbalizes concerns related to hospitalization	Anxiety level does not interfere with the ability to progress with cardiac rehabilitation	Anxiety level does not interfere with transition to home environment

continues

Exhibit 9–A–1 continued

Rule in on intermediate care unit ❑ Transfer from: CVICU ❑ local hospital ❑ Transfer time: _____ **Map ID No. 05006** **Page 2**	**PATIENT NAME** _____ **HOSP. NO.** _____

Critical Path (Care plan is on reverse side)	
Post MI Day 1	**Post MI Day 2**
Date		

	Post MI Day 1	**Post MI Day 2**
Consults	Cardiovascular Health Assessment Management Preventive Services (CHAMPS) Referral	CHAMPS Referral ❑ (if not done previous day)
Tests	Fasting Lipid Profile (if not done past 6 months) UA (if not done on admission) e2 PTT (if on heparin) pt/INR (if on Coumadin)	e2 (if previously abnormal or if Pt on diuretic) PTT (if on heparin) pt/INR (if on Coumadin) Echocardiogram or Cine CT (or IVG if MI not anterior Q-wave) (circle one) _____ EF _____ % (date)
Treatments/ Interventions	VS per unit routine Assessment per routine Telemetry Level II IV wt qd I & O cp protocol *Consider:* oxygen	VS per unit routine Assessment per unit routine Telemetry level I ❑ IV Daily wt. I & O cp protocol

Medications	ECASA qd *PRNs:*
	 Colace BID NTG Sublingual
	 NS lock Flush BID Laxative of Choice
	                        Antacid of Choice
	                        Tylenol
	                        Benadryl
	
	 Consider:
	 Heparin IV or Subq ❑
	 Beta blocker PO ❑
		ECASA qd . *PRNs:*
		 Colade BID NTG Sublingual
		 NS lock Flush BID Laxative of Choice
		                        Antacid of Choice
		                        Tylenol
		                        Benadryl
		
		 Consider:
		 DC Heparin IV/subq ❑
		 Substitute Coumadin for ECASA ❑

Diet	Modified cholesterol *Consider:* NAS	Modified cholesterol *Consider:* NAS

Activity	Explain activity progression ❑
	 MET LEVEL _____
		CHAMPS or PT BID
		     lab ❑ / ❑
		     floor ❑ / ❑
		 MET LEVEL _____

Teaching	*On admission Review:* ❑ unit routine ❑ call for cp ❑ diagnosis ❑ activity	Review lay map ❑ ❑ ❑ MI packet ❑ class schedule ❑ lay map
Discharge Planning	*On admission:* Assess home situation and family support ❑	

INITIALS/ _____/_____   _____/_____

SIGNATURE/ _____/_____   _____/_____

continues

Exhibit 9–A–1 continued

Map ID No. 05006 Page 3 • File most recent sheet of this number ON BOTTOM	DATE HOSP. NO. NAME BIRTH DATE ADDRESS SS# If not imprinted, please print date, hosp. no., name and location

Teaching Key:	3—Independent in carrying out activity and/or verbalizing information 2—Requires assistance to carry out activity or verbalize information 1—Unable to carry out activity or verbalize information	Initials = Information given, no further questions N/A = Not applicable # = Family only

Right margin tabs: A–20 | B Clin. Notes | C Laboratory | D X-Ray Exam | E Consultation | F Spec. Exam | G Therapy | H Pathology | I Diagnosis

CRITICAL PATH

Date	Post MI Day 3	Post MI Day 4	Post MI Day 5	Post MI Day 6
Consults				Phase II Referral local ❏ _____ UIHC ❏
Tests	PTT (if on heparin) pt/INR (if on Coumadin) *Consider:* Cardiac cath (if LV EF < 40%) _____ (date) CxR Pa/Lat (if not done on admit)	PTT (if on heparin) pt/INR (if on Coumadin) *Consider:* Submaximal GXT ❏ _____ (date)	PTT (if on heparin) pt/INR (if on Coumadin) *Consider:* CBC e2 Cardiac cath if ischemic GXT	pt/INR (if on Coumadin)
Treatments/ Interventions	VS per unit routine Assessment per unit routine Telemetry level I IV Daily wt. I & O cp protocol	VS per unit routine Assessment per unit routine DC Telemetry ❏ _____ (date) DC IV ❏ DC Daily wt. ❏ DC I & O ❏ cp protocol	VS per unit routine Assessment per unit routine cp protocol	Same as Day 6
Medications	ECASA qd (or Coumadin) Colace BID Saline lock flush BID Angiotensin converting enzyme (ACE) inhibitor if LV EF < 40% ❏ *PRNs* NTG sublingual Laxative of choice Antacid of choice Tylenol Benadryl PO *Consider:* DC heparin IV/subq ❏ Beta blocker ❏ Substitute Coumadin ❏	Same as Day 4 *PRNs:* Same as day 4 *Consider:* DC heparin IV/subq ❏ Beta blocker ❏ Substitute Coumadin ❏	Same as Day 4 *PRNs:* Same as day 4 *Consider:* DC heparin IV/subq ❏ Beta blocker ❏ Substitute Coumadin ❏	Same as Day 4 *PRNs:* Same as day 4 *Consider:* DC heparin IV/subq ❏ Beta blocker ❏ Substitute Coumadin ❏
Diet	Modified cholesterol *Consider:* NAS	Modified cholesterol *Consider:* NAS	Modified cholesterol *Consider:* NAS	Modified cholesterol *Consider:* NAS
Activity	CHAMPS or PT BID lab ❏/❏ floor ❏/❏ MET LEVEL _____	CHAMPS or PT BID lab ❏/❏ floor ❏/❏ MET LEVEL _____	CHAMPS or PT BID lab ❏/❏ floor ❏/❏ MET LEVEL _____	CHAMPS or PT BID lab ❏/❏ floor ❏/❏ MET LEVEL _____
Teaching	Review lay map ❏ ❏ **TOPICS** *date/initials/key Risk factors _____ Heart anatomy _____ Angina vs MI _____ Medications _____ Stress Management _____ Diet _____ Activity _____ Taking NTG _____ Smoking cessation _____	Review lay map ❏ ❏	Review lay map ❏ ❏ **TOPICS** *date/initials/key echo _____ GXT _____ cardiac cath _____ PTCA _____ CABG _____ Stent _____ Fasting Lipid Profile _____ Other _____	Review lay map ❏ ❏
Discharge Planning	Notify social worker of potential nursing home placement ❏ N/A ❏ Notified Hospital care needed ❏ N/A ❏ Notified ❏ Reviewed home care needs Order special equipment ❏ N/A ❏ Notified _____ Discharge referral form ❏ N/A ❏ Started ❏ Complete			Send pt home with B19b1 and B19b1a ❏
INITIALS/	____/ _____	____/ _____	____/ _____	____/ _____
SIGNATURE/	____/ _____	____/ _____	____/ _____	____/ _____
POSITION/	____/ _____	____/ _____	____/ _____	____/ _____

Source: Copyright © The Center for Case Management. Content Copyright 1996, The University of Iowa Hospitals and Clinics, Iowa City, Iowa, used with permission.

Exhibit 9–A–2 Heart Attack Patient Care Goals and CareMap®

		HEART ATTACK PATIENT CARE GOALS		
	Day 2	**Day 3 & 4**	**Day 5 & 6**	**Discharge Goal**
Activity	Your heart rate and blood pressure are within safe limits as you increase activity. You should not do activities other than those you have been instructed to do.	You will increase your activity without symptoms of stress on your heart. Heart rate and blood pressure will be kept within safe limits by keeping your activity within set guidelines.	Your activity is increased and you may go to CHAMPS* activity area.	You receive a written activity program to do at home and going to a local rehab program is encouraged. You will receive instructions on when to return to work. You know which activities are safe to do at home and work.
Warning Signs of a Heart Attack	Immediately report symptoms like those which brought you to the hospital to a care provider. Your warning signs are: _____	You know the symptoms that may warn of a heart attack. Report symptoms immediately to a care provider.	You know when and how to take nitroglycerin. Report the warning signs immediately to a care provider.	You know which symptoms to report to a doctor and which symptoms need immediate medical care. You are free of symptoms related to your heart disease.
Knowledge	Your knowledge and understanding of heart disease are assessed. You get written information to help you learn.	You go to available classes on diet, heart disease, activity, medication, and stress. Information and support to help stop smoking is provided (if needed).	Same as Day 3 & 4. You know your risk factors related to heart disease. You set goals to help lower your risk of another heart attack.	You receive a list of your medicine(s), including why and when to take each one. You know how to follow a low cholesterol diet. You stop smoking or set a stop date.
Anxiety/Stress	Your questions about your hospital stay, tests, and treatments are answered.	Help is available for money concerns. You receive emotional support as needed. Your family is involved in care and decisions when possible.	You learn skills to help you lower your stress.	You have goals for recovery. You have support and resources for your care at home.

*CHAMPS (Cardiovascular Health Assessment, Management, and Prevention Services)

continues

Exhibit 9–A–2 continued

HEART ATTACK PATIENT CAREMAP® DAY-TO-DAY ACTIVITIES		
	CareMap® Day 2	**CareMap® Day 3**
Date:		
Consultations/ Appointments	♥ CHAMPS (Cardiovascular Health Assessment Management Preventative Services) staff will be called to start your exercise sessions. ♥ Help may be offered to quit smoking. ♥ A social worker can visit you and your family for personal or money questions. ♥ A dietician will see you and your family as often as needed before discharge to teach you about cholesterol (fat) and sodium (salt) use.	
Tests	♥ Blood samples may be drawn during your hospital stay to check kidney function, medication levels, and cholesterol. ♥ Additional tests may also be needed based on your condition. ♥ The following are several tests that may be done during your hospital stay: • Heart tracing (EKG)—looks at the heart's rhythm • Chest x-ray—x-ray of heart size and lungs • Ultrasound (echo)—for information on the heart muscle, chambers, and valves • Nuclear medicine test—looks at pumping strength of the heart at rest and/or heart artery narrowings • Cardiac catheterization—looks at heart vessels • Treadmill test—(or at a follow-up appointment)	
Treatments	♥ Vital signs (blood pressure, pulse, breathing rate, temperature) may be checked every 4 hours while you are awake. ♥ You may be weighed every evening. ♥ You may have a heart monitor on. ♥ You may have an IV in. ♥ Your liquid intake and urine may be saved and measured.	Same as Day 2 except: ♥ Vital signs (blood pressure, pulse, breathing rate, temperature) may be checked 2 to 3 times a day.
Medications	♥ The medications you were taking before may be the same or may be changed. ♥ You will take a tablet daily as a mild blood thinner. ♥ You may also take a stronger intravenous blood thinner if needed. ♥ If needed, medications may be given for gas, constipation, generalized aches and pains, and sleep. ♥ Nitroglycerin may be given for your warning signs of: _____	♥ Same as Day 2.
Diet	♥ Cholesterol (fat) and sodium (salt) restricted diet.	♥ Same as Day 2.
Activity	♥ A physical therapist or nurse may introduce you to the CHAMPS program and begin your exercise sessions.	♥ Your activity may be increased as tolerated, based on your heart rate and blood pressure response to exercise.
Discharge Planning	♥ Please talk to your nurse about home situation and family support. Please let your nurse know if you need special help after discharge.	
Teaching	♥ We will show you your room and call-light system. ♥ Please call us as soon as your warning signs occur. ♥ We will give you a packet of learning materials. ♥ You may go to classes about your heart. Please check with a staff member about day and time.	♥ Cardiac risk factors may be reviewed with you: weight, smoking, family history, high cholesterol, diabetes, high blood pressure, sedentary lifestyle, and stress. ♥ Heart teaching and medications will be reviewed with you.

continues

Exhibit 9–A–2 continued

	CareMap® Day 4	CareMap® Day 5
Date:		
Consultations/ Appointments	♥ CHAMPS (Cardiovascular Health Assessment Management Preventative Services) staff will be called to start your exercise sessions. ♥ Help may be offered to quit smoking. ♥ A social worker can visit you and your family for personal or money questions. ♥ A dietician will see you and your family as often as needed before discharge to teach you about cholesterol (fat) and sodium (salt) use.	
Tests	♥ Blood samples may be drawn during your hospital stay to check kidney function, medication level, and cholesterol. ♥ Additional tests may also be needed based on your condition. ♥ The following are several tests that may be done during your hospital stay: • Heart tracing (EKG)—looks at the heart's rhythm • Chest x-ray—x-ray of heart size and lungs • Ultrasound (echo)—for information on the heart muscle, chambers, and valves • Nuclear medicine test—looks at pumping strength of the heart at rest and/or heart artery narrowings • Cardiac catheterization—looks at heart vessels • Treadmill test—(or at a follow-up appointment)	
Treatments	♥ Vital signs (blood pressure, pulse, breathing rate, temperature) may be checked every 4 hours while you are awake. ♥ You may be weighed every evening. ♥ You may have a heart monitor on. ♥ You may have an IV in. ♥ Your liquid intake and urine may be saved and measured. ♥ Vital signs (blood pressure, pulse, breathing rate, temperature) may be checked 2 to 3 times a day.	♥ You may no longer need to wear the heart monitor except for exercise. ♥ Your IV may be removed. ♥ Your liquid intake may no longer be measured and you may no longer need to save urine.
Medications	♥ The medications you were taking before may be the same or may be changed. ♥ You will take a tablet daily as a mild blood thinner. ♥ You may also take a stronger intravenous blood thinner if needed. ♥ If needed, medications may be given for gas, constipation, generalized aches and pains, and sleep. ♥ Nitroglycerin may be given for your warning signs of: _____ ♥ New heart medications may be started based on your test results.	♥ The medications you were taking before may be the same or may be changed. ♥ You will take a tablet daily as a mild blood thinner. ♥ You may also take a stronger intravenous blood thinner if needed. ♥ If needed, medications may be given for gas, constipation, generalized aches and pains, and sleep. ♥ Nitroglycerin may be given for your warning signs of: _____
Diet	♥ Cholesterol (fat) and sodium (salt) restricted diet.	♥ Cholesterol (fat) and sodium (salt) restricted diet.
Activity	♥ Your activity may include walking in the hallways and/or while in CHAMPS Lab using an exercise treadmill, Airdyne bike, or recumbent bike.	♥ Your activity may include walking in the hallways and/or while in CHAMPS Lab using an exercise treadmill, Airdyne bike, or recumbent bike.
Discharge Planning		A nurse may ask if you need a visiting nurse at home, or if you need special equipment ordered for home.
Teaching	♥ Cardiac risk factors may be reviewed with you: weight, smoking, family history, high cholesterol, diabetes, high blood pressure, sedentary lifestyle, and stress. ♥ Heart teaching and medications will be reviewed with you.	♥ Continues review of heart teaching and medications.

continues

Exhibit 9–A–2 continued

	CareMap® Day 6	CareMap® Day 7
Date:		
Consultations/ Appointments	♥ CHAMPS (Cardiovascular Health Assessment Management Preventative Services) staff will be called to start your exercise sessions. ♥ Help may be offered to quit smoking. ♥ A social worker can visit you and your family for personal or money questions. ♥ A dietician will see you and your family as often as needed before discharge to teach you about cholesterol (fat) and sodium (salt) use.	
Tests	♥ Blood samples may be drawn during your hospital stay to check kidney function, medication level, and cholesterol. ♥ Additional tests may also be needed based on your condition. ♥ The following are several tests that may be done during your hospital stay: • Heart tracing (EKG)—looks at the heart's rhythm • Chest x-ray—x-ray of heart size and lungs • Ultrasound (echo)—for information on the heart muscle, chambers, and valves • Nuclear medicine test—looks at pumping strength of the heart at rest and/or heart artery narrowings • Cardiac catheterization—looks at heart vessels • Treadmill test—(or at a follow-up appointment)	
Treatments	♥ You may no longer need to wear the heart monitor except for exercise. ♥ Your IV may be removed. ♥ Your liquid intake may no longer be measured and you may no longer need to save urine.	♥ You may no longer need to wear the heart monitor except for exercise. ♥ Your IV may be removed. ♥ Your liquid intake may no longer be measured and you may no longer need to save urine.
Medications	♥ The medications you were taking before may be the same or may be changed. ♥ You will take a tablet daily as a mild blood thinner. ♥ You may also take a stronger intravenous blood thinner if needed. ♥ If needed, medications may be given for gas, constipation, generalized aches and pains, and sleep. ♥ Nitroglycerin may be given for your warning signs of: _____	♥ The medications you were taking before may be the same or may be changed. ♥ You will take a tablet daily as a mild blood thinner. ♥ You may also take a stronger intravenous blood thinner if needed. ♥ If needed, medications may be given for gas, constipation, generalized aches and pains, and sleep. ♥ Nitroglycerin may be given for your warning signs of: _____
Diet	♥ Cholesterol (fat) and sodium (salt) restricted diet.	♥ Cholesterol (fat) and sodium (salt) restricted diet.
Activity	♥ Your activity may include walking in the hallways and/or while in CHAMPS Lab using an exercise treadmill, Airdyne bike, or recumbent bike.	♥ Your activity may include walking in the hallways and/or while in CHAMPS Lab using an exercise treadmill, Airdyne bike, or recumbent bike.
Discharge Planning	♥ Homegoing prescriptions may be written. Tell your nurse if you want them filled here or at home. ♥ Be thinking about who will give you a ride home at discharge.	♥ A doctor may give you a copy of hospital summary and needed follow-up appointments.
Teaching	♥ Continues review of heart teaching and medications.	♥ CHAMPS staff will review your home exercise program with you and set up for local cardiac rehabilitation. ♥ A pharmacist or nurse will review your medication schedule with you. ♥ Your nurse will review your homegoing instructions.

Source: Copyright © The Center for Case Management. Content Copyright 1996, The University of Iowa Hospitals and Clinics, Iowa City, Iowa, used with permission.

Exhibit 9–A–3 CareMap® Tracking Record

	DATE
	HOSP. NO.
	NAME
	BIRTH DATE
Page 1	ADDRESS
	SS#
• File most recent sheet of this number ON BOTTOM	If not imprinted, please print date, hosp. no., name and location

CareMap Name _____ Acute Myocardial Infarction (CVICU) _____   CareMap ID # _____ 05006 _____

CareMap Day/Sequence Date/Initials	Key Event	Variance/Reason	Action
Level I	1. C.I. ≥ 2 2. SaO_2 ≥ 90% 3. Anginal pain controlled 4. Fluid volume stable 5. Hemodynamics stable with or without pressors	❏ No Variance ❏ Variance: Dates of variance _____ / _____ ❏ Reason(s): _____ _____ / _____ _____ _____ / _____ _____ _____ / _____ _____ _____ / _____ _____	
Level II	1. Hemodynamics stable 2. C.I. ≥ 2 Weaning pressors 3. SaO_2 ≥ 90% without vent 4. Anginal pain controlled 5. Cardiac tests completed	❏ No Variance ❏ Variance: Dates of variance _____ / _____ ❏ Reason(s): _____ _____ / _____ _____ _____ / _____ _____ _____ / _____ _____ _____ / _____ _____	
Level III	1. C.I. ≥ 2 without IV pressors 2. SaO_2 ≥ 90% 3. Transfer to step down unit	❏ No Variance ❏ Variance: Dates of variance _____ / _____ ❏ Reason(s): _____ _____ / _____ _____ _____ / _____ _____ _____ / _____ _____ _____ / _____ _____	

INITIALS/ _____ / _____ _____ / _____ _____ / _____

SIGNATURES/ _____ / _____ _____ / _____ _____ / _____

POSITION/ _____ / _____ _____ / _____ _____ / _____

Side tab: A–21 B Clin. Notes C Laboratory D X-Ray Exam E Consultation F Spec. Exam G Therapy H Pathology I Diagnosis

continues

Exhibit 9–A–3 continued

	PATIENT NAME
Page 2	HOSP. NO.
• File most recent sheet of this number ON BOTTOM	If not imprinted, please print date, hosp. no., name and location

Uncomplicated MI ❑ Yes ❑ No
Reason(s):
❑ CHF—not resolving in 24 hr
❑ Cardiogenic shock
❑ Pericarditis/Dressler's Syndrome
❑ Pulmonary or systemic arterial thromboembolism
❑ Transfer CVICU

❑ Sustained and/or hemodynamically significant atrial or ventricular arrhythmia
❑ Resting myocardial ischemia
❑ Conduction abnormalities requiring transvenous/transcutaneous pacing
❑ Cardiac cath due ONLY to recurrent ischemia
❑ Other _____

CareMap Day/Sequence Date/Initials	Key Event	Variance/Reason	Action
Day 5	**Telemetry discontinued**	❑ No Variance ❑ Variance Reason(s):	
Discharge	**Patient discharge on aspirin or warfarin**	❑ No Variance ❑ Variance Reason(s): ❑ Documented allergy to aspirin AND warfarin not used relative to normal LV function ❑ Bleeding disorder ❑ Hb < 11 g/dl ❑ Platelets < 100,00 mm3 ❑ Ticlopidine therapy ❑ Other _____	
Discharge	**If LV EF < 40%, discharged on angiotension converting enzyme inhibitor** (any inhibitor & dose is acceptable)	❑ No Variance ❑ Variance Reason(s): ❑ Documented allergy or adverse drug effect ❑ Cr > 2.5 mg/dl ❑ Sustained systolic BP < 85 mm Hg ❑ Treatment with hydralazine and isosorbide dinitrate, isosorbide mononitrate, or nitroglycerin ❑ Other _____	
Discharge	**Discharged on beta-adrenergic blocker** (any beta blocker and dose acceptable)	❑ No Variance ❑ Variance Reason(s): ❑ Documented allergy ❑ Sustained systolic BP < 85 mm Hg ❑ Sustained P < 55 bpm ❑ Hx 2nd or 3rd degree AV block ❑ Obstructive lung disease including asthma & emphysema ❑ Diabetes mellitus ❑ CHF after 2nd hosp. day ❑ Severe LV dysfunction with LV EF < 40% by isotope **or** contrast angiogram **or** severe LV dysfunction on echocardiogram ❑ Hx depression ❑ Claudication ❑ Other _____	
Discharge	**Fasting lipid profile drawn during hospitalization or within 6 months prior to hospitalization**	❑ No Variance ❑ Variance Reason(s): ❑ Documented arrangement to obtain within 1 month of d/c ❑ Lipid disorder requiring therapy identified before index admission ❑ Other _____	
Discharge	**Risk stratification performed or planned outpatient**	❑ No Variance (cath performed ❑ and/or gxT performed ❑) ❑ Variance Reason(s): ❑ Baseline EKG abnormalities ❑ Musculoskeletal or exercise intolerance ❑ Other functional tests performed _____ ❑ gxT planned outpatient ❑ Other _____	

INITIALS/ _____ / _____   _____ / _____   _____ / _____

SIGNATURES/ _____ / _____   _____ / _____   _____ / _____

Source: Copyright © The Center for Case Management. Content Copyright 1996, The University of Iowa Hospitals and Clinics, Iowa City, Iowa, used with permission.

■ Part III ■
Surgical Cardiovascular

■ 10 ■

The Fast-Track Critical Pathway for Surgical Outpatients in Perianesthesia Care

Annabelle Borromeo, Pamela Windle, and Hermie Robles

Critical pathways are patient management tools that promote effective and efficient care. The impetus for the development of critical pathways in the Post Anesthesia Care Unit (PACU) and Day Surgery (DS) was the need to maintain quality patient care in light of decreasing health care resources. The Traditional Perianesthesia Critical Pathway or Traditional Pathway, for short, was first developed and implemented in 1992 at the St. Luke's Episcopal Hospital in Houston, Texas. The Traditional Pathway for the care of the outpatient surgical patient accounted for an average length of stay (LOS) in the PACU of 2 hours and an average LOS in DS of 2.5 hours. The average LOS of Surgical Outpatients (FY 1995–1996) from admission to PACU to discharge from DS was 4.5 hours. The average annual number of surgical outpatients per year was 5296. Average annual cost for the average LOS was approximately $2,500,000. Patient flow using the Traditional Pathway in place at the time is illustrated in Figure 10–1.

The Traditional Pathway required a minimum stay of 1 hour in the PACU, after which the patient was transferred to DS where a minimum of another hour was needed to prepare the patient for discharge. The rule of thumb for determining postanesthesia LOS was based on total surgical time; recovery time in PACU was generally half of total surgical time. If the surgical procedure lasted 2 hours, the patient recovered for at least 1 hour in PACU, and prepared for discharge in DS for another hour. The Traditional Pathway was essentially a protocol driven by minimum time requirements rather than an assessment of patient condition and readiness for discharge. At the time the Traditional Pathway was first designed, there was much concern about completely stabilizing the patient after receiving anesthetics during surgery. Many of these drugs were not eliminated from the patient's body quickly due to the prolonged half-life. Thus LOS was increased in anticipation of complications relating to the use of anesthetic agents.

Today, advances in perianesthesia medicine and nursing have made the Traditional Pathway obsolete. These advances include newly discovered short-acting anesthetics such as propofol, which the body quickly eliminates, thus producing fewer postoperative complications such as drowsiness and nausea and vomiting; new research-based preoperative protocols that require prophylactic interventions to decrease postoperative complications; and better preoperative screening and evaluation. These improvements in care management have made it possible to shorten LOS without jeopardizing the patient's safety and well-being. Thus, the Fast-Track Critical Pathway for Surgical Outpatients was developed (See Exhibit 10–1).

WHAT IS FAST-TRACKING?

Fast-tracking is a concept of shortening LOS that was first introduced in the cardiac surgical population. Fast-tracking studies showed faster recovery and earlier discharge from both the intensive care unit and the hospital without apparent increased risk to the patient.[1–6] The studies focused on improved anesthetic techniques and agents that provide shorter sedation and earlier extubation leading to shorter LOS. Not only was LOS shorter, but quality indicators such as intubation time, respiratory infections, wound infections, laboratory procedures, and patient satisfaction improved as well.[5] A basic assumption of fast-tracking is that the patient is not placed at any additional risk and the focus is on improving recovery and well-being.[4]

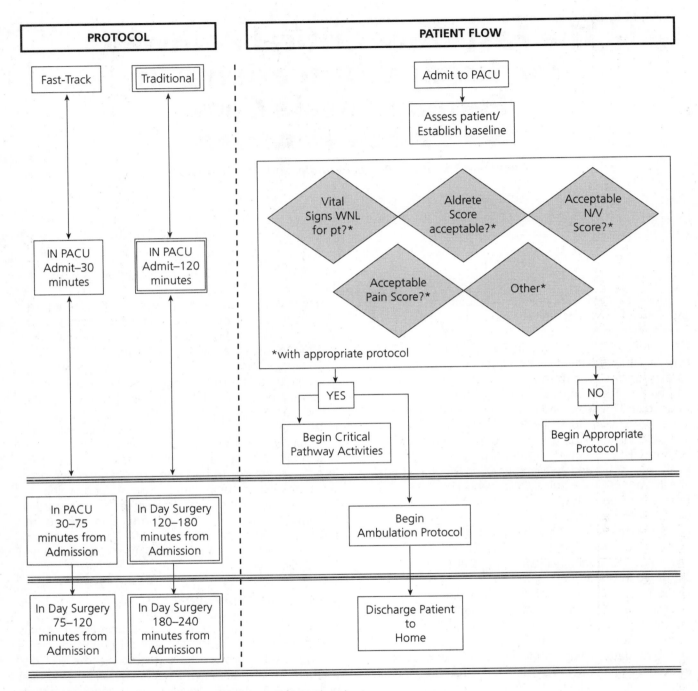

Figure 10–1 Fast-Track and Traditional Pathways Patient Flowchart

DEVELOPMENT AND IMPLEMENTATION OF THE PERIANESTHESIA FAST-TRACK CRITICAL PATHWAY

A Fast-Track Collaborative Practice Team (CPT) was formed consisting of nursing staff representatives from the PACU and DS Units, Anesthesia, Physical Therapy, Radiation Therapy, Radiology, and Laboratory Depart-

ments, and the Blood Donor Center. Interdisciplinary CPTs are assembled to help ensure high quality, cost-effective, and research-based clinical care throughout the continuum of hospitalization.[7] Insight into processes and practice patterns that either hinder or enhance the achievement of quality outcomes is more comprehensive when team composition is diverse. Important characteris-

Exhibit 10–1 Fast-Track Critical Pathway for Surgical Outpatients

DAY SURGERY/PACU SURGICAL OUTPATIENT CRITICAL PATHWAY

	Preadmission		Preop	Admission (30 Min)	Phase I Recovery		Phase II Recovery
	0–30 Minutes	60 Minutes	0–60 Minutes		60 Minutes	75 Minutes	0–60 Minutes
Initials							
Registration	Sign in at registration desk Complete insurance and anesthesia forms Complete registration with admitting clerk		Sign in at registration desk Complete insurance and anesthesia forms (if non pre-admit) Complete registration with admitting clerk Complete any additional business office paperwork (if pre-admit)				
Nursing	Medical-surgical history Psychosocial assessment Obtain baseline vital signs (BP, PR, RR, temperature, O_2 saturation) Height/weight pedi FHT		Review nursing history and verify information Check preoperative vital signs Review chart for completeness (consent, H&P, lab, clearances, etc) Obtain needed equipment (SCD pumps), old charts, etc. TED hose if ordered	**Check post-op orders** **General** VS q 15 min./post-op assess/check dressing/comfort measures/orientation/warm blanket(s)/O_2/pulse oximeter **Spinal/EPI** VS q 15 min./orient to routine/O_2/pulse oximeter/check dressing/complete assessment/check dermatone level **MAC/BLK** VS q 15 min./orient to routine/check level of anesthesia/complete assessment/check dressing/O_2/pulse oximeter	**GEN** Routine DSC discharge protocol **Spinal/EPI** Continue monitoring Continue post-op orders Pedi family visitation Wean off O_2	**Spinal/EPI** Routine DSC discharge protocol VS q 30° Complete sensation/movement for lower extremities	Assessment for local Reassessment/vital signs Discharge par score Discharge
Consults	Anesthesia interview Cardiology PRN Other consults		Anesthesia interview (if non pre-admit) Other consults PRN	**GEN** Anesthesiologist for vent. management Respiratory therapist for vent. patients MD/surgeon/consult PRN **MAC/BLK** Sign/out anesthesiologist	**GEN** S/O anesthesiologist to DSC (op)	**EPI** S/O anesthesiologist to home **Spinal** S/O anesthesiologist to DSC	Anesthesia PRN Physical therapy Home health
Lab		Complete lab work as ordered by surgeon/anesthesiologist Blood donor center	Complete additional lab tests (if required)	As ordered	Follow-up on pending results		Follow-up on pending results

continues

Exhibit 10–1 continued

| | Preadmission | | Preop | Admission (30 Min) | Phase I Recovery | | Phase II Recovery |
	0–30 Minutes	60 Minutes	0–60 Minutes		60 Minutes	75 Minutes	0–60 Minutes
Initials							
EKG		As ordered	As ordered	As ordered			
Radiology		Give request and direct to radiology (if ordered)	As ordered	As ordered	Follow-up on pending results		Follow-up on pending results
Nutrition			NPO	**GEN/Spinal/EPI** NPO; IV maintenance as ordered	**Spinal/EPI/GEN** Ice as ordered; IV maintenance; PO fluids	**Spinal/EPI** PO fluids if not contraindicated or if ordered	PO fluids; Discontinue IV when tolerating PO
Education	View preoperative video; Review preoperative and postoperative instructions; must include NPO, responsible adult/ride, arrival time (as determined by surgeon)		Orient to unit/routines; Confirm compliance with preoperative instructions; Assess learning needs and readiness to learn; Instruct patient according to appropriate teaching protocols; Identify barriers and reasons (ie, anesthesia, age, neuro status, language, other); Post instruction for non pre-admit; Demonstrate understanding of patient teaching	**GEN/Spinal/EPI/MAC/BLK** Assess learning needs and readiness to learn; Instruct patient according to appropriate teaching protocols; Identify barriers and reasons (ie, anesthesia, age, neuro status, language, other); **MAC/BLK** Demonstrate understanding of patient teaching	**GEN** DB&C; Demonstrate understanding of patient teaching; **Spinal/EPI** Inform patient re: return of sensation and movement	**Spinal/EPI** Provide other instructions specific to surgical procedure/anesthesia; Demonstrate understanding of patient teaching; Reinforce teachings as needed	Reorient to unit routine; Review discharge instructions (verbal and written) with patient and/or significant other
Discharge Planning	Initiate discharge planning: Responsible adult for 24 hours; Evaluate support system/coping mechanisms		Confirm discharge transportation for outpatient; Confirm responsible adult for 24 hours after discharge	**MAC/BLK** Discharge assessment/instructions	**GEN/EPI** Discharge assessment/instructions	**Spinal** Discharge assessment/instructions	Verify ride home; Re-evaluate support system; Referrals as indicated (ie, home health, social services)
Medications			Start IV TKO (unless otherwise ordered); Administer antibiotics and other meds (if indicated)	**GEN/Spinal/EPI/MAC/BLK** Pain management; Other meds	**GEN/Spinal/EPI/MAC** IV/IM meds; Epidural analgesia; PO meds		Analgesics/antiemetics PRN; Take-home medications
Activity			Ad lib in unit (unless sedated)	**Spinal/EPI** Bed rest; **MAC/BLK/GEN** Bed rest, HOB ↑, begin ambulation protocol	**GEN/Spinal/EPI** HOB ↑ per surgeon/anesthesia; Turn PRN		Recliner → ambulatory; or bed → recliner → ambulatory; Encourage ambulation

Source: Copyright © St. Luke's Episcopal Hospital.

tics that are valued in CPTs are clinical competence of team members, credibility, commitment, trust, and structured meeting times.[7] Members of the Fast-Track CPT carried out the following steps in developing and implementing the Fast-Track Critical Pathway that is illustrated in Figure 10–1:

1. Critically examine the current protocol and identify areas where the process
   - could be improved by decreasing variation in implementation
   - could be safely shortened or eliminated without compromising the outcome of quality patient care
   - simply "broke down"

   The PACU staff determined that the average 2-hour LOS in the PACU traditionally required to stabilize the patient postoperatively could be safely shortened due to the introduction of the newer, shorter-acting anesthetics. The process change mandated a shorter PACU LOS (maximum 75 minutes) and total hospital stay (maximum 120 minutes), provided the patient had no complications. The new LOS standards were based on the average half-life of the new anesthetic drugs. The difference in patient flow between the Traditional Pathway and the Fast-Track Pathway is illustrated in Figure 10–1 and Table 10–1.

2. Examine the pathway and ensure clarity of process. Clarify the roles of the team members and establish assessment criteria to determine patient readiness to move through the pathway.

   The Traditional Pathway did not clearly delineate the team members' activities, especially the role of the PACU nurse, during the 2-hour stay in PACU. The pathway was strictly based on time frames without considering the patient's condition and readiness for progression toward transfer or discharge. Nurses utilizing the Traditional Pathway assumed that the priorities included monitoring for postoperative complications, checking vital signs, keeping the patient warm, and teaching about the treatment plan. However, no guidelines were available to assist the nurse to perform consistently. The Fast-Track Pathway delineates the expected Critical Pathway Activities (See Exhibit 10–2) that serve as a "routine" to minimize care variability. The Critical Pathway Activities were developed by consensus after surveying the PACU staff members. The Ambulation Protocol and Protocol for Abnormal Vital Signs (Figures 10–2 and 10–3) are examples of those detailed activities.

3. The pathway should provide enough detail to decrease variability of practice while still allowing the nurse to practice sound clinical judgment.

   The Traditional Pathway was severely lacking in clarity and detail about what to do for common postoperative complications like abnormal vital signs, pain, and nausea and vomiting. Unwritten protocols for the most common complications were transmitted mostly by word-of-mouth, which accounted for wide variability in practice patterns that naturally led to variance in outcomes. Fast-Track protocols for the most common postoperative complications are shown in Figures 10–3 through 10–5.

4. Enlist staff "buy-in" by consulting the staff and developing the pathway using a multidisciplinary approach.

Table 10–1 Differences between the Fast-Track and Traditional Critical Pathways

	Fast-Track	Traditional
Protocol	Patient is moved aggressively through the system (PACU—Day Surgery—Discharge) using both patient criteria and fast-track timelines. ***Patient is in and out of the system within a span of 2 hours or less.***	This protocol is how patients are traditionally moved through the system. Patient is moved passively through the system following the Critical Pathway. The Critical Pathway mandates a minimum 2-hour stay in the PACU for monitoring and stabilization purposes, therefore, ***the patient generally stays within the system longer than 2 hours.***
Mobilization and ambulation	Began in the PACU ***30 minutes*** after admission to PACU, first by raising the HOB, until the patient is finally seated on a recliner chair. This progression of actiities is dependent on patient assessment for readiness. Thus, the term "Fast-Tracking."	Began in Day Surgery approximately ***2 hours after admission to PACU.*** No attempt at mobilizing the patient is done in PACU. Basically same procedure as in "Fast-Track" protocol but performed in Day Surgery approximately 2 hours after admission to PACU.
Mode of transfer from PACU to Day Surgery	Recliner Chair	PACU Bed

Exhibit 10–2 Critical Pathway Activities

1. Begin efforts to awaken patient
   1.1 Call patient's name
   1.2 Introduce self to patient
   1.3 Orient to unit and time
   1.4 Inform patient about surgery being over
2. Instruct patient about goals of care
   2.1 Stabilization of patient condition
   2.2 Monitoring for anesthesia effects
   2.3 Deep breathing and coughing
   2.4 Pain management
3. Judicious use of medication for relief of postoperative complications such as nausea and vomiting, and pain
4. Discontinue oxygen when the following goals are achieved:
   4.1 Oxygen saturation > 92%
   4.2 Stable vital signs
   4.3 Pain Scale of 6 or less
5. NPO until patient is seated on recliner chair, and only upon patient request
6. HOB up 30–45 degrees depending on patient comfort
   6.1 Raise HOB up to 45 degrees, use bed measure
   6.2 Ask patient if he/she is comfortable
   6.3 If patient is uncomfortable, lower HOB slowly until patient feels comfortable
7. Evaluate for appropriateness of initiation of Ambulation Protocol
8. Evaluate for appropriateness of Transfer to Day Surgery from PACU
9. Evaluate for appropriateness of Discharge to Home from Day Surgery

The PACU nursing staff was initially apprehensive about attempting mobilization and ambulation in the PACU. These activities were traditionally the domain of the Day Surgery staff. The PACU/DSC Professional Staff Nurse Council held staff meetings to allow the staff to verbalize their concerns as well as to give the Nurse Manager the opportunity to enlighten the staff on the goals of the Fast-Track Pathway.

Not only should nursing and other ancillary nursing staff members be involved, but physicians as well as other members of the health care team must be included in the change process as well. The multidisciplinary approach to problems ensures that possible variances from the unique perspective of each discipline can be identified and prevented.

5. Start with a small pilot study.

Any change in process is better received if it is first tested using a small sample and if there is a perceived deadline to testing the change. The PACU/DS staff first did a dry run of the Fast-Track pathway on several patients. Initial minor problems that came up were identified and solved, thus preventing escalation into major problems. One problem that was identified and solved was the undue strain on the back of the staff assigned to transport the patient on the recliner chair from PACU to Day Surgery Center (DSC). The problem was found to be due to the recliner chair construction. The recliner chairs were then modified based on the staff members' recommendations, and transport from PACU to DSC was no longer a problem. Also, to protect the staff members, patients who weighed more than 200 lbs were transported to DSC on the PACU bed. In addition to identification of problems, the pilot study also gave the staff members a chance to verbalize their concerns. Further opportunities for fine-tuning the pathway were also identified and incorporated.

THE FAST-TRACK CRITICAL PATHWAY

The goal of the Fast-Track Critical Pathway is to move the patient aggressively from Preadmission to the PACU and DSC to achieve a LOS target of 2 hours or less, while still maintaining quality care. The basic premise is that the goals of shortened LOS, decreased complications, and lowered cost of treating these complications may be achieved if mobilization protocols are initiated in the PACU instead of in Day Surgery. The differences between the Traditional and Fast-Track pathways are illustrated in Figure 10–1. The pathway is divided into 4 phases: preadmission, preoperative, phase I recovery, and phase II recovery.

Surgical outpatients are preadmitted 1 to 3 days before scheduled surgery. Preadmission accounts for a maximum of 60 minutes of the pathway. Preoperative activities commence the day of surgery and account for 60 minutes of the pathway. Phase I recovery, which occurs in the PACU, is allotted 30 to 75 minutes and the variability is accounted for by the type of anesthetic used. As a general rule, patients who receive general anesthesia recover faster than patients who have spinal or epidural anesthesia. Phase II recovery, set in the Day Surgery Center, is carried out over a 1-hour period. The patient is then discharged to home in the care of a responsible adult. Follow-up activities (eg, postop patient satisfaction telephone survey for outpatients) are carried out within 48 hours of the patient's discharge from the Day Surgery Center.

Preadmission

The first hour of preadmission involves signing-in and registration activities, obtaining the patient's history, and performing physical and psychosocial assessments. It is during this time that consultations with members of the anesthesia department and other departments occur. Coordination with the medical staff can potentially be

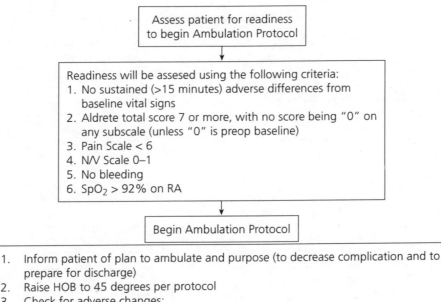

Figure 10–2 Ambulation Protocol

the greatest source of variance during this time. Physicians' schedules are not always predictable. This problem can be prevented by close coordination with the chiefs of the anesthesia department and other consulting departments prior to implementation of the pathway.

The first portion of the pathway during the first hour is essentially devoted to obtaining information from the patient to determine what makes the patient unique; the second portion is devoted to giving information to the patient. During this phase, the nurses share information on the surgical procedure through a variety of means (eg, videos, handouts, discussion), review preoperative and postoperative instructions, and initiate discharge planning activities. These activities include a review of the critical pathway, the goals of the critical pathway and the need for a responsible adult during the first 24 hours after discharge.

During the second hour of preadmission, coordination with the laboratory, EKG and radiology departments is required. Blood specimens are drawn for laboratory tests, and EKG and x-rays are obtained. The nurses obtain the blood specimens, ancillary personnel obtain the EKG, and the patient is transported to the radiology department for the necessary x-rays. After all the tests are done, the patient is discharged and is asked to return early morning the day of surgery.

Preoperative

The preoperative pathway begins when the patient registers early in the morning on the day of surgery. Admitting and business office staffs ensure that all the necessary forms are completed. The nurse reviews the patient's history, prepares the patient for surgery, draws any additional

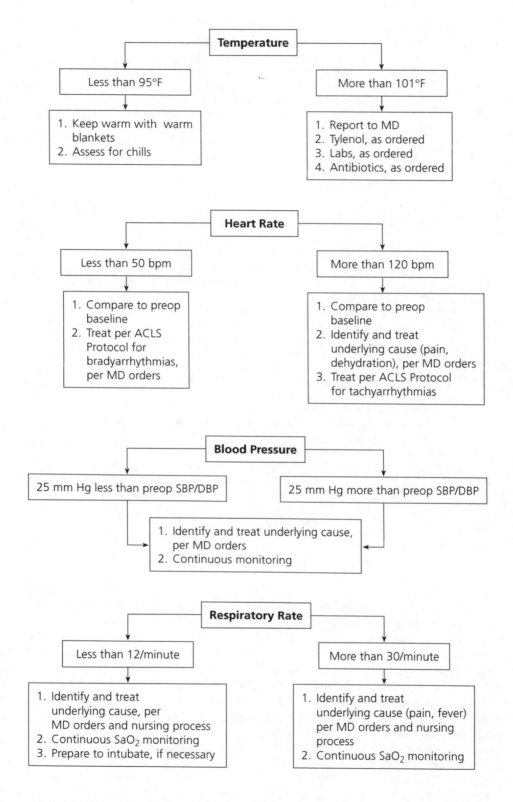

Figure 10–3 Protocol for Abnormal Vital Signs. (Vital signs must always be compared to preop baseline before they are classified as abnormal. Vital sign trends signifying movement toward the above absolute abnormal values must be monitored.)

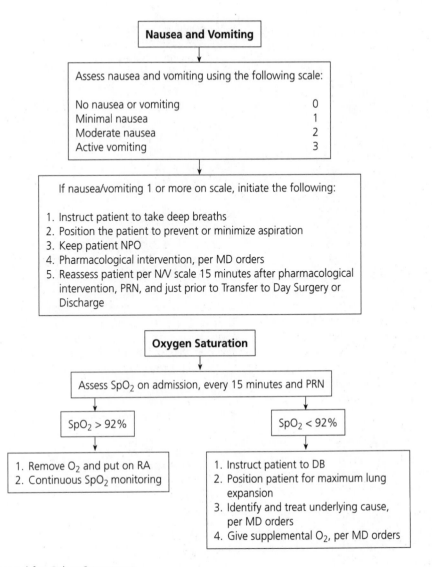

Figure 10–4 Protocol for Other Symptoms

lab work, starts the IV, and orders the necessary equipment. The anesthesia physician on call interviews the patient and once again goes over the anesthesia procedure. Any additional consults are also done at this time.

Information given during the preadmission process is reviewed with the patient. The patient is encouraged to ask questions. The patient also performs a return demonstration of proper coughing and deep breathing techniques. The nurse confirms the availability of discharge transportation and a responsible adult to assist the surgical outpatient up to 24 hours postdischarge. The patient is transported to the operating room on call.

Phase I Recovery

The target LOS in phase I recovery, which occurs in PACU, is a minimum of 30 minutes to a maximum of 75 minutes, depending on the type of anesthetic used. In general, patients who have had spinal or epidural anesthesia stay in the unit longest due to more stringent discharge criteria. The patient is required to regain sensation and movement of both lower extremities before being transferred to DSC.

Upon admission to PACU, quick head-to-toe assessment of general body systems, level of consciousness,

Ask patient to rate postoperative pain
on a scale of 0 to 10 with "0" being no pain and "10" being
the worst pain the patient has ever felt.

↓

If pain scale > 6, initiate the following:

1. Assess pain
2. Treat pain according to MD orders
3. Teach patient jaw relaxation technique
4. Begin PCA orders and teach patient how to
   use PCA properly for pain control

Figure 10–5 Protocol for Pain

and the surgical incision and dressings, if applicable, is performed. Comfort measures, such as warm blankets, are immediately instituted, if appropriate, and the patient is assessed for the presence of postoperative pain and other possible postoperative complications. If the patient has had spinal/epidural anesthesia, the dermatome level is also assessed. All consults are coordinated during the first 30 minutes. Respiratory care, x-rays, lab work, supplies, and patient equipment are obtained.

During the first 30 minutes of stay in the PACU, the patient is constantly monitored for comfort, stability, and readiness for ambulation using the ambulation protocol in Figure 10–2. The patient is considered ready for ambulation when the following criteria are met:

1. no sustained (> 15 minutes) adverse differences from baseline vital signs
2. Aldrete Scale total score of 9 or more, with no score being "0" on any subscale (unless "0" is preoperative baseline) (See Exhibit 10–3)
3. pain score < 6 (on a scale of 1 to 10)
4. nausea/vomiting scale 0 to 1 (See Figure 10–4)
5. no bleeding
6. SpO$_2$ > 92% on room air

The ambulation protocol is initiated during the latter part of the first 30 minutes in the PACU. The patient is evaluated for the presence of adverse changes in vital signs and SpO$_2$ level, pain or discomfort, and nausea and vomiting during the ambulation and mobilization process. The appropriate protocols are initiated according to the presenting symptoms (Figures 12–3 through 12–5). The patient is considered ready for transfer to the Day Surgery Center if the following criteria are met:

1. patient score of 7 or more on the Aldrete Scale (range 0 to 10)

2. able to tolerate sitting up on a recliner chair for at least 10 minutes without adverse changes from pre-operative baseline in:
   - vital signs
   - level of consciousness
   - pain scale (6 or less)
   - SpO$_2$ (92% or more on room air)
3. able to walk a minimum of 3 feet without changes from preoperative baseline
4. able to tolerate transfer from PACU to Day Surgery on a recliner chair

Phase II Recovery

Phase II recovery activities are carried out in the Day Surgery Center. One hour is allotted for these activities. The patient is admitted to Day Surgery and admission temperature and blood pressure are obtained. The nurse notifies family members in the waiting room and invites them to see the patient. The family is instructed on the logistics for discharge (eg, patient pick-up, approximate time of discharge, patient clothes and belongings). Any ordered treatments are administered at this time. Pre-

Exhibit 10–3 Protocol for Aldrete Scoring

Each patient will be scored on the Aldrete Scale on Admission, 30 minutes after Admission, 60 minutes after Admission, and upon discharge from PACU.

Aldrete Scale involves scoring the patient on the following criteria:

Activity:	Able to move 4 extremities	2
	Able to move 2 extremities	1
	Able to move 0 extremities	0
Respiration:	Regular, able to DB and C freely	2
	Dyspnea, limited or obstructed breathing	1
	Apneic	0
Circulation:	BP +/– 0–20 mm Hg preop	2
	BP +/– 20–25 mm Hg preop	1
	BP </> 25 mm Hg preop	0
LOC:	Awake, alert	2
	Drowsy, but arousable	1
	Unarousable, nonresponsive	0
SpO$_2$:	> 92% on room air	2
	Requires support O$_2$ to maintain > 92%	1
	< 92% even with O$_2$ support	0
Highest Possible Score:		**10**

Patient must have a score of 7 or more, and no score of "0" (unless preop baseline), to begin Ambulation Protocol or Transfer to Day Surgery from PACU.

Source: Reprinted with permission from J.A. Aldrete and D. Kroulik, A Post Anesthesia Recovery Scote, *Anesthesia Analgesia*, Vol. 49, pp. 924–928, © 1970, Williams & Wilkins.

scriptions are ordered, and discharge instructions are reviewed. The patient's IV line is discontinued and instructions on the care of the IV site are discussed. Physical therapy orders are also carried out (eg, gait training, crutch walking). Finally, the patient is asked to void. The patient is then discharged to home in a wheelchair if the following discharge criteria are met:

1. pain scale < 6
2. nausea/vomiting (N/V) scale 0 to 1
3. vital signs stable
4. absence of bleeding
5. verbalized understanding of discharge instructions

OUTCOMES MEASURED

The following outcomes are measured throughout the surgical outpatient's stay and 24 hours postdischarge:

1. PACU Length of Stay
2. Day Surgery Length of Stay
3. Total Length of Stay
4. Time to Recline from Admission to Unit
5. PACU Cost of Intervention
6. Day Surgery Cost of Intervention
7. Home Cost of Intervention
8. Total Cost of Intervention
9. Patient Satisfaction

If the patient "falls off" of the pathway for Fast-Track, that is, if the patient is not discharged within 2 hours of admission to PACU, reasons for the variance are documented in the progress notes. For example, waiting for a ride, patient due to void, waiting for physical therapy, or doctor's orders (doctor request to increase LOS, or waiting for antibiotic or Decadron infusion). The following are documented in the progress notes:

- vital sign changes and actions taken to correct abnormal vital signs
- complaints of pain and actions taken to relieve pain
- complaints of nausea and vomiting and relief interventions
- any other complaints and/or symptoms and relief interventions
- any other positive or negative comments about perianesthetic experience

THE PILOT STUDY

A pilot study was conducted to determine the effects of fast-tracking surgical outpatients on the outcomes of LOS, incidence of postoperative complications, cost of treatment of postoperative complications, and patient satisfaction. The following hypotheses were tested:

1. There will be no difference between the Fast-Track (FT) Group and the Traditional (T) Group in LOS.
2. There will be no difference between the FT Group and the T Group in incidence of postoperative complications.
3. There will be no difference between the FT Group and the T Group in cost of treatment of postoperative complications.
4. There will be no difference between the FT Group and the T Group in patient satisfaction.

Method

A quasi-experimental study was conducted. Every other outpatient who met inclusion criteria was randomly selected to participate in the study. Inclusion criteria included patients classified as "outpatient," who received general anesthesia during surgery, with ages ranging from 16 to 90 years. All types of surgical procedures were included, except those that required extended stay per surgeon's orders, such as some laparoscopic cholecystectomy, laminectomy, surgeries prohibiting weight-bearing postoperatively, and patients weighing more than 200 lbs. Selected patients were then cared for using the Traditional Pathway. The care was given by a select group of PACU nurses familiar with the research protocol. After a sufficient number of subjects were enlisted to achieve a power of .80, the same method was used to assign an equal number of patients to the FT Group.

The following data were collected for the PACU and Day Surgery Center stays: time of admission to PACU/DSC, time to recline in chair in PACU (if FT Group) or in Day Surgery (if T Group), postoperative complications in PACU/DSC, time of discharge to DSC, and time of discharge from Day Surgery.

A follow-up telephone call was done 24 hours after the postoperative stay to determine patient satisfaction with the postoperative experience and presence and treatment cost of complications.

A total of 168 patients were enrolled in the study (Traditional n = 84, Fast-Track n = 78). Six patients in the FT Group were dropped because they could not be contacted by telephone 24 hours after the postoperative stay.

Statistical Analysis

After screening the data for violations of assumptions, *t*-tests were performed to examine the differences in LOS (hypothesis 1), cost of treatment (hypothesis 3), and patient satisfaction (hypothesis 4) between the two groups. The chi-square test was used for incidence of postoperative complications (hypothesis 2).

All *t*-tests conducted in this study, except for the patient satisfaction test, were equal variance *t*-tests. To avoid the problem of inflated error rates because of the number of *t*-tests applied (three), the Bonferroni adjustment was used. Only results significant at .017 are reported.

Results

Hypothesis 1: There will be no difference between the Fast-Track Group and the Traditional Group in LOS.

The *t*-test revealed no significant differences in the LOS between the two groups in PACU LOS, Day Surgery LOS, or total LOS.

Hypothesis 2: There will be no difference between the FT Group and the T Group in the incidence of postoperative complications.

To determine whether this difference was significant, the chi-square test was used. There was a significantly lower incidence of postoperative complications in the FT Group (n = 78) compared to the T Group (n = 84) in the PACU; χ^2 (21, N = 162, df = 8, p = .006). No significant differences were found between groups for incidence of post-operative complications in Day Surgery; χ^2 (4, N = 162, p = .35, ns). The distribution of postoperative complications in PACU, DSC, and home is illustrated in Table 10–2.

Hypothesis 3: There will be no difference between the FT Group and the T Group in the cost of treatment of postoperative complications.

The *t*-test revealed no significant differences in the cost of treatment of postoperative complications between the two groups. The means of the collected variables, including cost of treatment, are listed in Table 10–3.

Hypothesis 4: There will be no difference between the FT Group and the T Group in patient satisfaction with the postoperative experience.

There was significantly greater satisfaction with the postoperative experience in the FT Group than the T Group (*t* = –3.192, df = 133, p = .002).

Discussion

The study revealed no statistically significant difference in the LOS of surgical outpatients who were cared for with either of the two pathways. The practical significance of the study is apparent, however, when the potential cost savings are determined. Each minute the patient spends in the PACU/Day Surgery Center costs approximately $2.41.[8] The average LOS difference between the FT and T protocols is 23 minutes. The patient stays in the hospital an average of 23 minutes longer if the Traditional Pathway is used. Total average savings for each FT patient is $53.43. Each month, the average number of outpatients in our facility is 90. Total potential average monthly savings is $4808.70 or approximately $57,704.40 a year.

There is a statistically significant decrease in the incidence of postoperative complications in the FT Group. This finding may be due to patient and nurse focusing on ambulation and mobilization activities rather than on postoperative complications. There was no difference in cost of treating postoperative complications. Patients on the FT Pathway were more satisfied with the care received postoperatively compared to patients in the Traditional Pathway. Contrary to popular belief, the FT Pathway does not necessarily exert a negative impact on patient satisfaction.

CONCLUSION

The Traditional Critical Pathway for Surgical Outpatients, developed by a multidisciplinary staff at St. Luke's Episcopal Hospital in Houston, Texas, needed to be evaluated in view of rapid changes in perianesthesia care. These changes, including the introduction of anesthetics with short half-lives, served to accelerate the need for change. The Traditional Pathway was characterized

Table 10–2 Distribution of Postoperative Complications in PACU, Day Surgery, and Home

Postoperative Complications	PACU		Day Surgery		Home	
	TG	FT	TG	FT	TG	FT
Pain	46	31	24	16	26	22
Nausea/Vomiting (N/V)	2	3	3	4	5	6
Shivering	2	3	0	0	0	0
Pain and N/V	6	6	1	0	4	2
Drowsiness	1	5	0	0	0	0
No complications	20	29	55	58	49	48

Table 10–3 Means of Collected Variables (N = 162)

Variable	Traditional (n = 84)	Fast-Track (n = 78)
Age	43	44
PACU Length of Stay (in minutes)	94	78
Day Surgery Length of Stay (in minutes)	84	77
Total LOS (in minutes)	178	155
Time to Recline from Admission to Unit (in minutes)	75	66
PACU Cost of Intervention	$10.12	$7.36
Day Surgery Cost of Intervention	$1.78	$2.13
Home Cost of Intervention	$.57	$.47
Total Cost of Intervention	$12.48	$9.96
Patient Satisfaction (Scale of 1 to 3)	2.63	2.87

mainly by activities that were time driven, rather than based on an assessment of the patient condition at any given time. The Traditional Pathway also did not clearly delineate staff responsibilities. Variation in process became the norm, not the exception.

A Collaborative Practice Team (CPT) was formed to evaluate the relevance and usefulness of the Traditional Pathway. The CPT developed a new pathway, called the Fast-Track Pathway for Surgical Outpatients, that reflected new developments in anesthesia practice. The purpose of the Fast-Track Pathway was to achieve the outcome of efficient and quality care. The Fast-Track Pathway is characterized by early mobilization practices and set protocols for dealing with the most common postoperative complications.

A pilot study was conducted, which showed no significant difference in the LOS between patients in the Traditional Group and the Fast-Track Group. However, the practical significance of the difference is evident when the cost savings were tallied in favor of the Fast-Track Group. Patients in the Fast-Track Group were significantly more satisfied than those in the Traditional Group.

Critical pathways need to be evaluated periodically for practicality, usefulness, and relevance. We recommend that pathways be reviewed every 2 or 3 years to ensure that the pathway incorporates the latest research findings and developments. At St. Luke's Episcopal Hospital, we are currently investigating the effects of complementary healing modalities, healing health care environments, and other therapeutic modalities that might exert an effect on the outcomes of LOS, cost of stay, and patient satisfaction with care. The results of these and other research studies will be incorporated in the next critical pathway review.

REFERENCES

1. Aps C. Fast-tracking in cardiac surgery. *Br J Hosp Med.* 1995;54: 139–142.

2. Lazar HL, Fitzgerald C, Heeren T, Aldea GS, Shermin RJ. Determinants of length of stay after coronary artery bypass graft surgery. *Circulation.* 1995;92(suppl 9):II20–II24.

3. Mounsey JP, Griffith MJ, Heaviside DW, Brown AH, Reid DS. Determinants of length of stay in intensive care and in hospital after coronary artery surgery. *Br Heart J.* 1995;73:92–98.

4. Engleman RM. Mechanisms to reduce hospital stays. *Ann Thorac Surg.* 1996;61(suppl 2):S26–S29.

5. Riddle MM, Dunstan JL, Castanis JL. A rapid recovery program for cardiac surgery patients. *Am J Crit Care.* 1996;5:152–159.

6. Anderson J. Fast track for ACB patients. In: Houston S, Cole L, Wojner A, Luquire R, eds. *Outcomes Management: A User's Guide.* 2nd ed. Houston, TX: Center for Innovation, St. Luke's Episcopal Hospital; 1996:149–152.

7. Grady GF, Wojner, AW. Collaborative practice teams: the infrastructure of outcome management. *AACN Clin Iss.* 1996;7(1):153–158.

8. Borromeo AR, Windle PE. Benchmarking for unrelieved pain in a post anesthesia care unit. *Best Pract Benchmarking Health Care.* 1997;2(1):20–23.

■ 11 ■

Developing a Pathway and Improving Outcomes for Carotid Endarterectomy

Rella Adams, Dana Danielson, Cynthia Hinojosa, Eric Six, Denise Tucker, and Pam Warner

Valley Baptist Medical Center (VBMC) is a 444-bed acute care facility located in South Texas. With today's rapidly changing health care environment, hospital leaders, staff, and physicians place quality outcomes as a top priority, while addressing reimbursement and length of stay issues. A major task related to this process was to develop diagnosis-specific pathways and protocols.

One of the authors, Dr. Six, a practicing neurosurgeon and medical director at VBMC, had a special interest in evaluating cost, length of stay, and outcomes of patients undergoing carotid endarterectomy. Direct physician involvement is the key to maximizing cost reductions and implementing a majority of the cost containment opportunities. This physician's input during the early stages of this project not only initiated the development of the Endarterectomy Restorative Care Pathways (RCPs®), but served as the motivational force during the entire project.

Because of the medical director's interest in comparing costs for all carotid endarterectomy patients, an initial patient-specific and physician-specific computer analysis was done of all carotid endarterectomy cases in 1995. This analysis revealed that the length of stay and charges per case were above national best practice benchmarks. Also, when a comparison was made among all neurosurgeon, and cardiovascular surgeons at VBMC, it was found that some patients' lengths of stay were longer and the hospital charges per case on some patients were significantly greater than others. Therefore, the goals in this project were to maintain or improve the existing clinical outcomes, while continuing to reduce the length of stay and significantly reduce cost per case for carotid endarterectomies. This chapter discusses the steps used to meet these goals.

PATHWAY/PROTOCOL DEVELOPMENT

A process improvement committee oversees multidisciplinary groups of health care professionals and physicians. Together they comprise continuous quality improvement (CQI) teams that develop pathways. The CQI teams are responsible for the review, selection and development of the pathway. The goals and objectives are to

1. develop and implement RCPs® and associated material such as patient educational guidelines for a selected population
2. improve methods and timeliness of diagnosis and treatment benefiting a population of patients
3. minimize variation in patterns of practice, standardizing whenever possible
4. evaluate variation in outcomes of health care service
5. focus on resource-intensive services

DEVELOPMENT OF RCPs®

Steps in the development of RCPs® are as follows:

1. Choose a diagnosis or diagnosis related group (DRG). The basis for making this selection may be related to case volume, potential opportunities for improvement when compared to best practice, or, as in the case of carotid endarterectomy, the amount of physician interest in the project. According to Dr. Six, from a physician's perspective, changing the way one practices medicine is a difficult learning process. The impetus for change is the challenge of maintaining or improving clinical outcomes and, at the same time, reducing the cost of

care. The stimulus comes from reviewing individual statistics when compared to national best practice.

2. Form a multidisciplinary CQI team. The team must include all appropriate care-delivery disciplines in order to encourage buy-in and practice change. Membership for the carotid endarterectomy team included physicians, nurses, case managers, social services/discharge planners, pharmacists, and representatives from laboratory and radiology services. Not all individuals need to be permanent members of the CQI team. For example, pastoral care and rehabilitation services staff attended one meeting to outline their roles in the care.

3. Assess, compare, and map current practice. The development of the RCP® for carotid endarterectomy began with a review of current practice, which was the standard of care. Because physician acceptance is crucial to the success of any pathway, the physicians performing this procedure were approached. Since both neurosurgeons and cardiovascular surgeons perform the endarterectomies, it was challenging to write one pathway that would serve the physicians, the clients, and the hospital staff.

All four surgeons have varied practice patterns. These were analyzed and all variations were discussed in order to reach a comparison of their practices. The desired goal was finally realized in a draft of one standardized RCP® for carotid endarterectomy (Appendix 11–A, Exhibit 11–A–1).

Once the basic outline of current practice was set into the format, members of the multidisciplinary team met to review the product. Reducing the lengths of stay cannot be accomplished by arbitrarily sending patients home earlier. Preparation and changes in behavior have to occur beginning with the physician's office. The office staff work with the patients during the preadmission process. The patients then are referred to the case managers in the inpatient setting, returning to the physician's office staff for postoperative follow-up.

Reducing the length of stay requires increased vigilance by both physician and nursing staff during the continuum of care in order to detect any potential complications, whether they are physical, psychological, and/or social. Secondly, reducing utilization requires a careful review of laboratory tests, pharmacologics, operating room time and equipment, and even practices within anesthesia. Surprising discoveries were made during this project.

- Medications and laboratory tests were ordered as "routine" by nursing staff, from a standard order set. If there is one message learned from this experience, it is that the new environment requires more attention to detail. The ultimate impetus for both hospitals and physicians to make needed changes is that third-party payers have this information already. They will use it when offering contracts for managed care and when negotiating for fee-for-service contracts.

- Oxygen was given to all postoperative patients in the recovery room. This was a standard practice by anesthesia.

- Antibiotics could be substituted by pharmacy with additional cost not being considered.

In the process of defining the RCP®, these items were analyzed and changed appropriately by the physicians through the interdisciplinary meetings.

4. Review, revise, improve RCP®. At this stage, the final RCP® was developed. As stated, all opportunities to improve had been considered and incorporated. Each discipline reviewed the product one last time prior to final print.

5. Development of complementary documents. The carotid endarterectomy RCP® is accompanied by a patient education pathway (Exhibit 11–A–2 in Appendix). This teaching document outlines each anticipated step in the acute care episode. At the time the patient is scheduled for surgery, the physician's office staff also schedule a preoperative appointment for the patient to come to the hospital to prepare for the upcoming inpatient procedure. During this Pre-Admission Teaching and Testing (PATT) visit, the patient is registered, interviewed, and assessed, and all necessary documents are completed. The patient meets the case manager who follows the patient's progress throughout the continuum. Any necessary laboratory, radiology, and EKG tests are completed at this time. The case manager evaluates all the preoperative diagnostic data and ensures physician notification of any necessary medical or cardiac clearance/consultation. Needless delays on the scheduled date of surgery have been avoided by this process.

A major accomplishment during this phase is the preoperative teaching. The patient receives and reviews the teaching pathway. Family members and/or home caregivers are encouraged to be present. This educational process has proven to be key in effectively reducing the length of stay and is vital to patient satisfaction. In the PATT process, teaching is done at a time when the patient can assimilate the information, rather than expecting the patient to absorb the teaching on the day of surgery. As a consequence, an emotional bond develops between the patient and the case manager that extends throughout the continuum of care.

During the PATT process, the patient and family tour the nursing unit and meet their nurses. Patients benefit greatly from this orientation process. Many patients who arrive somewhat nervous and anxious leave feeling informed and secure, knowing a contact person is available to answer questions. Occasionally there are patients who do not complete this PATT process and, as a result, differences are noted in the postoperative period. These patients are less prepared to accept their roles in the recovery and rehabilitation phase. Physicians state that they are immediately able to recognize those patients who have not attended PATT.

The overall goal of the PATT appointment is to meet the holistic needs of the patient and the family. Ongoing monitoring of patient satisfaction has revealed a higher level of satisfaction from those patients and families who participate in the PATT program.

6. Develop Performance Improvement Indicators/ Monitor Variance. The final responsibility of the CQI team is to identify performance improvement indicators to measure benefits of the RCP®. The case manager is responsible for monitoring the RCP® throughout the continuum of the hospital stay and reviews the charts to screen for variances. If variances to the RCP® are noted, the case manager documents and reports the variance. He or she is available to offer advice, clinical expertise, and educational opportunities for the physicians, nursing staff, and patient and family. On a day-to-day basis the primary care nurse updates the RCP®. The nursing staff are prompt in notifying the case manager when a variance occurs. Likewise, the physician and the staff nurse can also be notified of a variance by the case manager. Other disciplines, such as respiratory therapy may be involved in the case because of a comorbid condition. They are aware of the RCP® and the desired outcomes. The same type of notification would hold true for any of the ancillary services.

The RCP® is readily available for documentation by any of the team members. When a major variance or complication of carotid endarterectomy, such as stroke, myocardial infarction, cranial nerve injury, or wound hematoma occurs, the RCP® is adjusted accordingly and is managed closely. Additional input may be required from physical therapy, occupational therapy, speech and language pathology, food services, or social services to address the concerns of ancillary services. Multidisciplinary team conferences are held as necessary with the case manager serving as the coordinator. The case manager then communicates to the team, the patient, or the family.

PERFORMANCE IMPROVEMENT AND OUTCOMES MEASUREMENT

The medical board's performance improvement committee, endorsed by the board of trustees, systematically monitors, evaluates, and prioritizes the organizationwide performance and outcomes management program at VBMC. Multidisciplinary CQI teams form, as needed, to systematically improve process and outcome. The development and monitoring of the Carotid Endarterectomy RCPs® has been the result of this process.

Generic, as well as physician-specific, data are collected using performance indicators that measure process and outcomes. The outcomes measurement is accomplished using a computerized process called Integrated Quality and Resource Management (IQRM). Using this system, comprehensive financial and health care information can be collected, analyzed, trended, and prioritized. The information is adjusted for severity of illness and can be analyzed at the macro and micro level.

A comprehensive diagnosis-specific analysis typically begins with severity-adjusted market comparison of cost, length of stay, mortality, and common complications. The analysis is then sorted into physician-specific clinical and financial comparisons. Further benchmark comparison at the line-item level, such as the number of diagnostic tests or the type of pharmaceuticals ordered, provides the foundation for increasing efficiency of care and improving clinical outcomes.

This process was used for the initial analysis of all carotid endarterectomy cases in 1995 and was instrumental in providing comparative data following the implementation of the RCP® in 1996. Clinical outcomes (Figure 11–1), such as postoperative infection, myocardial infarction, stroke, cranial nerve injury, bleeding, or death are monitored through the case managers and the IQRM database and have always been well below national average.

The utilization of resources and the financial indicators revealed the greatest opportunity for improvement. The overall length of stay averaged 4.76 days, which was 1.76 days over the best practice benchmark. Figure 11–2 reveals that the physician-specific length of stay varied greatly without the standardized RCP®. This graph also depicts the improvement realized by use of the RCP®, PATT process, and case management.

Figure 11–3 depicts the physician-specific case mix index. Under the Medicare Prospective Payment System (PPS),

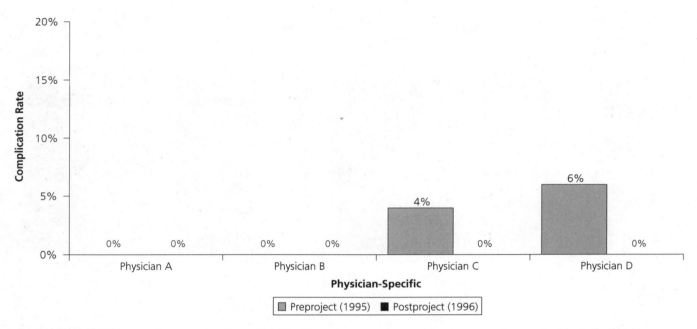

Figure 11–1 Physician-Specific Complications Carotid Endarterectomy. *Source:* Copyright © Valley Baptist Medical Center.

each discharge is assigned a DRG based on the patient's condition and other information pertaining to the treatment received. The DRGs serve to group cases with similar resource requirements. Using this scale, Figure 11–3 depicts that all carotid endarterectomies had a similar case mix index and, therefore, should utilize the same amount of resources and have similar lengths of stay.

When a variance occurs or a physician-specific trending report exceeds acceptable benchmarks, the findings are reviewed with the attending practitioner through a peer review process. This action has been effective in modifying practice patterns to bring outcomes to desirable expectations. This proves that outcomes assessment and management reduces length of stay, uses resources

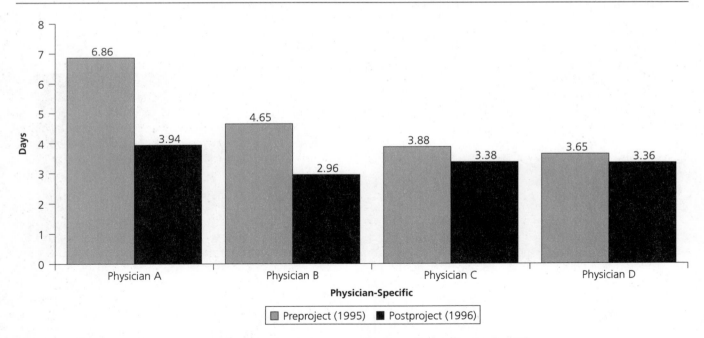

Figure 11–2 Length of Stay Carotid Endarterectomy. *Source:* Copyright © Valley Baptist Medical Center.

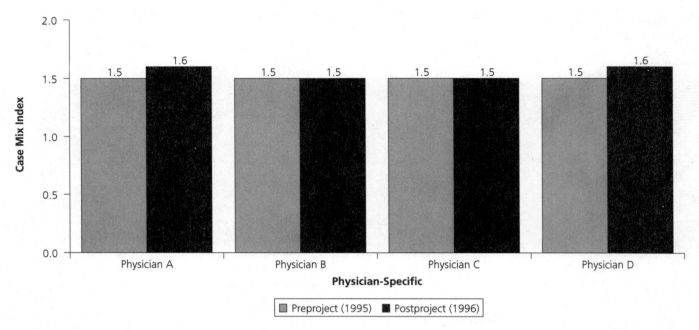

Figure 11–3 Physician-Specific Case Mix Index Carotid Endarterectomy. *Source:* Copyright © Valley Baptist Medical Center.

efficiently, and reduces cost. Figure 11–4 reveals that an overall cost savings of $93,624 was realized with this project, while quality of care was maintained or enhanced.

An ongoing patient satisfaction survey process allows an evaluation of implemented changes. The patients and families consistently report high ratings for the preopera-

tive preparation and the entire hospital experience. Feedback prior to the implementation of the RCP® was also satisfactory but the specific comments now being offered lead the entire team to conclude that positive changes have been made. All caregivers, including the physicians, voice improved satisfaction with the new processes. Patients, families, and the entire health care team

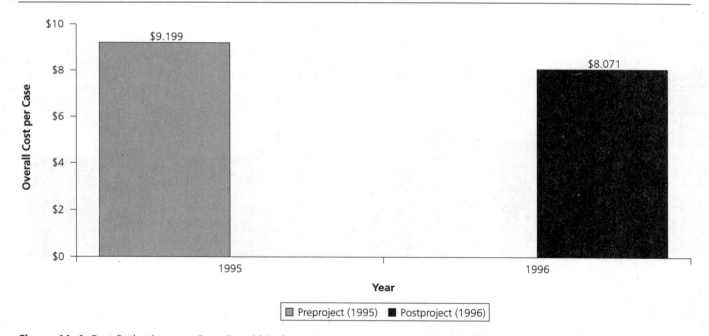

Figure 11–4 Cost Reductions per Case Carotid Endarterectomy. *Source:* Copyright © Valley Baptist Medical Center.

are able to follow an uninterrupted continuum of care for carotid endarterectomy procedures.

CONCLUSION

RCPs® are intended to be used as tools for quality, utilization, and outcome management. RCPs® also provide templates that offer approaches to the diagnosis, management, or prevention of specific diseases and conditions. They are not intended to substitute for, or mandate, the professional judgment of any physician involved in the care and treatment of a patient, nor are they intended to be construed or considered as standards of care. They are clinical practice guidelines.

The hospital staff and physicians realized that the Endarterectomy RCP® offered an opportunity to improve utilization of resources and quality of patient care at VBMC. Its final draft was shared with all the hospital staff, medical staff, appropriate committees, and the Board of Trustees, who endorsed its implementation.

Patient satisfaction, utilization, and financial indicators will continue to be monitored. In fact, length of stay for endarterectomy has dropped another day in 1997 to 1.7 and the cost was reduced by another $2000 per case. In the near future, VBMC's computerized repository will allow administrators, physicians, and hospital staff to perform detailed analysis on both financial and clinical data within minutes of a request. Future opportunities for improvement will be easily prioritized and monitored.

What is the future of RCPs®? Most likely, they will become Phases of Care (POC). A POC is the product of an integrated medical system. It includes all the services provided to a group of patients with medical/psychological problems across the continuum of care over a specific time period. This type of integrated system would be driven by data from computer systems. The data would enable caregivers to locate patients within the system, analyze their needs, and appropriately assign them to services along the continuum of care in order to offer timely, cost-effective, quality outcomes.

■ Appendix 11–A ■

Clinical and Patient Pathways with Acronyms

Acronyms/Definitions:

DRG—Diagnosis Related Grouping
Preop—Preoperative
Postop—Postoperative
CBC—Complete Blood Count
Lab—Laboratory
Protime—Prothrombin Time
PTT—Partial Thromboplastin Time
EKG—Electrocardiogram
Meds—Medications
Neuro—Neurological
↑—Increase
MD—Medical Doctor
IV—Intravenous
pt—Patient

Exhibit 11–A–1 Restorative Care Path

VALLEY BAPTIST MEDICAL CENTER
Harlingen, Texas
DX: CAROTID ENDARTERECTOMY
RESTORATIVE CARE PATH
KEY: Initials = Completed; N/A = Not Applicable; 0 = Variance
Page 1 of 2

DRG: _____

Emergency Call: _____

Special Considerations: _____

Allergies: _____

Case Manager: _____

☐ Emergency Room   ☐ Direct Admit
☐ Pre-Admit Testing & Teaching   ☐ Day Surgery

	PREOP	POSTOP EVENING OF SURGERY	Day 2:	Day 3:			
	Date:	Date:	Date:	Date:	Initial	Signature/Status	Initial	Signature/Status
	RN Review	RN Review	RN Review	RN Review				EXPECTED OUTCOMES
CONSULTS							
Consulting Physicians	___ Identify family doctor	___ Inform family doctor that patient on floor postop					___ Consults completed
	___ Preop medication administered by anesthesia						___ Follow-up appointments made
LAB Tests	___ Lab: CBC, Electrolytes Other						___ Lab tests within normal limits
	___ Lab results on chart preop		___ pro time/PTT				
Daily Lab							
EKG	___ EKG—interpreted copy on chart preop						
X-Rays	___ Chest X-ray report on chart preop						
OPERATIONS/ SPECIAL PROCEDURES							
OTHERS							

RN

continues

VALLEY BAPTIST MEDICAL CENTER
Harlingen, Texas
DX: <u>CAROTID ENDARTERECTOMY</u>
RESTORATIVE CARE PATH
KEY: Initials = Completed; N/A = Not Applicable; 0 = Variance
Page 2 of 2

	Initial	Signature/Status	Initial	Signature/Status
				RN

DRG: Emergency Call: Case Manager:

Special Considerations: Allergies:

□ Emergency Room   □ Direct Admit
□ Pre-Admit Testing & Teaching   □ Day Surgery

	PREOP Date:	POSTOP Date: EVENING OF SURGERY	Day 2: Date:	Day 3: Date:	EXPECTED OUTCOMES
Assessment & Evaluation	RN Review ___ Permit complete ___ List of home meds ___ List allergies ___ Complete database ___ Case manager assessment	RN Review ___ Blood pressure parameters: ___ to ___ ___ Neuro vital signs every 15 minutes until stable, then every 1 hour ___ Keep temperature less than 99° rectally	RN Review ___ Blood pressure parameters: ___ to ___ ___ Neuro vital signs every 4 hours (when arrives on floor) ___ Keep temperature less than 99° rectally	RN Review ___ Blood pressure parameters: ___ to ___ ___ Neuro vital signs every 4 hours ___ Keep temperature less than 99° rectally	___ Neuro vital signs stable ___ I&O within expected norm ___ Bowel sounds present ___ Incision line clean and dry free of redness ___ Pain or edema
Activity/Safety		___ ↑ head of bed 30° at all times ___ Coma position until fully awake ___ Turn every 2 hours ___ ↑ head of bed to sitting position slowly, dangle if tolerated Potential Physical Injury Level: II	___ ↑ head of bed 30° at all times ___ Chair → Dangle ___ Ambulate in unit before transfer to ___		___ Independent ambulation
Physical Therapy		___ Type ___ ___ MD Conference ___ Location	Potential Physical Injury Level: II ___ Physical Therapy as needed ___ Type ___ ___ MD Conference ___ Location	Potential Physical Injury: II ___ Type ___ ___ MD Conference ___ Location	
Protective/Supportive Device	Location				
Methicillin Resistant Staphylococcus Aureus					
Treatments		___ Specific gravity ↑ 1 hour ___ I&O hourly ___ Asess need for Physical Therapy, Occupational Therapy, Speech Therapy ___ Drains intact compressed ___ Foley catheter patent	___ I&O every 4 hours ___ Ask about discontinuing drain ___ Discontinue Foley catheter— reinsert as needed	___ I&O every 4 hours ___ Drain discontinued ___ Voiding every shift	

continues

Exhibit 11–A–1 continued

	RN Review	RN Review	RN Review	RN Review	EXPECTED OUTCOMES
Dressing Change		Reinforce dressing as needed	Bulky dressing	Light dressing	
Occupational/ Speech Therapy					
Respiratory Therapy					
Diet		Nothing by mouth until fully awake, then clear liquid and progress to regular	Regular	Regular Diet	— Tolerating diet well
Feedings					
Meds/IV Fluids			Discontinue IV—Saline Lock		— Meds given in safe & timely manner.
Pain Management		Notify doctor of pain			
Patient/Family Education	Preop teaching; Initiate Patient Education Record	Postop teaching; Neuro Intensive Care Unit brochure explained to family	Orient to nursing unit		— Patient/Family demonstrates understanding of home instructions
Psychosocial	Have pt bring significant other to Patient Admission Testing & Teaching; Designate family spokesperson; Establish need for chaplain visits				— Review med prescriptions; Ensure written discharge instructions complete
Discharge Planning/ Continuum of Care	Discharge needs assessed	Rehabilitation consult as needed	Discharge planning complete; Home Health as needed		Discharge: ❑ Nursing Home Placement ❑ Home Health ❑ Other ❑ Home ❑ Pulmonary Rehab ❑ Transfer ❑ Records faxed ❑ Report called ❑ Discharge instructions given
Others					
7A–7P					
7P–7A					
Room #	Last Name	First Name MI		MD	Admit Date

Exhibit 11–A–2 Patient Teaching Pathway

PATIENT PATHWAY FOR CAROTID ENDARTERECTOMY SURGERY
VALLEY BAPTIST MEDICAL CENTER

Physician _____

Case Manager/Social Worker _____

Phone _____

This pamphlet tells you what treatments and care you can expect to receive on a day-to-day basis, while you are in the hospital. This plan of care may change according to your needs.

QUESTIONS I WANT ANSWERED!

1. _____

   _____

2. _____

   _____

3. _____

   _____

4. _____

   _____

5. _____

   _____

6. _____

   _____

7. _____

   _____

8. _____

   _____

9. _____

   _____

10. _____

    _____

---

continues

Exhibit 11–A–2 continued

	BEFORE SURGERY (PATT)	AFTER SURGERY (DAY 1–2)	DISCHARGE (DAY 2–3)
Activity	• You will be able to continue normal activity.	• The head of your bed should be up 30 degrees at all times. • You will be on bed rest until fully awake. Your nurse will assist you to slowly raise the head of your bed. The nurse will be monitoring your blood pressure and making sure that you are able to tolerate the activity. • You will increase your activity as you are able. Your nurses will assist you with any new or first-time activity.	• Once you are home, gradually increase your activity. Do not get overtired. No strenuous exercises. If there are any other special instructions, your doctor will let you know.
Treatments	• Your nurse will check your temperature, blood pressure, heart rate, and breathing.	• Your blood pressure, heart rate, temperature, and breathing rate will be checked often. • You will have a large bandage on your neck. It will be checked and changed according to your doctor's orders. You will have a drain at the incision site. Your nurse will also check and empty this as ordered by your doctor. • Your nurse will encourage you to do deep breathing exercises often to keep your lungs clear. Also, it is important to turn from side to side at least every 2 hours. • Your doctor may order oxygen for you after surgery. A Respiratory Care Practitioner will check you from time to time.	• Your nurse will continue to check your blood pressure, heart rate, temperature, and breathing. • Your doctor or nurse will remove your bandage or change it as needed. • A Respiratory Care Practitioner will check to see if oxygen is still required. The oxygen will be discontinued before you are discharged. • When you are dismissed, your nurse will give you a staple remover kit. You will need to take this to your next office visit with you.
Tests	• Your doctor has ordered blood tests to be drawn or other tests to be done. Your nurse will explain these to you.	• More blood tests may be done if ordered by your doctor.	
Medication	• Your nurse will start an intravenous line (IV) to give you antibiotics, other IV medications, and fluids as ordered by your doctor. • If your doctor has ordered that you are to continue your home medications before surgery, these medications will be given with a small sip of water.	• Your nurse will continue to check your IV on a regular basis. • Your home medications will be started by the nurse when ordered by your doctor. • If you are having pain, ask your nurse for the pain medication your doctor has ordered.	• If your doctor has prescribed any new medications for you to take when you go home, you will receive a medication information sheet about that medication.
Teaching	• Your doctor and nurse will explain the procedures that will be done, medications to be given, and any supplies that might be used. • Chaplain services are available at your request. • Instructions will be reviewed with you and your family. • You will be given an opportunity to ask questions. (List your questions on the back.)	• Feel free to ask questions at any time.	• Your nurse will review the discharge orders with you before you are dismissed. You are encouraged to have a family member or friend there at the time.
Diet	• You should not eat or drink anything after midnight.	• Your diet will be continued as ordered by your doctor. Usually you will start out with clear liquids to drink and advance to your regular diet as you are able.	• You can continue your diet as recommended by your doctor.
Discharge Planning	• The Social Worker/Discharge Planner and/or Case Manager may meet with you to discuss any discharge needs.	• A Social Worker/Discharge Planner and/or Case Manager may meet with you to discuss any discharge needs.	• If appropriate, the doctor will discharge you.

Source: Copyright © Valley Baptist Medical Center.

Coronary Intervention Collaboration Model: Cardiovascular Laboratory to Home

Celine Peters, Cynthia H. McMahon, and Linda C.H. Stennett

Mission Hospital Regional Medical Center (MHRMC), a 271-bed facility, established an outcomes management approach to patient care in 1995. The department's goal is to manage and enhance patient care for a defined patient population. The definition of outcomes management used at MHRMC is "a multidisciplinary health care delivery process designed to integrate all services, clinical and financial, while maintaining or increasing quality, throughout the continuum of care." Outcomes management differs from case management. Case management is the focus on managing the care of individual patients. Outcomes management focuses on aggregate patient populations, the experience of a defined population as a whole, with the purpose of making research-based decisions to patient care.[1] The intent of the department is to bring disciplines together that "touch" the patient and as a team make recommendations to care. The purpose of outcomes management is to coordinate services, provide cost-effective and efficient care and increase or maintain the quality of care.[2] This chapter will describe the process of outcomes management with a cardiac patient population. Strategies to achieve clinical and financial outcomes will also be discussed.

PATHWAY

A profile of the high-risk diagnoses seen at the hospital was developed as an initial step in process development. These high-risk diagnoses included those that met one of the following criteria: high volume, high charges, or high length of stay. Coronary Interventional Procedures met the criteria for high volume and high charges. Further investigation of this diagnosis was prompted by the cardiologists.

They were interested in obtaining preprinted percutaneous transluminal coronary angioplasty (PTCA) orders, a tool used in the outcomes management (OM) process and through the cardiology subsection committee requested to evaluate care of the PTCA patient population.

The cardiovascular outcomes manager spearheaded the outcomes management process. Outcomes management is the art of managing patient outcomes. As the process evolved, it became evident that other procedures occur in the heart catheterization laboratory besides PTCA. The patient population that was targeted included those receiving PTCA, stents, atherectomy, and intravascular ultrasound. The Interventional ICD-9 codes examined were 36.01, 36.02, 36.05, and 36.06 as primary diagnoses. The title of the pathway then changed to incorporate all interventional procedures performed in this specialty laboratory, and the pathway was titled "Coronary Intervention Pathway." It was developed through an interdisciplinary process that focused on the continuum of care within the acute care setting. The theory of continuum of care extends beyond the traditional models of medicine, which has historically focused on the acute phase of care. Many hospitals throughout the development of pathways have focused on the preprocedure and postprocedure phases of care for intervention patients. Few have extended that continuum into the procedure phase of care.

The patient care problems identified for this patient population were wide variations in care among physicians, incongruity of care among units, inconsistency in patient education materials, absence of a means to track outcomes, and current methodology for data collection on these patients.

The purpose identified by the interdisciplinary team was to develop a clinical pathway using all procedural phases for the patient receiving an interventional procedure. The team consisted of the following disciplines: cardiologists, nurses from the coronary intensive care unit, the cardiopulmonary unit, the cardiovascular laboratory, and cardiac rehabilitation, and also included the cardiovascular outcomes manager, pharmacist, discharge planner, chaplain, dietitian, and cardiac liaison nurse.

The pathway design included categories of care, interventions, the expected outcomes, variances from expected outcomes, and an area for comments related to the variances. The continuum was divided into three phases of care with each having its own clinical pathway with separate expected outcomes. Phase I included admission through the interventional procedure itself and discharge from the cardiovascular laboratory to the coronary intensive care unit. Phase II is the immediate postprocedural phase with expected outcomes in the intensive care unit that achieve physical and hemodynamic stability. The patient remains in this phase until outcomes are achieved. Phase III focuses on the care on a telemetry unit to home.

The development of the intervention clinical pathway also included an emphasis on data collection not only for variances to expected outcomes, but also to track quality indicators that are typically measured on this patient population. These included indicators such as: emergent to the operating room, postprocedure hematoma, and contrast reactions. For each variance and quality indicator, a definition was agreed upon by the interdisciplinary team. Nationally recognized definitions were used when available.

Preprinted physician orders were developed to minimize the variability in treatment protocols among physicians and are initiated in the cardiovascular laboratory. The orders provide the interventions required that are listed on the pathway. Educational materials for patients and families were also developed to provide consistent preprocedure and discharge instructions.

CARE/CASE MANAGEMENT

The outcomes health care delivery model used at MHRMC is shown in Figure 12–1. It describes the dynamic process of patient care. The patient, population, or community is at the core or focus of the health care process. The key communicator is the health care team (HCT) that actively interacts with the patient and family during the five phases of the health care continuum. The results of these interactions provide the clinical and financial outcomes that are evaluated quarterly and provide further recommendations to the plans of care. Critical in the evaluation of outcomes is the social, environmental, cultural, and spiritual dimensions of care that describe

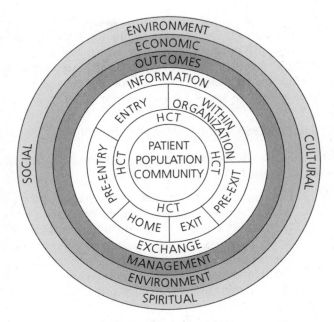

Figure 12–1 Health Care Delivery Model. *Source:* Copyright © Mission Hospital Regional Medical Center.

the uniqueness of a targeted population and impact the method in which information is exchanged. The health care delivery team is the multidisciplinary team that recommends changes to care. These recommendations are disseminated to health care peers including the bedside nurse. It is the bedside nurse who is seen as the patient's true case manager since this individual cares for the patient a minimum of 8 hours per day. The goal is to have the bedside nurse place all patients on the clinical pathway that is kept on the nursing unit. This person oversees the patient interventions with outcomes and documents care given on the clinical pathway in a "documentation by exception" format.

ROLE OF THE OUTCOMES MANAGER

The outcomes manager is a clinical nurse specialist (CNS) who is responsible for pathway development, implementation, and subsequent reports. The outcomes managers are assigned by patient populations and currently include the following service lines: cardiovascular, medical, neurological, surgical, oncology, and women and children. The role of the outcomes manager incorporates the traditional role of the CNS as clinical practitioner, consultant, educator, and researcher with the added role of financial analyst. The outcomes manager establishes, implements, and communicates clinical pathways in collaboration with the multidisciplinary team, reviews targeted clinical and financial outcomes, and tracks variances to care. These duties fulfill the roles of educator, consultant, researcher, and financial analyst. The clinical practitioner may case-manage high-risk patients, those that are complicated with multiple specialty consults, complex psychologically or functionally challenged, or those with an unusual or unfamiliar diagnosis.[3]

THE PROCESS

The outcomes management process includes the following steps.[4]

1. Define the patient population and multidisciplinary team members. Targeted patient population were those at high risk, high volume, high length of stay, or high charges. Validate the decision to evaluate a specific population with physicians and payers.
2. Conduct a literature review on the patient population and perform retrospective and concurrent chart review to have knowledge related to current care.
3. Benchmark current practice with national, state, and local standards to create a baseline for future comparisons.
4. Develop clinical, charge, patient, and family education pathways in collaboration with the multidisciplinary team.
5. Target clinical and financial indicators to monitor quarterly. These are based on recommended changes to patient care.
6. Implement, evaluate, revise, validate, reevaluate to begin to predict changes for the aggregate population.

The first step is to decide on a patient population to evaluate the care process. This is derived using a list of high-risk diagnoses seen at the institution. To validate the selection of a high-risk diagnosis, a list from payers may be compiled to pursue joint venture projects that will benefit the payer and allow the health care provider to increase quality of care. Once the diagnosis is determined, multidisciplinary team members are identified.

Multidisciplinary team members collaborate to improve practices and outcomes for the given patient population. Typically this occurs when knowledge is gained from the retrospective and concurrent chart review regarding the care of the patient population. There is a minimum of three physicians representing three different group practices invited as members of the team. Each team member is made aware of the time commitment. There is a maximum of three meetings, each lasting 1 hour. Sub–task-force meetings serve to accomplish the tasks, and findings are presented at the regular team meetings.

The second step in the outcomes process is the clinical pathway development, a tool of outcomes management.

The chart review notes current practices in care and becomes the foundation. A literature review searches for current technologies and care that may be included in the clinical pathway. Literature review should always include clinical practice guidelines from the Agency for Health Care Policy and Research.[5] Established standards of care, patient care plans with interventions specific to the population, and existing patient education teaching plans are included in the literature review.

In the third step of the process, comparisons of the organization's volume, charges, length of stay, and outliers can be made with national, state, and local standards. A baseline benchmark report affords insight into progress on the clinical and financial outcomes for each subsequent quarter reporting.

The purpose of clinical pathway development is to specify actions or intentions that will take place with the patient and family to achieve desired outcomes. The development of clinical pathways achieve collaborative communication between disciplines, appropriate bedside care, positive clinical outcomes, and decreased resource consumption. A clinical pathway is to be used as a guideline only in the plan of care. The health care team has the right to deviate from the plan when the condition of the patient warrants.

Charge pathways are developed for several reasons. They are used to determine total charges incurred per day and charges for each cost center. They identify opportunities to decrease resource consumption and determine appropriate length of stay. Lastly, they validate frequency of test or medication orders to be added to the clinical pathway. Patient and family education pathways communicate the expectations of the patient care plan. Prehospitalization and posthospitalization pathways match the plan of care (clinical pathway) and length of stay.

The fourth step in the outcomes process is the multidisciplinary team's recommended clinical indicators, to be reported quarterly. These indicators represent key variances to care that impact patient recovery and length of stay. They are unique to a patient population. The clinical indicators are listed as clinical variances on the clinical pathway and the bedside nurse evaluates patient care based on the variances.

Financial outcomes are the targeted charge savings per patient population. Each clinical change from current practice has a resultant financial dollar impact. Therefore, all charges are summed to reflect an aggregate dollar savings for the population, based on volume per year, per quarter.

Lastly, the team and designated hospital staff implement the clinical pathway, physician orders, and patient and family education pathways. Variance definitions ensure that all staff interpret variances in the same manner, preserving the integrity of clinical outcomes reports.

Quarterly, the outcomes manager reviews the coupled variance data for those that occur for greater than 20% of the population and this information is reported back to the team. Through this process, the team may make additional changes in practice, revisions are then implemented again, and clinical outcomes reviewed.

IMPLEMENTATION

Pivotal to the outcomes process is knowledge and understanding of the OM tools (clinical pathway, preprinted physician orders, and patient and family education materials) by all users. The following implementation strategies are used with each pathway. The efforts have produced high, nurse-compliance rates in placing a patient on the pathway once the learning curve was passed. The compliance rate for the coronary intervention pathway was 82% of the total volume from February to June 1997. The positive strategies are

1. Nursing staff participate on the multidisciplinary team.
2. Clinical pathway is part of the medical record.
3. Clinical pathway is a documentation-by-exception method of charting.
4. Staff education is conducted on all shifts and through self-learning modules.
5. Staff support is given throughout the implementation phase (2 weeks with a 1-month postimplementation follow-up).

The coronary intervention pathway was implemented February 1997 (Appendix 12–A). Licensed personnel in the cardiovascular laboratory document during the procedural phase of the clinical pathway and the bedside nurse documents care on the nursing unit. A future direction will be the documentation by all appropriate health care disciplines on the clinical pathways. Therefore, it will further foster an interdisciplinary approach to patient care problems.

The greatest advantage of pathway use to patient care delivery has been the consistency of practice patterns among medical, nursing, and support staff in the care of the interventional patient. As a result of this consistency and predictable patterns, we have realized increased efficiencies of care as well. Efficiency is measured through length of stay and charges. The pathway, by including the procedural phase of care, has been extremely effective in communicating expected variances that were recognized into the postprocedural phase.

Perhaps the greatest disadvantage is a reflection of its greatest strength, that of consistency of care. At times, the clinical pathway is perceived as a "cookbook" of medicine, particularly among the physicians. An argument against the development of a clinical pathway is that it would lead to a lower level of quality of care when care is not individualized. However, the pathways are used on a daily basis by the physicians themselves through the preprinted orders. Quarterly, the variances, complications, and quality indicators are reviewed by the physicians through their appropriate medical staff committee structure and the presented data are compared to baseline. The data reflect that quality of care is enhanced, not compromised.

Acceptance of the clinical pathway is high; the bedside nurse is willing to implement a pathway. However, patient care time commitments are barriers to completing the form despite initial enthusiasm. The outcomes managers and unit director's support has assisted nursing staff through the change process. Opportunities to discuss pathway changes to facilitate compliance, clarification of clinical variance definitions, and feedback regarding patient care sustain staff interest in the project. Future plans are to delete the multidisciplinary care plan if a patient is placed on a clinical pathway that clearly states the plan and problems of care. Patients do not currently review their coronary intervention pathway.

VARIANCES

The multidisciplinary team members recommend clinical indicators for a specific diagnosis. Variance indicators for the coronary intervention pathway are expected outcomes not met. Each variance is assigned a code letter or number, which is kept consistent among all patient populations in which the variance is appropriate. Daily, each identified variance is documented on the clinical pathway by code and is followed by an explanation of the action taken to alleviate the variance and facilitate achievement of the desired outcome. The nurse will use these variances at the bedside to revise the plan of care, if necessary. Exhibit 12–A–1 in Appendix 12–A is a list of variance definitions to assist staff in variance determination.

The variance and procedure codes documented on each phase of the pathway are seen in Exhibit 12–A–2. Since the pathways are a part of the medical record, clinical pathways are printed on NCR (no carbon required) paper. This allows the top sheet to remain in the chart and the second to be sent to the outcomes management department. The collected data are entered into an outcomes management database. Quarterly reports generated with the database are interfaced into the hospital information system to keep data entry to a minimum. In the future, the intent is to have the pathways available at the point of care. Targeted time frame is 1 year.

The cardiovascular outcomes manager quarterly reviews all pathways and analyzes variances using the chart or consulting the staff member. The variances are entered into a database, and reports are generated to determine compliance to the pathway (number of pathways/total number of patients) and frequency (number of patients with a specific variance/total number of patient on pathway). The standard (or goal) of the clinical pathways is to see no more than 20% of the population experiencing variances. Therefore, clinical variances are reevaluated by the multidisciplinary team if the occurrence is greater than 20%.

The multidisciplinary team may recommend variance changes or add clinical variances. It is foreseeable that changes in patient care through improved technology and changing practice patterns will generate additional ones. The intent is to also remain current with nationally recognized outcomes. If any of the above situations occur, a change in outcomes measures will occur.

OUTCOMES

Clinical Outcomes

A sample of clinical variances are summarized in Table 12–1 for the time period February to June 30, 1997. The focus is on consistent documentation and education concerning the definition of variances and outcomes by the bedside nurse. The director of outcomes management and cardiovascular outcomes manager present the quarterly report at the quality management committee and nursing staff meetings, respectively. Trending comparisons are made by previous quarter.

Financial Outcomes

Targeted financial savings for a given diagnosis are determined by the multidisciplinary team members. Each

change in care made by the team is calculated into dollars and charges are used as a reflection of cost. For instance, if the length of stay is decreased by 1 day, average charges by each cost center at baseline are identified on discharge day. Therefore, sample pharmacy charges may be $100 on the last day, these charges ($100) are multiplied by the number of patients that would be affected per year, and that number is calculated to reflect a charge savings by pharmacy. For instance, 500 patients per year seen with a given diagnosis; 125 are already being discharged in an appropriate time frame. That means savings are possible for 375 patients whose stay can be decreased by 1 day. The $100 charge from pharmacy seen on the last day is multiplied by 375, yielding targeted charge savings from pharmacy per year of $37,500. The $37,500 is divided by 4 to give pharmacy charge savings per quarter. Each cost center is calculated the same way. For specific medication changes, charges for ideal drugs are subtracted from charges of other drugs to more closely reflect the anticipated or targeted savings. The changes to care made for the coronary intervention pathway is summarized in Table 12–2. Actual charge savings versus targeted charge savings are reported quarterly to quality management and can be seen in Table 12–3.

Reports

Reports are generated on clinical and financial outcomes and compliance to the clinical pathway. These reports include comparisons to baseline. By year's end, the outcomes management department will be able to report on actual resource reduction through a decision support system. Reviewing these measures has proven successful with the payers. They have a vested interest in reducing unnecessary resource consumption that does not impact the quality of care. Historically, with other pathways, this information is appreciated by the payer, and goals to enhance care are identified. Reports are separately prepared to contain only a payer group's respective patients.

Table 12–1 Coronary Intervention Clinical Outcomes February to June 30, 1997

	Baseline (1996, 6 months)	1997 Year-to-Date (6 months)
Patient Volume	217	201
Length of Stay (LOS) in Days		
with Outliers (twice average LOS)	3.11	2.88
Length of Stay without Outliers	2.9	2.42
Number of Patients Admitted		
through Emergency Room	69	33
Clinical Indicators (Based on		
130 pts/5 months on clinical		
pathway)		
Chest pain ≥ 3 on pain scale		9 (7%)
Hematoma		11 (8%)
Reaction to contrast medicine		2 (1.5%)

Table 12–2 Coronary Intervention Targeted Financial Charge Savings

Recommendations	Charge Savings
Decrease in medications	$403
Decrease in Central Supply Charge	$108
Decrease in Length of Stay	$432
Decrease in Laboratory Charges	$624
Total Charge Savings/Patient	$1567
Target Patients	150
Total Targeted Charge Savings/Year	$235,050
Total Targeted Charge Savings/Quarter	$58,762

Table 12–3 Coronary Intervention Financial Outcomes

	Baseline (1996, 6 months)	*1997 Year-to-Date (6 months)*
Volume	217	201
Total Charge Savings	N/A	$136,728

CONCLUSION

The coronary intervention pathway established at MHRMC was intended for all patients with a coronary intervention performed in the heart catheterization laboratory. It was developed through a multidisciplinary team approach following the outcomes management process. The outcomes management department at MHRMC oversees the development, implementation, and evaluation of clinical pathways. The role of the outcomes manager, an advanced practice nurse, includes that of researcher, clinical practitioner, and financial analyst. It is through the accountability of these roles that the outcomes manager analyzes the clinical and financial outcomes of care with comparisons to baseline values.

Since the coronary intervention pathway was implemented in February 1997, the clinical outcomes reflect information collected from a total of 130 patients. There were no clinical variances to care that occurred greater than 20% of the population. To date, the length of stay is reduced by 8% from baseline (3.11 to 2.88 days).

Financial charge savings realized since February are $136,728. This met the targeted charge savings per quarter. Staff satisfaction with the clinical pathway is difficult to measure at this time since it is still considered a relatively new product and not fully integrated into the daily regimen of patient care.

Use of pathways has helped the hospital organization conserve hospital resources, increase efficiency of services, and communicate consistency of care across a selected aggregate population. To date, the organization has 11 other established pathways, with positive financial savings. Institution of the clinical pathway and system of reporting outcomes based on the pathway is supported and valued at Mission Hospital Regional Medical Center.

REFERENCES

1. Zander K. Part II, Identifying patient population for case management. *New Definition*. 1994;(9)4:1.

2. Moss MT, O'Conner S. Outcomes management in peri-operative services. *Nurs Econ*. 1993;(11)6:364–369.

3. Gibson SJ, Martin SM, Johnson MB, Blue R, Miller DS. CNS directed case management. *J Nurs Adm*. 1994;24(6):45–51.

4. Windle PE, Houston S. Commit. *Nurs Manage*. 1995;26(9):64AA–64HH.

5. The Agency for Health Care Policy and Research. *Clinical Practice Guidelines*. Silver Spring, MD: AHCPR Publications Clearinghouse; 1994.

■ Appendix 12–A ■

Clinical Pathway, Summary, and Definitions of Variances

Exhibit 12–A–1 Definitions of Variances

A. **CHEST PAIN**—3 on pain scale despite medication administration. (Does not apply to balloon inflation period during coronary intervention.)

B. **SYSTOLIC B/P**—less than 90 or greater than 180 mmHg

C. **BRADYDYSRHYTHMIAS**—sinus bradycardia with symptoms
Junctional rhythm
2nd or 3rd degree heart block

D. **VENTRICULAR DYSRHYTHMIAS**—all rhythms originating from the ventricle with ↓ or without symptoms (premature ventricular contractions excluded, ventricular tachycardia must be 10 or more beats to qualify).

E. **HEMATOMA**—hematoma of procedural puncture site equal to or greater than the size of a fifty-cent piece.

F. **BACK OR GROIN PAIN**—4 on pain scale despite medication administration.

G. **BLEEDING**—heavy bleeding at procedural puncture site necessitating unusual compression time or frequent dressing change.

H. **TEMPERATURE > 100 DEGREES**

I. **MD ALTERS PATHWAY**—MD orders listed interventions to be done with different frequency or not at all.

J. **SKIN INTEGRITY COMPROMISED**—Blisters or abrasions of skin resulting from equipment use, patient transfers and positioning, or immobility.

K. **PROCEDURE RELATED INFARCT OR EXTENSION OF MI**—myocardial infarction or extension that occurred during the interventional procedure and as a result of the intervention. Creatinine phosphokinase total must be greater than 300 with MB fraction elevation above normal range.

L. **DISSECTION**—Angiographically significant and flow-limiting dissection of vessel observed during or soon after the interventional procedure.

M. **FAILURE TO CROSS LESION/DELIVER OR DEPLOY STENT**—Inability to mechanically cross the coronary artery lesion or deliver and deploy stent during the interventional procedure.

N. **REACTION TO CONTRAST MEDIUM**—allergic reaction to the medium used for contrast during the coronary interventional procedure.

O. **OTHER**—Other variances. When this is used the actual variance must be identified.

Source: Copyright © Mission Hospital Regional Medical Center.

Exhibit 12–A–2 Post Coronary Intervention Clinical Pathway Summary

EXPECTED LOS: 24 HOURS PTCA; 36 HOURS STENT

NURSING DIAGNOSIS:
1. Comfort: Alteration in, Pain
2. Hemodynamics, altered
3. Dysrhythmia potential
4. Breathing pattern, ineffective
5. Anxiety
6. Tissue perfusion: alteration in
7. Cardiac Output: decreased

DEFINITIONS:
Coag = Coagulation Panel
CBC = Complete Blood Count
ACT = Activated Clotting Time
Hgb = Hemoglobin
Hct = Hematocrit
VS = Vital Signs
q = every
IV = Intravenous
LR = Lactated Ringers
NS = Normal Saline
Chol = Cholesterol
ASA = Aspirin
NTG = Nitroglycerine
WNL = Within Normal Limits
AAO = Alert & Oriented
HOB = Head of Bed
O_2 = Oxygen
IC = Intracoronary
NBP = Non-Invasive Blood Pressure

CODES/VARIANCE:
A. Chest Pain ≥ 3
B. Systolic B/P < than 90 or > 180 mm Hg
C. Bradydysrhythmias
D. Ventricular Dysrhythmias
E. Hematoma
F. Back or Groin ≥ 4
G. Bleeding
H. Temperature > than 100°
I. MD Alters Pathway
J. Skin Integrity Compromised
K. Dissection
L. Procedure-related infarct or extension
M. Failure to cross lesion/deliver stent
N. Reaction to contrast medium
O. Other

INDICATORS	CV CATH LAB INTERVENTIONS	CORONARY INTENSIVE CARE INTERVENTIONS	CARDIOPULMONARY UNIT INTERVENTIONS
Consults	Interventional Cardiologist. Surgical Standby & Referral (as indicated).	Interventional Cardiologist. Pastoral Care prn.	Interventional Cardiologist. Discharge Planner. Cardiac Rehabilitation. Pastoral Care prn.
Tests	Preprocedure labs: Chemistry, Coag, CBC, Blood Type and Screen. ACT prior to heparinization and as indicated during procedure.	EKG as ordered and prn. ACT before and 1 hr post heparin drip discontinuation while sheath in.	EKG prn chest pain. Cardiac Enzymes, CBC, Chem 7, and Hgb and Hct as ordered.
Activity/Skin and Tissue Integrity	Activity as directed during procedure.	Bed rest, affected leg immobilized and HOB 30° max elevation while sheath in. Walk 8 hr after sheath removal.	Ambulate.
Neurovascular Cardiac Respiratory	VS q 5 min or as indicated. Continuous ECG monitoring. Continuous arterial line, Swan Ganz, Pulse oximetry, and NIBP monitoring.	VS q 15 min × 4; q 30 min × 4; then q 2 hr. Assess A-line waveform, groin site, distal pulses, mentation, rhythm, and pain with VS. Systems assessment q 4 hr.	VS q h4 (routine) and prn. Monitor site, distal pulses, mentation, rhythm, and pain with VS.
Fluids Nutrition Elimination	IVs: LR or NS—rate as indicated by patient status. Foley catheter—if indicated.	Low Fat/Low Cholesterol Level II diet. Finger foods first meal. IV or saline lock as ordered. Urinary cath in/out if indicated.	Low Fat/Low Cholesterol Level II diet. IV as ordered—saline lock when feasible.
Medications	ASA/TICLID as needed. Pain evaluation q 5 min and prn. Sedation analgesia as ordered. Heparin and NTG infusions. Antiarrhythmics and other NTG as indicated. BP treated as directed. Prophylactic antibiotic as ordered.	Heparin drip and NTG drip (wean as ordered). Ancef every 8 hrs until sheath removed. Administer ASA, sleeper, sedation, analgesics, and O_2 as ordered.	Sleeper, sedative, analgesics, and O_2 as ordered.
Teaching	Pain Scale.	Assess home needs/discharge plan. Reinforce diagnosis information, activity restrictions, postprocedure care, and to report bleeding or chest discomfort to nurse. Self-care and discharge instructions.	Assess home needs/discharge plan. Reinforce diagnosis information, activity restrictions, postprocedure care, and to report bleeding or chest discomfort to nurse. Self-care and discharge instructions.

*Clinical pathways are guidelines to care and may be modified to individual needs.

continues

Exhibit 12–A–2 continued

Nursing Diagnosis:	1. Comfort: alteration in, Pain	3. Dysrhythmia potential	5. Anxiety	
	2. Hemodynamics, altered	4. Breathing pattern, ineffective	6. Tissue perfusion: alteration in	7. Cardiac Output: decreased

INDICATOR	INTERVENTION	EXPECTED OUTCOMES	VAR. CODE	Initials	NOTES
Consults/ Referrals	Interventional Cardiologist. Surgical Standby (as indicated) Surgical Referral (as indicated)	Patient/family aware of risks of procedure and available standby Surgeon and heart team available as indicated			_____ _____ _____
Tests	Preprocedure labs: Chemistry, Coag Panel, CBC, Blood Type/Screen, ACT prior to heparinization and as indicated during procedure.	Labs WNL. Blood products available (*if indicated) Blood adequately anticoagulated during procedure			_____ _____ _____ _____
Neurovascular Cardiac Respiratory	VS q 5 min or as indicated Continuous ECG monitoring Continuous arterial line monitoring Swan Ganz monitoring Pulse oximetry monitoring NBP monitoring	VS and cardiac rhythm WNL as indicated with associated monitoring			_____ _____ _____ _____ _____ _____
Fluids Nutrition Elimination	IVs: LR or NS—rate as indicated by patient status Foley catheter—if indicated	Maintained patient stability during procedure Maintenance of patient comfort with adequate urine output			_____ _____ _____ _____
Medications	Antiarrhythmics as indicated ASA/TICLID as needed Pain evaluation q 5 min or as indicated Additional sedation as indicated Additional analgesia as indicated _____ MSO4 _____ FENTANYL _____ DEMEROL Prophylactic antibiotic if indicated Heparin infusion as indicated ____ NTG _____ IV ____ SPRAY _____ IC BP treated as indicated _____ PROCARDIA _____ NTG	Patient free of lethal arrythmias Patient averted from thrombotic event Maintenance of patient comfort level during procedure Freedom from infection Patient relieved of hypertension/coronary artery spasm Ischemic chest pain managed to < 3 on pain scale after medication administration			_____ _____ _____ _____ _____ _____ _____ _____ _____ _____ _____ _____ _____ _____ _____ _____ _____ _____
Teaching	Pain Scale.	Patient able to quantify pain Patient/Family understand activity restrictions			_____ _____

Signature/Initials

_____   _____

_____   _____

Distribution: Original, Chart
Yellow, Outcomes Management

**POST CORONARY INTERVENTION
CLINICAL PATHWAY—CV CATHLAB**

PHASE I DATE: _____

VARIANCE CODES:
A. Chest Pain ≥ 3
B. Systolic B/P < 90 or > 180 mm Hg
C. Bradydysrhythmias
D. Ventricular Dysrhythmias
E. Hematoma
F. Back or Groin Pain ≥ 4
G. Bleeding
H. Temperature > 100°
I. MD Alters Pathway
J. Skin Integrity Compromised
K. Dissection
L. Procedure-related infarct or extension
M. Failure to cross lesion/deliver stent
N. Reaction to contrast medium
O. Other

PROCEDURE CODES:
1. PTCA Uncomplicated
2. PTCA Procedure Upgraded
3. PTCA/Stent
4. Stent
5. Rotational Coronary Atherectomy
6. Directional Coronary Atherectomy
7. Intra-Aortic Balloon Pump
8. Repeat Procedure—Planned
9. Repeat Procedure—Unplanned
10. Emergency Procedure
11. Reopro
12. Valvuloplasty
13. Intracoronary Thrombolytic: TPA
14. Intracoronary Thrombolytic: UK
15. Transfusion
16. To OR emergently

ADDRESSOGRAPH

continues

Exhibit 12–A–2 continued

Nursing Diagnosis:	1. Comfort: alteration in, Pain	3. Dysrhythmia potential	5. Anxiety
	2. Hemodynamics, altered	4. Breathing pattern, ineffective	6. Tissue perfusion: alteration in 7. Cardiac Output: decreased

INDICATOR	INTERVENTION	EXPECTED OUTCOMES	VAR. CODE	Initials	NOTES
Consults/ Referrals	Interventional Cardiologist Pastoral Care prn	Patient/family aware of postprocedure plan of care Patient/family emotional/spiritual needs addressed			_____ _____ _____
Tests	EKG as ordered and prn ACT before and 1 hr post heparin drip discontinued while sheath in	EKG within normal limits Steady decline in value to less than 160 No bleeding complications			_____ _____ _____
Activity/Skin and Tissue Integrity	Bed rest, affected leg immobilized while sheath in HOB 30° max. elevation while sheath in Walk 8 hr after sheath removal	No complications r/t skin immobility No bleeding or hematoma at site			_____ _____ _____
Neurovascular Cardiac Respiratory	VS q 15 min × 4; q 30 min × 4 then q 2 hr Assess A-line waveform, groin site, distal pulses, mentation, rhythm, and pain with VS Systems assessment q 4 hr	VS WNL A-line Patent—good waveform Drsg. dry and intact Distal pulses palpable AAO × 3 Absence of dysrhythmias and pain			_____ _____ _____ _____ _____
Fluids Nutrition Elimination	Low Fat, Low Cholesterol Level II Finger foods first meal IV or Saline lock as ordered. Urinary cath In/out if indicated Encouraged po fluids	Able to tolerate diet Maintain adequate hydration Urine output adequate			_____ _____ _____ _____ _____
Medications	Heparin drip (wean as ordered) NTG drip (wean as ordered) Ancef IVPB until sheath removed Administer ASA, sleeper, sedation, analgesics, and O_2 as ordered	Freedom from ischemic pain Remains comfortable Freedom from infection Exhibits restful sleep			_____ _____ _____ _____
Teaching	Reinforce diagnosis information, activity restrictions, postprocedure care, and to report bleeding or chest discomfort to nurse	Patient/family understanding of activity restrictions and when to notify nurse			_____ _____ _____

Signature/Initials

_____   _____

_____   _____

Distribution: Original, Chart
Yellow, Outcomes Management

**POST CORONARY INTERVENTION
CLINICAL PATHWAY—CICU to TRANSFER to
CARDIOPULMONARY**

PHASE II DATE: _____

VARIANCE CODES:
A. Chest Pain ≥ 3
B. Systolic B/P < 90 or > 180 mm Hg
C. Bradydysrhythmias
D. Ventricular Dysrhythmias
E. Hematoma
F. Back or Groin Pain ≥ 4
G. Bleeding
H. Temperature > 100°
I. MD Alters Pathway
J. Skin Integrity Compromised
K. Dissection
L. Procedure-related infarct or extension
O. Other

ADDRESSOGRAPH

continues

Exhibit 12–A–2 continued

Nursing Diagnosis: 1. Comfort: alteration in, Pain   3. Dysrhythmia potential   5. Anxiety
2. Hemodynamics, altered   4. Breathing pattern, ineffective   6. Tissue perfusion: alteration in   7. Cardiac Output: decreased

INDICATOR	INTERVENTION	EXPECTED OUTCOMES	VAR. CODE	Initials	NOTES
Consults/ Referrals	Interventional Cardiologist Pastoral Care prn Discharge Planner Cardiac Rehabilitation	Patient/family aware of postprocedure plan of care and plan for discharge Patient/family emotional/spiritual needs addressed			_____ _____ _____ _____
Tests	EKG prn chest pain Cardiac Enzymes, CBC, Chem 7, and Hgb and Hct as ordered	EKG WNL Lab values WNL			_____ _____ _____
Activity/Skin and Tissue Integrity	Ambulate	No complication r/t Skin immobility No bleeding from PTCA site Normal reconditioning			_____ _____ _____ _____
Neurovascular Cardiac Respiratory	VS q 4 hr (routine) and prn Monitor site, distal pulses, mentation, rhythm and pain with VS	VS WNL Distal pulses palpable Alert/Oriented × 3 Absence of dysrhythmias/pain			_____ _____ _____ _____
Fluids Nutrition Elimination	Low Fat, Low Cholesterol Level II IV as ordered-heparin lock when feasible	Maintain adequate intake and urine output Nutrition adequate to meet metabolic needs			_____ _____ _____
Medications	Sleeper, sedative, analgesics, and O₂ as ordered	Freedom from ischemic pain Remains comfortable Freedom from infection Exhibits restful sleep			_____ _____ _____ _____
Teaching	Assess home needs/discharge plan Reinforce diagnosis information post-PTCA Self-care and discharge instructions	Patient/family understand when to notify nurse Verbalize discharge instructions			_____ _____ _____ _____

Signature/Initials

_____   _____

_____   _____

Distribution: Original, Chart
Yellow, Outcomes Management

POST CORONARY INTERVENTION
CLINICAL PATHWAY—CARDIOPULMONARY to DISCHARGE

PHASE III DATE: _____

VARIANCE CODES:
A. Chest Pain ≥ 3
B. Systolic B/P < 90 or > 180 mm Hg
C. Bradydysrhythmias
D. Ventricular Dysrhythmias
E. Hematoma
F. Back or Groin Pain ≥ 4
G. Bleeding
H. Temperature > 100°
I. MD Alters Pathway
J. Skin Integrity Compromised
K. Dissection
L. Procedure-related infarct or extension
O. Other

ADDRESSOGRAPH

Source: Copyright © Mission Hospital Regional Medical Center.

■ 13 ■

The Coronary Artery Bypass Graft Clinical Pathway: Changing Paradigms in Patient Outcomes

Mary Porter Schooler

When clinical pathways were introduced at the University of Kentucky (UK) Hospital, they were objective evidence of the dramatic shifts occurring in the way this institution approaches health care. Where once the plan of care was limited to a single health care discipline, the plan of care outlined on the pathway became multidisciplinary in focus with health care disciplines being held accountable for their respective outcomes. Instead of being an open-ended plan of care, the pathway directs that interventions and outcomes are to be met within a specified time frame. Even though past nursing documentation had been centered around expected patient outcomes, outcomes on the pathway take on a greater significance because they are monitored, they determine patient progression on the pathway, and their exceptions are analyzed.

This chapter describes the Coronary Artery Bypass (CABG) clinical pathway. As the first pathway implemented at UK Hospital, its format, content, and use prompted many changes in multidisciplinary care for and considerations of this patient population.

Multidisciplinary groups called clinical group management teams (CGMT) develop, implement, and revise clinical pathways. Usually, a case manager serves as team leader and the physician representative as process owner of the team. Organized according to a patient population or product line, each CGMT reports to the clinical process and outcomes management (CPOM) steering committee. The CPOM initiative is devoted to improving clinical and financial outcomes throughout the institution.

The CABG pathway originated from the cardiac CGMT. Development, implementation, and evaluation of clinical pathways is an ongoing project of this team. The CABG clinical pathway was chosen as the first pathway to be launched by the team because this patient population is fairly predictable in their course of care and recovery, easily identifiable for the retrieval of financial data, and in need of improvement in services as determined by the chief of the cardiothoracic service.

CASE MANAGEMENT

Case management makes a significant contribution to the clinical pathway process. Composed of masters'-prepared nurses within the department of nursing, the nursing case management model at UK Hospital is organized around patient population groups. Most of the case managers focus on patient groups by working with specific physician services. In general, the case managers' role consists of coordinating the discharge planning process, managing patients across the continuum of health care services, providing clinical expertise, interpreting clinical and financial data, and participating in program development activities of the institution. Daily activities of the case managers may include participating in physician rounds, leading discharge planning rounds, assessing patient readiness for discharge, recommending and obtaining inpatient and outpatient services, consulting with other case managers and multidisciplinary staff regarding patient care options, following discharged patients by phone or in the outpatient clinic, and updating multidisciplinary staff regarding patient status and discharge plan. Whereas staff RNs are responsible for planning and delivering care for a given shift and for individual patients, the case manager focus is on the "big picture" of the patients' care and that of an entire group of patients. The way the case managers operationalize their individual roles varies according to the needs of their patient population and the nature of their physician team.

In addition to leading the CGMTs in the development and implementation of pathways, the case managers use the clinical pathway, along with other members of the multidisciplinary team, to monitor and track the patients' progression through the health care continuum and to direct appropriate care. Where interferences with the achievement of outcomes are noted (at UK, interferences or variances are referred to as "exceptions"), the case manager assists members of the multidisciplinary team in designing whatever interventions are necessary to get the patient back on track. Case managers assist in the analysis of clinical and financial data derived from the pathway. Based on these data, the case manager leads and/or participates in efforts to make any system or clinical practice changes that would improve future pathway outcomes.

PATHWAY DEVELOPMENT

Since the introduction of the nursing case management model at UK in 1990, several case managers began experimenting with the use of clinical pathways. Variations in pathway format and documentation methods were tried without much success.

After 4 years of struggle, clinical pathways took on an added significance in the institution and a multidisciplinary focus with the adoption of the CPOM process. Viewed as a mechanism to improve clinical and financial outcomes, the commitment to clinical pathways accelerated.

The Role of the Cardiac CGMT

The CABG clinical pathway was one of the early pathways that the cardiovascular case manager attempted to develop and introduce. At the inception of the cardiac CGMT, the development and implementation of the CABG clinical pathway was a primary goal. To start this process, the cardiovascular case manager and the chief of cardiothoracic surgery formulated a draft of the clinical pathway that reflected an integration of CABG protocols cited in the literature with the current standards of care for CABG patients at UK Hospital. Thus, the pathway was a combination of both existing and desired multidisciplinary clinical practice for CABG patients (Exhibit 13–A–1 in Appendix 13–A).

Health care disciplines represented on the pathway were invited to add their input. Each made recommendations regarding intervention, outcomes, and their timing on the pathway. Multidisciplinary staff participating in the development were staff nurses, attending physicians (surgeons and anesthesiologists), clinical nurse managers, physical therapists, respiratory therapists, nutritionists, pharmacists, quality assessment/utilization management (QA/UM) staff, cardiac perfusionists, and financial analysts. Both staff and supervisory levels of health care providers were involved.

Multidisciplinary input was also critical in deciding which disciplines were responsible for the documentation on the pathway and for being held accountable to pathway-specific outcomes. Key factors discussed in this process were clarifying which disciplines not only had access to the information determining whether or not an outcome was met, but also which discipline was recognized as being the professional expert for making judgments about a given outcome. For example, "Wound status allows discharge" is primarily the clinical judgment of a physician whereas "Indicates desired pain relief" is primarily an assessment/evaluation function of a nurse. Abbreviations of the discipline(s) follow each outcome on the pathway.

The cardiac CGMT also developed a CABG patient/family pathway and a follow-up pathway. Written in lay language but formatted as a clinical pathway, the patient/family pathway highlights the interventions, outcomes, and expected progression of the CABG patient. It is given to the patient/family at the time of admission. The follow-up pathway is utilized as a tool for documenting outcomes verbally reported by patients during phone calls made by nursing staff after discharge.

Initially, the CABG pathway was considered a "fast-track" pathway. The terminology was used in order to emphasize the importance of early extubation, early mobilization, early discharge teaching, and early discharge planning. The name was later dropped in order to avoid giving patients and families the perception that they would be pushed through their recovery process beyond what was safe and reasonable. However, the concepts of faster progression and more timely occurrence of interventions remained priorities for the team.

Another issue before the team was how to manage patients with multiple comorbidities and those with anticipated complicated postoperative courses on a pathway. For example, should patients with renal failure, cardiac assist devices, diabetes, pulmonary disease, or ventricular dysfunction be excluded from the pathway? Would these patients be more likely to "fall off" the pathway secondary to a prolonged postoperative recovery? Would these patients contaminate pathway data so that data analysis would be difficult and inaccurate conclusions be drawn? After much discussion, it was decided to have no exclusion criteria on the pathway. All CABG patients would be placed on the pathway. Patients whose care demands and status were no longer reflected on the pathway would be removed from the pathway and placed on a corresponding CABG patient care plan or other care plan (a nursing-developed care plan) that was most relevant for their needs. The impact of outliers on the data would be considered in the data analysis phase. Reasons for removal from the pathway could be monitored.

After implementation of the CABG pathway, the cardiac CGMT recognized that it would be redundant to develop a separate pathway for valve procedures. The postoperative care is very similar to open-heart CABG procedures and often valve and CABG procedures are combined. As the result, the CABG pathway was revised to include interventions and outcomes related to the care of cardiac valve patients. The CABG pathway was renamed the CABG/valve clinical pathway. Patients in the valve-related DRGs (diagnosis-related groups) are separated from the CABG DRGs for data reporting and analysis purposes.

The Role of the CPOM Steering Committee

While the cardiac CGMT worked on the development of the CABG/valve pathway, the CPOM steering committee made decisions regarding pathway implementation procedures. A standardized pathway format for the institution was agreed upon. Plans for education and data management were outlined by this group.

The CPOM steering committee provides oversight to the clinical pathway process throughout the institution. All pathways must be approved by this group before they are introduced into the hospital system. The steering committee also facilitates any institutional changes that are necessary to ensure the continuation of the pathway system.

USING THE UK CLINICAL PATHWAY SYSTEM

All multidisciplinary staff are to use the clinical pathway to guide and direct patient care activities. All multidisciplinary staff are to document on the clinical pathway. Physicians and other disciplines who normally document in the medical record progress notes are to continue that documentation practice in addition to recording on the pathway.

The physicians primarily use the CABG/valve pathway through their standardized orders. To facilitate that pathway interventions, and thus outcomes, are met, standardized physician orders based on the clinical pathway were written. These orders are for the immediate postoperative period in the ICU and for care in the telemetry/step-down unit upon transfer.

Social services and dietetic personnel assigned to cardiothoracic patients use the clinical pathway to prioritize patients in need of their assessment and intervention. The pathway outlines the optimal timing for their interventions, a timing that they determined themselves. Respiratory therapists utilize the pathway in making a post-ICU evaluation of the patients' respiratory status. Physical therapy intervention on the pathway is activated through the standardized orders for instruction on a postdischarge exercise and activity program.

Pathway Documentation

The implementation of the CABG pathway ushered in a new era of charting by exception. While bedside flowsheet documentation continues, charting by exception replaces the previous narrative documentation system.

Interventions and outcomes on the pathway are initialed and dated when met, or circled when unmet. An exception note consisting of the exception, the plan, and the outcome is written for each unmet pathway outcome on a clinical pathway documentation record. If there are no exceptions on pathway outcomes, an exception note is not required.

For data analysis purposes, each exception note is given an exception code. Exception codes consist of numbers assigned to a list of possible reasons for the exception. Grouped according to system, clinician, or patient reasons, the codes and their corresponding descriptor are printed at the top of the documentation record. Exceptions are coded for data analysis purposes. Any untoward patient event not addressed on the pathway may be documented in a narrative format as an event note on the documentation record. An example of this would be if a patient fell out of bed.

A signature record for health care provider signatures completes the set of the pathway and documentation record forms. Forms in the pathway system used by all patient populations, such as the documentation record and the signature record, are ordered by patient clerical services' staff. Clinical pathways associated with a specific diagnosis are ordered by the patient service coordinators assigned to nursing divisions. All pathways and related forms are stored at each clerical station. During admission, the patient's clinical pathway and its associated documents are kept with the bedside chart.

Clinical pathways replace the nursing kardex/patient care plan. Even though multidisciplinary staff are to document on the clinical pathway, the patient's registered nurse is ultimately responsible for that documentation. With the assistance of other team members and the patient, the RN determines when outcomes are met and individualizes the pathway to best reflect the patient's abilities, needs, and status.

OUTCOMES

Changing the focus from process to outcomes in health care is a major paradigm shift for health care providers. In an era of decreased resources and a demand for quality services, the achievement of optimal clinical and financial outcomes is of great importance. Whether contracts are obtained or health care services are selected, either by an individual consumer or by a managed care organization, increasingly depends upon the kinds of health care outcomes that can be produced.[1]

On the clinical pathway, both multidisciplinary interventions and outcomes are listed for each category of patient care and for each time interval. Outcomes are printed in bold type to distinguish them from interventions and to emphasize their significance.

Outcome Selection

Before the clinical pathway is ready for use, the CGMT decides which outcomes on the pathway are to be tracked. Once again, each multidisciplinary member has an opportunity to have input on what outcomes are considered critical for patient progression, quality, and optimal resource utilization. Because of their involvement in selecting the outcomes, members of the team readily agree to be held accountable to them.

The cardiac CGMT chose 13 different pathway-specific outcomes (Table 13–1). Some of these outcomes only appear in one pathway time interval, eg, "Extubation within 8 hours of arrival to CT-ICU" (CT-ICU Phase I), while other outcomes occur in multiple time intervals, eg, "Absence of atrial fibrillation/flutter" (CT-ICU Phase I through Telemetry

Phase 5). Using a utilization management software package, hospital QA/UM staff create a data entry screen listing each selected outcome and the time interval on the pathway in which it is to be evaluated.

In addition to pathway-specific outcomes, the team decides what other related outcomes provide important information for the patient group/diagnosis represented by the pathway. Exhibit 13–1 outlines the other outcomes monitored by the cardiac CGMT.

Data Retrieval

For the purpose of data retrieval, each pathway is assigned a code consisting of a letter followed by a two-digit number. The letter represents the CGMT that generated the pathway and the number refers to the order of development The CABG/valve pathway is B01. The "B" is the letter assigned to the cardiac CGMT and "01" indicates that the CABG/valve pathway is the first pathway created by this team.

At the time the patient is placed on a pathway, clerical assistants enter the patient into the hospital's clinical data/order entry system according to the pathway code, the status of the pathway (Status 1 = on the pathway; Status 2 = removed from the pathway), and the date this occurs. If the patient is removed from the pathway prior to discharge, the pathway status and date are changed in the computer system. Daily census reports generated by the hospital's information management department identify which patients have been or are on a pathway, the name and the code of the pathway, the status of the pathway, and the effective date of the pathway status. As a result, all members of the multidisciplinary team have the potential to rapidly determine on a daily basis which patients are currently on a pathway.

When the patient is removed from the pathway or discharged, the pathway and its associated forms are filed behind the progress notes in the patient's chart. QA/UM staff receive a daily report of all patients discharged from the hospital who had been placed on a pathway. They

Table 13–1 Pathway-Specific Outcomes

Pathway Phase	Pathway Outcome
OR	Antibiotic within 2 hours of incision
CT-ICU Phase 1	Extubated within 8 hours of arrival to CT-ICU
CT-ICU Phase 1	Absence of atrial fibrillation/flutter
CT-ICU Phase 2	Up in chair with meals
CT-ICU Phase 2	Transfer to telemetry
CT-ICU Phase 2	Absence of atrial fibrillation/flutter
CT-ICU Phase 2	Beta blocker
Telemetry Phase 1	Ambulate in room
Telemetry Phase 1	Absence of atrial fibrillation/flutter
Telemetry Phase 1	Tolerates solid food
Telemetry Phase 2	Absence of atrial fibrillation/flutter
Telemetry Phase 2	Ambulates 150 ft TID
Telemetry Phase 3	Absence of atrial fibrillation/flutter
Telemetry Phase 4	Tolerates 50% of 2 meals
Telemetry Phase 4	D/C pacer wires
Telemetry Phase 4	Discharge not delayed due to wound
Telemetry Phase 4	Absence of atrial fibrillation/flutter
Telemetry Phase 5	Discharge not delayed due to wound
Telemetry Phase 5	Discharged

Source: Copyright © University of Kentucky Hospital.

Exhibit 13–1 Other Related Pathway Outcomes

Discharge disposition
Returns to the OR
Returns to ICU
Readmissions within 14 days
Complications rate (by ICD-9 code)
Ventilator time
Costs

Source: Copyright © University of Kentucky Hospital.

access the medical record and review the pathway outcomes. Data regarding whether the pathway-specific outcomes are met, unmet, unknown, or not applicable are entered into the utilization management database. If an outcome is not met, the exception code recorded on the documentation record is entered as well. Patient satisfaction data are obtained through a patient satisfaction phone survey administered by the patient representative office after discharge. Patient representatives identify pathway patients via a daily hospital census report.

Communicating Pathway Outcomes

The frequency of data analysis depends upon the volume of patients placed on a given pathway. Outcomes are analyzed and reported on a 6-month basis for the CABG/valve pathway. The volume of patients is too small for meaningful data analysis to occur at a more frequent time interval. Because there are fewer valve cases, data for the valve DRGs will be analyzed on a yearly basis.

Data from the hospital's clinical data system and the utilization management database are downloaded into the Transition Systems, Incorporated (TSI) database. Patient satisfaction survey results are entered into a separate database. Reports are then generated from TSI by the clinical data management and analysis staff on both the pathway-specific and other related outcomes listed in Table 13–1 and Exhibit 13–1.

The cardiac CGMT has a process for communicating outcomes data. The senior clinical analyst reviews the detailed outcomes report with the case manager. This information is shared with all multidisciplinary members of the cardiac CGMT. In the next phase of data dissemination, the detailed reports are sent to the cardiothoracic surgery physicians and to the cochairpersons of the CPOM steering committee.

Highlights of the findings are presented to the entire CPOM steering committee. System, clinician, and patient issues contributing to the outcome data as well as any future action plans for impacting the outcomes are discussed with the steering committee. This group may respond with recommendations for future CGMT activities or data analysis procedures.

Once the physicians and CPOM committee have reviewed the data, members of the cardiac CGMT are responsible for communicating the data and proposed action plans to their multidisciplinary group. This mechanism helps to ensure that all levels of staff are updated on the outcomes from the CABG/valve patient population.

CABG/Valve Pathway Outcomes

The cardiac CGMT has completed two 6-month cycles of outcomes analysis since the CABG/valve pathway was implemented in January 1996. The two time periods are January 1996 through June 1996 and July 1996 through December 1996. At the time of this writing, the CGMT is awaiting reports from the third cycle, January 1997 through June 1997.

The primary focus of data analysis has been on DRG 106 (CABG with cardiac catheterization) and DRG 107 (CABG without cardiac catheterization). Data obtained are compared with the 6-month time period prior to implementation and to any previous 6-month cycle since pathway implementation.

Length of Stay Outcomes

For DRG 106, median total length of stay (LOS) increased by 2 days during the first 6 months after implementation of the pathway, then decreased by 1 day the following 6-month period (Figure 13–1). However, the postoperative average length of stay for ICU and tele-

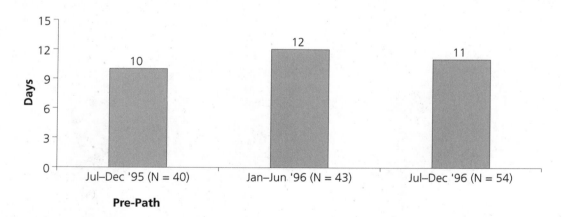

Figure 13–1 DRG 106 Median LOS. *Source:* Copyright © University of Kentucky Hospital.

metry care decreased. This patient group is commonly admitted to the hospital with a variety of acute diagnoses. The need for heart surgery is determined after admission but the patient's condition does not permit discharge from the hospital prior to surgery.

Nevertheless, based on clinical experiences, the perception among the team members was that the number of days between the decision to operate and the actual surgery could be decreased. While the average preoperative LOS for DRG 106 ranged from 4.0 to 4.5 days since pathway implementation, data from the clinical data management and analysis for July 1996 through November 1996 for DRG 106 showed that some patients had from 6 to 9 days between the day of their cardiac catheterization and the day of their operative procedure.

The cardiac CGMT decided to further investigate what factors were responsible for prolonging LOS between cardiac catheterization and surgery. The team wanted to ascertain if the factors involved were truly related to the patient's condition or from system or clinician sources. The QA/UM team members began gathering data on all cardiac catheterization patients who had surgery in the same admission. If the LOS between cardiac catherization and surgery was greater than 48 hours, the reasons for the delay would be noted. Thirty consecutive patients meeting the 48-hour criteria were followed. The final report on these data is pending.

For DRG 107, the average total LOS decreased significantly from 10.1 to 6.64 days for the first 6 months after pathway implementation. On the other hand, the average total LOS increased in the second 6-month interval to 11.4 days. Further data analysis revealed that there were three outlier cases in the second 6-month period, none in the first and only one in the 6 months prior to implementation. In addition, the preoperative LOS remained below what it had

been prior to pathway implementation. It was believed that the higher acuity and longer LOS of these three outliers skewed the mean LOS. Therefore, the median total LOS was used to make comparisons (Figure 13–2).

DRG 107 patients are commonly those patients who are admitted for elective surgery. Part of the initial decrease in the average LOS for this DRG was attributed to a decrease in the patients' preoperative LOS. A preoperative workup/hotel plan had been developed by the cardiac CGMT as a part of the overall initiative to improve LOS outcomes. According to this protocol, patients undergo their preoperative evaluation during the day or days preceding the surgery. If the patients meet certain distance criteria, they are offered, at the hospital's expense, lodging at a local motel on the night prior to their surgery.

Financial Outcomes

The average cost per case for DRG 106 remained relatively unchanged from prepathway implementation through the following two 6-month periods of data analysis. The average cost for DRG 107 decreased by approximately $4000 per case during the first 6 months after implementation as compared to the 6 months preceding implementation. However, the average cost per case increased by $9000 from the first 6 months to the second 6 months after pathway implementation. The median cost for DRG 107 remained approximately the same from prepathway through the following two 6-month cycles.

Increases in the average were once again attributed to the three outlier cases for DRG 107 in the period of July 1996 through December 1996. In addition, while a dramatic change was not noted in the DRG 106 cases, new attending physician staff and resulting practice changes during this time period may have contributed as well.

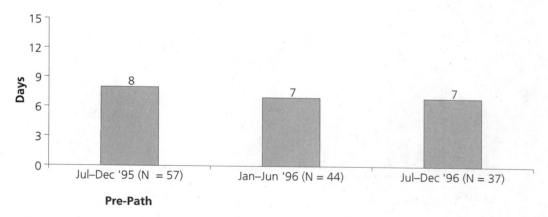

Figure 13–2 DRG 107 Median LOS. *Source:* Copyright © University of Kentucky Hospital.

Patient Satisfaction Outcomes

Overall patient satisfaction scores from the patient satisfaction survey ranged from 90 to 92 out of a possible 100 for time periods since pathway implementation. Since this type of survey had not been consistently done for this patient population prior to pathway implementation, there is no benchmark for comparison. In general, the scores were considered positive.

Pathway-Specific Outcomes

For the first 6-month period of evaluation, there were so many missing pieces of information from pathway documentation inconsistencies and inaccuracies that no credible data could be reported regarding pathway-specific outcomes. Overall, the percentage of "unknowns" entered into the utilization management database was high. This emphasized the fact that data accuracy and integrity are highly related to the quality of the documentation produced by clinicians and others with direct access to the medical record.

Based on these findings, efforts to educate staff regarding documentation policies were intensified. As the result, for some of the pathway-specific outcomes, fewer "unknowns" were reported for the July–December time interval. The percent of patients who met the outcome "Extubation within 8 hours of arrival to CT-ICU" could be reported because there were 0% and 6% "unknowns" for DRG 106 and 107, respectively (Figure 13–3). Low percentages of "unknowns" also allowed more accurate reporting of the data on the "Absence of atrial fibrillation/flutter" outcomes (Figure 13–4).

Prolonged postoperative intubation and atrial dysrhythmias prolong LOS, increase cost, and contribute to patient discomfort. These outcomes continue to be ones of great concern to the cardiac CGMT. As a result of these data, new action plans are being instituted by the cardiac CGMT. Revised protocols for early extubation have been developed and are in the process of being implemented. The use of beta blockers postoperatively is considered daily during physician rounds on postoperative patients. Atrial dysrhythmias, when they occur, are treated much more aggressively.

Beyond Data Analysis

Reviewing outcomes and data are a critical piece of the pathway, but the process of quality improvement does not end here. As indicated above, different strategies are initiated in an attempt to achieve more optimal outcomes. The pathway is revised to reflect any changes in action plans. Outcomes may be added or moved to different points on the pathway. Hopefully, adjustments made in practice and/or with the pathway will be reflected in the next cycle of data reports.

PATHWAY IMPLEMENTATION ISSUES

Conceptually, the majority of multidisciplinary staff at all levels support the intent of the pathway to decrease cost and improve quality. However, as with any major change, implementing the CABG/valve pathway has not been a smooth experience. Adjusting to a new documentation system has been frustrating for staff. Switching from open-ended, narrative documentation to only charting by exceptions has been a source of great discomfort. Physicians are not participating in pathway documentation, primarily due to the duplication of documentation on the pathway and in the progress notes. Future plans are to integrate clinical pathways into an automated bedside documentation system that will streamline the process.

All staff in the institution have a long way to go in adopting an outcomes focus in relation to patient care. Implementation of the clinical pathway did not create an

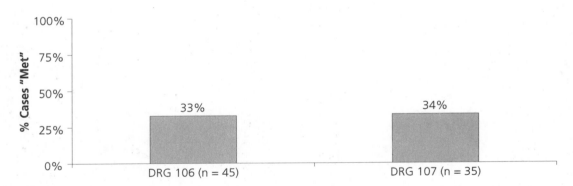

Figure 13–3 Extubation within 8 hours of arrival to ICU (July–December 1996). *Source:* Copyright © University of Kentucky Hospital.

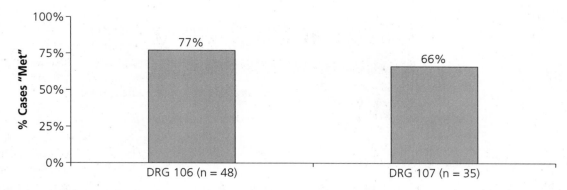

Figure 13–4 Absence of Atrial Fibrillation/Flutter (July–December 1996). *Source:* Copyright © University of Kentucky Hospital.

instantaneous change. However, as more clinical and financial data are produced, both directly and indirectly related to the clinical pathway, paradigms and behavior are beginning to make the transition.

CONCLUSION

The CABG/valve clinical pathway was developed and implemented by the cardiac CGMT, a multidisciplinary team of health care providers with expertise in the care of the cardiac surgical patient population. Utilized by case managers and multidisciplinary staff, the pathway outlines the optimal interventions and outcomes necessary for timely progression through the postoperative recovery period until discharge.

Data management systems in the hospital support the pathway process and generate reports on outcomes. Clinical and financial outcomes monitored by the cardiac CGMT include LOS, costs, ventilator time, and atrial dysrhythmias. The team analyzes data on many other outcomes as well. However, the data are only as accurate as the documentation and data entry from which they are derived.

Information about outcomes are disseminated to multidisciplinary staff. Clinical practice and system changes are implemented in order to improve outcomes data. The clinical pathway is revised to reflect these action plans.

Introduction of the clinical pathway has created an enormous change for the hospital not only in the mechanics of its use, but in its emphasis on outcomes-based practice. Nevertheless, the clinical pathway is a critical tool for the University of Kentucky Hospital's survival in today's health care environment.

REFERENCE

1. Wojner AW. Outcomes management: an interdisciplinary search for best practice. *AACN Clin Issues.* 1996;7(1):133–145.

■ Appendix 13–A ■
Clinical Pathway

Exhibit 13–A–1 CABG/VALVE Pathway

UNIVERSITY OF KENTUCKY HOSPITAL CHANDLER MEDICAL CENTER LEXINGTON, KENTUCKY	"Clinical pathways are guidelines for consideration which may be modified according to the individual patient's needs"	*See standard protocol ___ Continue on extended pathway	Patient Name Medical Record # Date of Birth

CLINICAL PATHWAY

DIAGNOSIS	ADMISSION DATE	PRIMARY NURSE	ATTENDING PHYSICIAN
CABG, VALVE, CABG/VALVE			

PATIENT PROBLEM LIST
1. Altered tissue perfusion
2. Ineffective airway clearance
3. Fluid volume excess
4. Decreased PO intake
5. Impaired skin integrity
6. Knowledge deficit
7. Postoperative pain
8. Potential intraoperative injury
9. Cardiac dysrhythmias

EXCLUSION CRITERIA: None

PARAMETERS: SDA = Same day admit　　Dry Weight

INTERVAL POD	PREOP PHASE	OR	CT-ICU PHASE I (CT-I)	CT-ICU PHASE II (CT-II)	TELEMETRY PHASE I	TELEMETRY PHASE 2	TELEMETRY PHASE 3	TELEMETRY PHASE 4	TELEMETRY PHASE 5
ASSESSMENT	H&P (SDA) Preanesthetic evaluation (SDA) Nursing database Vital signs Q 8 hours Check BP both arms (SDA) Pain management Goal ____ (SDA) Normal cardiovascular assessment* (MD/NUR)	Perioperative nursing assessment Anesthesia monitoring per protocol* CI > 2.0 (MD) Adequate BP, filling pressures x 4 (MD) Adequate ventilation and oxygenation: PaO₂ WNL, PaCO₂ WNL (MD)	Hemodynamics Q 1 hr Cardiovascular assessment* Q 4 hr Pain assessment Q 1 hr CI > 2.0 (MD/NUR) Pedal pulses x 4 (MD/NUR) HR 60–100 (MD/NUR) SVO₂ > 60% (MD/NUR) Absence of dysrhythmias (MD/NUR) Absence of atrial fibrillation/flutter (MD/NUR) EMV 10–11t (MD/NUR) Extubation within 8 hrs of arrival to CT-ICU (MD/RT/NUR) Breath sounds clear, equal bilaterally (NUR) RR normal, unlabored (NUR) Indicates desired pain relief (NUR) Extubated–transfer to CT-ICU phase II (MD/NUR)	Hemodynamics Q 1–4 hr Cardiovascular assessment* Q 8 hr Pain assessment Q 4 hr Pain management goal ____ CI > 2.0 (MD/NUR) Pedal pulses x 4 (MD/NUR) Absence of dysrhythmias (MD/NUR) Absence of atrial fib/flutter (MD/NUR) EMV 15 (MD/NUR) SVO₂ > 60% (MD/NUR) Breath sounds clear, equal bilat (NUR) RR normal, unlabored (NUR) Indicates desired pain relief (NUR) Hemodynamically stable-transfer to Telemetry (MD/NUR) Respiratory status allows transfer (MD/NUR)	Cardiovascular assessment* Q 8 hr Assess incisions Q 8 hr Pain assessment Q 4 hr Vital signs Q 4 hr x 24 hr Absence of dysrhythmia (MD/NUR) Absence of atrial fib/flutter (MD/NUR) EMV 15 (MD/NUR) Breath sounds clear, equal bilat (NUR) RR normal, unlabored (NUR) Indicates desired pain relief (NUR) Ready to ambulate in hall-progress to Telemetry #2 (MD/NUR)	Cardiovascular assessment* Q 8 hr Assess incisions Q 8 hr Pain assessment Q 8 hr Vital signs Q 8 hr Absence of dysrhythmia (MD/NUR) Absence of atrial fib/flutter (MD/NUR) EMV 15 (MD/NUR) Breath sounds clear, equal, bilat (NUR) RR normal, unlabored (NUR) Indicates desired pain relief (NUR) Ambulating in hall-progress to Tele #3 (NUR) Atrial fibrillation rate <120 if present-progress to Telemetry #3 (MD/NUR)	Cardiovascular assessment* Q 8 hr Assess incisions Q 8 hr Pain assessment Q 8 hr Vital signs Q 8 hr Absence of dysrhythmia (MD/NUR) Absence of atrial fib/flutter (MD/NUR) Breath sounds clear, equal, bilat (NUR) RR normal, unlabored (NUR) Indicates desired pain relief (NUR) Atrial fibrillation rate <120 if present-progress to Telemetry #4 (MD/NUR) Pulmonary toilet per patient-progress to Telemetry #4 (NUR)	Cardiovascular assessment* Q 8 hr Assess incisions Q 8 hr Pain assessment Q 8 hr Vital signs Q 8 hr Normal cardiovascular assessment* (MD/NUR) Absence of atrial fib/flutter (MD/NUR) Breath sounds clear, equal, bilat (NUR) RR normal, unlabored (NUR) Indicates desired pain relief (NUR) Room air saturation >90% with activity-progress to Telemetry #5 (MD/NUR)	Cardiovascular assessment* BID Assess incisions Q 8 hr Pain assessment Q 8 hr Vital signs Q 8 hr Normal cardiovascular assessment* (MD/NUR) Absence of atrial fib/flutter (MD/NUR) Breath sounds clear, equal, bilat (NUR) RR normal, unlabored (NUR) Indicates desired pain relief (NUR)
TESTS/LABS	CXR (SDA) EKG (SDA) A9 (SDA) Hitachi (SDA) U/A (SDA) T&C 4 units PRBC (SDA) O₂ saturation on room air ABG if saturation <90% PT/PTT (SDA) Hemogram with differential (HEMD) (SDA) Glucose < 250 (MD/NUR)	ABG (5) HCT (5) K+ (4) VBG (1) Glucose (diabetics only)	CXR A9, HEMD, PT/PTT, ion CA O₂ saturation panel K+, ion CA Q4 hrs x 24 (Diabetics) glucose Q 4 H x 24 Continuous O₂ saturation EKG ABG on arrival ABG end of wean and after extubation ABG Q 6 H x 24 if unable to wean in 6–8 hrs Glucose < 250 (MD/NUR) SPO₂ > 90% (MD/NUR) ABGS WNL (MD/NUR) LABS WNL (MD/NUR)	A7 CXR Hemogram with differential (HEMD) CXR after chest tubes D/Cd O₂ saturation Q 8 hr EKG PT/INR SPO₂ > 90% Glucose < 250 LABS WNL (MD/NUR) Glucose < 250 (MD/NUR)	O₂ saturation Q 8 hr Finger stick BS QID if diabetic PT/INR SPO₂ > 90% with activity Glucose < 250 (MD/NUR)	O₂ saturation Q 8 hr PT/INR CXR SPO₂ > 90% with activity (MD/NUR)	O₂ saturation Q 12 hr PT/INR A7, CBC	D/C O₂ saturation EKG	PT/INR INR acceptable level for discharge (MD/NUR)

continues

Exhibit 13–A–1 continued

INTERVAL	PREOP PHASE	OR	CT-ICU PHASE I (CT-I)	CT-ICU PHASE II (CT-II)	TELEMETRY PHASE I	TELEMETRY PHASE 2	TELEMETRY PHASE 3	TELEMETRY PHASE 4	TELEMETRY PHASE 5
DATE									
POD									
CONSULTS	Social services notified (SDA); Social services here (SDA); Anesthesia (SDA); CT surgery (SDA); RN coordinator (SDA); Consider chaplain	Perfusion services	Respiratory therapy	RT transfer protocol; Consider physical therapy; Dietetics		Consider physical therapy	Physical therapy–"home exercise program"; Dietetics re-evaluation		
NUTRITION/ FLUID BALANCE	NPO after mn; Prudent diet	IV fluids; NPO; Adequate intravascular volume (MD); Absence of bleeding (MD); HCT > 21 (MD)	NPO; IV KVO; Weigh Q night; Hemodynamics stable (MD/NUR); UOP > 30 cc/hr (MD/NUR); Abdomen soft, nondistended (MD/NUR); Chest tube output < 100 cc/hr (MD/NUR)	Clear liquids; Advance to prudent diet; D/C IV fluid; Weigh Q night; Hemodynamics stable (MD/NUR); UOP > 30 cc/hr (MD/NUR); Tolerates clear liquids (dietetics/NUR); Abdomen soft, nondistended (MD/NUR)	Prudent diet; Weigh Q evening; 1800cc fluid restriction; Strict I & O; UOP >240 cc/8 hr (MD/NUR); Trending to dry weight (MD/NUR); Taking solid food (dietetics/NUR); Abdomen soft, nondistended (MD/NUR)	Prudent diet; Weigh Q evening; 1800cc fluid restriction; Strict I & O; UOP >240 cc/8 hr (MD/NUR); Trending to dry weight (MD/NUR); Tolerating solid food (dietetics/NUR); Abdomen soft, nondistended (MD/NUR)	Prudent diet; Weigh Q evening; 1800cc fluid restriction; Strict I & O; UOP >240 cc/8 hr (MD/NUR); Trending to dry weight (MD/NUR); Tolerates 50% of 2 meals a day (dietetics/NUR); Absence of N & V (NUR)	Prudent diet; Weigh Q evening; 1800cc fluid restriction; Strict I & O; UOP >240 cc/8 hr (MD/NUR); Trending to dry weight (MD/NUR); Tolerates 50% of 2 meals a day (dietetics/NUR); Absence of N & V (NUR)	Prudent diet; Weigh Q evening; 1800cc fluid restriction; Strict I & O; UOP >240 cc/8 hr (MD/NUR); Tolerates 50% of 2 meals a day (dietetics/NUR); Absence of N & V (NUR); Discharged on 1800cc fluid restriction × 2 weeks (MD/NUR)
ACTIVITY	Up ad lib		Bedrest with HOB 30 degrees; Turn Q 2 hr; Leg exercises	Dangle; Up in chair with meals; Sternal precautions; Up in chair (NUR)	Up in chair with meals; Sternal precautions; Ambulating in room (NUR)	AM care with assistance; Sternal precautions; Tolerates activity progression (NUR); Ambulates 60–90 feet TID (NUR)	AM care with assistance; Sternal precautions; Tolerates activity progression (NUR); Ambulate 150 ft TID (NUR)	Ambulate 200 ft TID or stairs 1/2–1 flight + 100 ft TID; Sternal precautions; Ambulates 200 ft TID (NUR)	Ambulate 200 ft TID or stairs 1/2–1 flight + 100 ft TID; Sternal precautions; Independent with ADLs (NUR); Ambulates 200 ft TID (NUR)
TREATMENTS	Hibiclens shampoo/ shower; Preop checklist	Cardiopulmonary bypass per protocol*; Deep line/A-line placement; Perioperative care plan; Perioperative nursing standards met (NUR)	Pacer:; Foley; Chest tube; Invasive lines; ACE wrap; Ventilator wean to extubate*; NG; Incision care per protocol*; Skin intact (NUR)	Pacer:; D/C Foley; D/C chest tubes; D/C invasive lines; D/C ACE wrap; Knee-high TED stockings; Mask/NC O2; Incentive spirometry Q 2 hr while awake; Incision care per protocol*; Nebulized treatments; Facial shave Q day (MALE); TCDB Q 1–2 while awake; Transfer to telemetry (NUR); Effective secretion clearance (NUR); Skin intact (NUR); Incision well approximated (MD/NUR)	Telemetry; Pacer standby; Incentive spirometry Q 2 hr while awake; Incision care per protocol*; D/C O2 if room air saturation > 90%; Knee-high TED stockings; Facial shave Q day (MALE); TCDB Q 1–2 while awake; Effective secretion clearance (NUR); Incision well approximated (MD/NUR)	Telemetry; Pacer wires capped; Wean O2 per protocol*; Incentive spirometry Q 2 hr while awake; Incision care per protocol*; Knee-high TED stockings; Facial shave Q day (MALE); TCDB Q 1–2 while awake; Effective secretion clearance (NUR); Incision well approximated (MD/NUR)	Telemetry; Wean O2 per protocol*; Incentive spirometry Q 2 hr while awake; Incision care per protocol*; Knee-high TED stockings; Facial shave Q day (MALE); TCDB Q 1–2 while awake; Effective secretion clearance (NUR); Incision clean, well approximated (MD/NUR)	D/C telemetry; D/C O2; Incentive spirometry Q 2 hr while awake; Incision care per protocol*; Knee-high TED stockings; Facial shave Q day (MALE); D/C pacer wires (MD/NUR); Effective secretion clearance (NUR); Incision clean, well approximated (MD/NUR)	Incentive spirometry Q 2 hr while awake; Incision care per protocol*; Knee-high TED stockings; Facial shave Q day (MALE); Effective secretion clearance (NUR); Incision clean, well approximated & free of drainage & infection (MD)
MEDICATIONS DRIPS BLOOD PRODUCTS	Coumadin stopped (SDA); Continue routine meds (SDA); Sleep medication; Heparin GTT; Anesthetic pre-meds; Cefuroxime to OR	Blood products per protocol*; Heparin; Papaverine; Incision irrigation; Cefuroxime; Anesthetic medications; Adequate amnestic, anesthetized muscle-relaxed state (MD); Antibiotic within 2 hrs prior to incision (MD/NUR)	IV antibiotics Q 8 hr x 3; IV H2 blocker; IV analgesic; IV pressors; Electrolyte replacement per standardized orders; Diuretic; ASA in 8 hours if no bleeding; Transfuse HCT < 21; Sliding scale insulin IV; IV volume expansion	Saline lock; PO analgesics; PO H2 blocker; D/C IV GTT; Stool softener; Diuretic; ASA; Electrolyte replacement per standardized orders; Beta blocker; Preop noncardiac meds & antihypertensives; Daily ASA therapy (MD/NUR)	Saline lock; PO analgesics; Diuretic; ASA; Stool softener; PO H2 blocker; Beta blocker; Preoperative noncardiac meds & anti-hypertensives; PO anticoagulation	Saline lock; PO analgesics; Diuretic; ASA; Stool softener; Beta blocker; D/C H2 blocker; Preoperative noncardiac meds & anti-hypertensives; PO anticoagulation	Saline lock; PO analgesics; Diuretic; ASA; Stool softener; Beta blocker; Preoperative noncardiac meds & anti-hypertensives; PO anticoagulation	D/C saline lock; PO analgesics; Diuretic; Stool softener; Beta blocker; Preoperative noncardiac meds & anti-hypertensives; PO anticoagulation	PO analgesics; ASA; Stool softener; Beta blocker; Preoperative noncardiac meds & anti-hypertensives; PO anticoagulation; Discharge on above meds (MD/NUR)

continues

Exhibit 13–A–1 continued

DATE									
INTERVAL	**PREOP PHASE**	**OR**	**CT-ICU PHASE I (CT-I)**	**CT-ICU PHASE II (CT-II)**	**TELEMETRY PHASE I**	**TELEMETRY PHASE 2**	**TELEMETRY PHASE 3**	**TELEMETRY PHASE 4**	**TELEMETRY PHASE 5**
POD									
TEACHING	Respirex Visual analog scale (SDA) Foot and ankle exercises Splinting Heart pillow Preop video Teaching booklet (SDA) Orient to routine ICU tour Review pathway with patient/family (SDA) Plan of care session per protocol* **Verbalize understanding of pre- and post-op care (NUR)**		Family teaching re: postop Unit policy Reorient patient Review pathway with patient/family Plan of care session per protocol* **Verbalize understanding of ICU environment/procedures (NUR)** **Participate with care (NUR)**	Activity Cough/deep breathing exercises Review visual analog scale Review pathway with patient/family Plan of care session per protocol* **Verbalize understanding of pain scale (NUR)** **Participates with care (NUR)** **Demonstrates adequate pulmonary toilet (NUR)** **Verbalize no recall of intraop procedure (MD/NUR)**	Floor routine* Pain management Review pathway with patient/family Plan of care session per protocol* **Verbalizes understanding of floor routine (NUR)**	Pain management Incision care Review pathway with patient/family Plan of care session per protocol* **Verbalizes understanding of incision care (NUR)**	Medications, home activity Risk factors, sexuality Reportable S&S, emotions F/U care Diet class/instruction Review pathway with patient/family Plan of care session per protocol* **Verbalize understanding of:** D/C medications (NUR) Home monitoring (NUR) Risk factors (NUR) F/U care (NUR) Diet guidelines (dietetics) Home activity (NUR) Emotions/sexuality (NUR) Fluid restriction (NUR)	Reinforce D/C teaching Nutrition guidebook Review pathway with patient/family Plan of care session per protocol* **Verbalize understanding of D/C instructions (NUR)** **Verbalize understanding of Coumadin and INR F/U**	Reinforce D/C teaching Review pathway with patient/family Plan of care session per protocol* **Verbalize understanding of D/C instructions (NUR)**
DISCHARGE PLANNING	Identify caretaker Functional level Home support Educational level Financial counseling Verify in-network home health provider					Consider D/C needs for home O₂	Consider home health Plan D/C transportation Verify support systems	Make home health referral D/C instruction sheet Discharge before 1100 Activate F/U pathway F/U appointment made **Discharge (MD/NUR)** **Wound status allows discharge (MD)**	D/C instruction sheet Discharge before 1100 Activate F/U pathway **Discharge (MD/NUR)** **Wound status allows discharge (MD)**

Source:

■ 14 ■

Open Heart Surgery Clinical Pathway: A Guide to Managing Patient Outcomes

Laura S. Savage

In the 21st century, cardiac care will be influenced by economic forces, clinical outcomes, and quality of care. Over the past 5 to 7 years, open heart surgery has changed dramatically. Patients were routinely hospitalized for 10 to 14 days after surgery. Currently, the average length of stay is 4 days or less. Coronary artery bypass grafting (CABG) accounts for a large part of the estimated $259.1 billion spent for cardiovascular therapy, including hospitalization, medications, disability, and medical and nursing services in 1997.[1]

The current health care environment drives concern for economics and cost containment. At the same time, clinical outcomes and quality care cannot be sacrificed. Reducing hospital stay has become a primary focus of insurers and administrators. To ensure adherence to quality outcomes, the Medical College of Virginia Hospitals (MCVH) implemented a clinical pathway for the management of open heart surgery (OHS) patients. We began with patients undergoing CABG, and soon expanded the path to include all OHS (valve replacement or repair and reoperative CABG).

Pathway development provides for efficient management of the cardiac surgery population. The process of developing a pathway allowed us to examine our current practice, assess for duplication of services, and streamline operations. Our goal was to enhance quality care based on patient outcomes. Standard order sets were developed to provide guidance to house staff who are charged with managing the day-to-day care of open heart surgery patients. The fast-track concept facilitates early discharge by focusing on patient outcomes as indicators of readiness for discharge. The fast track refers to patients who undergo open heart surgery without complications. Their postoperative course meets expected outcomes without varying from the guideline.

The advanced practice nurses for cardiovascular and critical care were instrumental in facilitating the development of the open heart surgery clinical pathway. A key player was the cardiac surgeon. The physician champion is essential to the successful implementation of clinical pathways. This pathway was selected as the first in the institution based on the fact that CABG is a fairly predictable procedure and the group of cardiac surgeons practice in a similar manner. In addition to the advanced practice nurses and the cardiac surgeon, the following disciplines collaborated on the development of the path: staff nurses, physical and occupational therapists, the social worker, dietician, cardiac anesthesia, and the community health coordinator.

PATHWAY

In 1993, an interdisciplinary group including intensive care, intermediate care and advanced practice nurses, cardiac surgeons, social workers, physical and occupational therapists, dieticians, and the cardiac anesthesiologist convened to examine open heart surgery practices. At the same time, an independent consultant firm was hired by the hospital to assess for cost-saving opportunities. Our focus was to enhance cardiovascular services by providing cost-effective, quality care. We undertook the analysis phase by examining our internal practices. We benchmarked our practice against academic medical center norms, regional markers, and our internal prac-

The Open Heart Surgery Clinical Pathway was initially developed by the following individuals: Donna George, RN, BSN, Nurse Manager, Medical College of Virginia Hospitals; Lisa Pettrey, RN, MN, Nurse Manager, Medical College of Virginia Hospitals; and John Spratt, MD, Cardiovascular Surgeon. A special thanks to Teresa Day, Decision Support Analyst, and Kory Kittle, Product Line Administrator, for their support.

tice. We limited our focus to patients with diagnosis-related groups (DRGs) 106 and 107, that is, coronary artery bypass grafting with and without cardiac catheterization, respectively. Figures 14–1 and 14–2 compare MCVH with the Richmond area. We identified an opportunity to reduce length of stay (LOS) for CABG patients.

The team approached this project utilizing principles of quality management. The method used by the team is known as the FADE process.[2] There are four phases that guide the team through problem solving: focus, analysis, develop, and execute. All aspects of cardiac care were initially assessed for cost-savings opportunities. These included evaluating the number of routine lab tests ordered, volume of cross-match orders for blood transfusion, number of chest x-rays and electrocardiograms (ECGs) ordered, LOS in the cardiac surgery intensive care unit (CSICU), LOS in the intermediate unit, and operating room (OR) supplies used, including surgeon-specific equipment. We addressed cost of services and supplies in the following areas: cardiac catheterization lab, anesthesia, OR, pharmacy and intravenous medications, and respiratory therapy.

A literature review was conducted to determine current practice patterns. At the time of the initial path development, a 5-day LOS was the national norm. Given the number of disciplines that interfaced with this patient population, the path was reviewed and feedback solicited from all disciplines. Each contributed specific clinical outcomes. All of this information was integrated into a pathway that began preoperatively and continued until the fifth postoperative day. The categories of care included tests, medications, diet, patient/family education, nursing assessments and interventions, house staff assessment and procedures, consults, and patient activities. As illustrated in Appendix 14–A, Exhibit 14–A–1, specific outcomes were identified for each day. In order to promote adherence to the pathway, each discipline contributed to the development of standard order sets. House staff were guided through the appropriate course of care based on outcomes. The nursing standard of care, Adult Patients Undergoing Cardiac Surgery, was utilized as the basis for nursing assessments and interventions.[3] Medications and laboratory tests were determined by the cardiac surgeons' preference and based on research findings and national benchmarking projects. Exhibit 14–A–2 details the Clinical Guideline for Open Heart Surgery.

Due to the anticipated shortened LOS, patient/family education became very important. A patient/family guideline was developed to educate the patient on day-to-day expectations. An example of the patient guide is shown in Exhibit 14–A–3. In order to increase cooperation and compliance, it is essential for the patient and family to be well-prepared and anticipate daily treatment goals. All patients receive this guide during their preoperative outpatient evaluation or when hospitalized. When patients have short hospital stays, discharge plan-

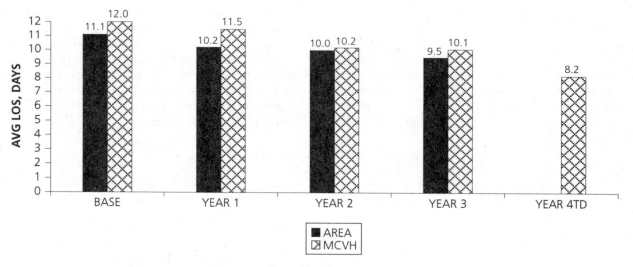

BASE PERIOD: JULY 1993 THROUGH FEBRUARY 1994
CHARGE AND LOS OUTLIERS AND DEATHS EXCLUDED
INPATIENTS DISCHARGED 07/01/93 THROUGH 06/30/97
RICHMOND AREA DATA FOR YEAR 3 COVER 03/01/96 THROUGH 06/30/96

Figure 14–1 Richmond, Virginia Area Length of Stay DRG 106 Coronary Bypass with Cardiac Catheterization. *Source:* Copyright © Medical College of Virginia Hospitals.

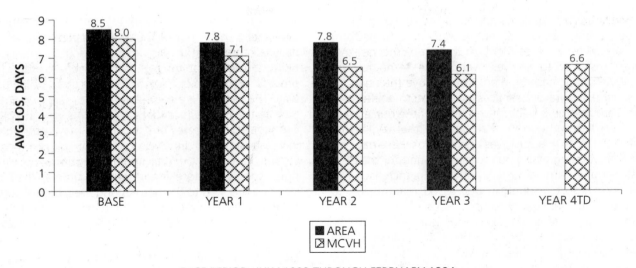

BASE PERIOD: JULY 1993 THROUGH FEBRUARY 1994
CHARGE AND LOS OUTLIERS AND DEATHS EXCLUDED
INPATIENTS DISCHARGED 07/01/93 THROUGH 06/30/97
RICHMOND AREA DATA FOR YEAR 3 COVER 03/01/96 THROUGH 06/30/96

Figure 14–2 Richmond, Virginia Area Length of Stay DRG 107 Coronary Bypass without Cardiac Catheterization. *Source:* Copyright © Medical College of Virginia Hospitals.

ning must begin on admission. In the case of same day admissions for surgery, discharge planning can begin prior to hospitalization. Patient/family education strategically progresses from reinforcement of pulmonary toileting techniques to issues of home care.

ROLES OF THE INTERDISCIPLINARY TEAM

The clinical pathway guides the coordination of care for multiple disciplines and monitors patient progress toward desired outcomes. Care of the postoperative cardiac surgery patient is truly collaborative. Each member of the team contributes to patient outcomes with the goal of achieving maximum function and independence at discharge.

In order to enhance the discharge process, daily interdisciplinary rounds were initiated in October 1996. The rounds are facilitated by the cardiovascular clinical nurse specialist. They are attended by the core team, which includes the primary/associate nurse, house staff, social worker, and nurse reviewer for quality improvement and utilization. The purpose of daily rounds is to efficiently plan care with the team and the patient in order to facilitate discharge planning.

Key issues that may impact discharge or the current plan of care are identified with the team. After rounds, the team members can better prioritize care. Rounds occur after the cardiac surgery house staff have conducted morning rounds and prior to the physician team round-

ing in the late afternoon. In this way, timely interventions and decisions can occur. As part of the interdisciplinary team, the quality improvement/utilization nurse assists the team by identifying issues related to acuity of care and intensity of service. The team is better able to move the patient to the appropriate level of care, thereby enhancing financial reimbursement. As a result of interdisciplinary rounds, the OHS path was reduced from 5 days to a 4-day LOS.

The plan of care is carried out and/or modified by the primary and associate nurses. Outcomes are measured on a daily basis by the cardiovascular clinical nurse specialist and primary and associate nurses. If the patient has not achieved the desired outcome for the day, this becomes the focus of care. For example, if the patient has not achieved the desired state regarding bowel function, this becomes the goal of the day. Prior to implementing the path, discharge was often delayed for alteration in elimination. Now, laxatives are a part of the care. Nurses are aware that this outcome (ie, successful bowel movement) must be accomplished by postoperative day number 3 or sooner.

The other disciplines play an integral role in ensuring an efficient discharge. The respiratory therapist evaluates each patient's pulmonary status upon transfer to the intermediate unit. Use of the incentive spirometer is reinforced as well as administration of inhalation therapy as indicated. Prior to implementing the path, the acuity of patients began to change. There was an increase in the number of CABG patients admitted the day of surgery. These patients required intensive staff time to perform their preoperative assessment. At the same time, patients were transferred to a lower level of care 24 hours after surgery. This increased patient acuity drove the need for supportive respiratory services in the intermediate care unit. Through the collaborative efforts of the nurse manager, cardiac surgeon, and respiratory therapy, a full-time respiratory therapist was assigned for these patients to assist with aggressive pulmonary toileting for the early postoperative cardiac surgery patient. The increased respiratory support resulted in decreased readmission to the CSICU for pulmonary complications.

The physical therapist is also an integral part of the cardiac rehabilitation team. All patients are evaluated and treated by a physical therapist, who designs an individualized activity/exercise program for the initial postoperative period. Patient progress is measured daily. Activity tolerance is a key indicator of readiness for discharge. In the future, low-level stress testing will be available for all OHS patients prior to discharge.

The dietician provides education to patients and families on the second and third postoperative days. A heart-healthy diet is prescribed for all. Individualized diet counseling occurs, and diets are tailored to the patient's specific medical needs. The community health coordinator, in conjunction with the advanced practice nurse, evaluates the patient's need for home health services. The social worker is also a key member of the team. Through the initial nursing assessment and daily interdisciplinary rounds, referrals are made to alert the social worker to important patient issues. All patients are screened for psychosocial needs.

CARE MANAGEMENT

Cardiac surgery patients on the pathway are managed by the primary and associate nurses and the cardiovascular clinical nurse specialist. Daily interdisciplinary rounds are conducted to evaluate patient progress toward clinical outcomes. The cardiovascular clinical nurse specialist serves as the case manager for open heart surgery patients. She facilitates daily rounds to ensure that patient outcomes occur as expected. Standardized order sets promote compliance with the path by house staff. In addition, each discipline is alerted to postoperative patients by receiving consults generated from the computerized order sets.

At MCVH, we are fortunate to have an automated order-entry system. This system contributes to the success of physicians managing patient care. Standardized order sets were developed by the interdisciplinary team. Order sets are organized along the continuum of care. These include the preoperative workup, immediate postoperative care, transfer out of the intensive care unit, and, finally, discharge from the hospital. Each set of screens guides the physician through the appropriate orders. This process facilitates compliance with the goals of the pathway. The surgeon is also free to enter orders that are not part of the OHS clinical pathway. In this way, he or she can individualize care. This, however, may contribute to increased cost and unnecessary testing. To assist the physicians, during the initial implementation, the clinical nurse specialists closely monitored all orders. The house staff are oriented to the path during their cardiac surgery rotation. With increased attention on cost, house staff have begun to evaluate their own ordering practices.

DOCUMENTATION OF THE CLINICAL PATHWAY

Fast-track documentation was developed by the cardiovascular clinical nurse specialist to facilitate compliance of the nursing staff to the pathway. The path documentation allows nurses to document interventions and clinical outcomes. It serves as a communication tool for the rest of the team for assessing patient progress. The format has been well accepted as it streamlines documentation. Exhibit 14–A–4 illustrates the first postoperative day. The caregiver initials each intervention as it is completed. Patient outcomes are documented and progress is tracked each shift. It is an interdisciplinary form on which each team member may document. At present, each discipline documents their plan and outcomes on the progress note. Ideally, as flowsheets are revised, plans are to incorporate path documentation into the bedside flowsheet. Currently, this document is maintained in the medical record.

Variations from the path are also noted on this form. Variances are utilized by unit-based quality improvement teams as well as the departmental quality committee. Documentation begins in the intensive care unit (ICU) as demonstrated in Exhibit 14–A–5. The OHS clinical pathway also serves as a teaching tool for house staff. They continue to document patient outcomes on the progress notes; however, the computerized order sets guide their daily practice. Through interdisciplinary rounds, house staff have been educated in how to incorporate outcomes into their daily documentation.

Initially, the management of patients on the critical pathway was facilitated by the cardiovascular clinical nurse specialist on the step-down floor and the critical care clinical nurse specialist in the CSICU. Currently, the cardiovascular clinical nurse specialist facilitates outcome management of all open heart surgery patients. Specific documentation was developed by the cardiovascular clinical nurse specialist to ensure compliance by the house staff. Education is conducted to orient new nursing staff to documentation and the concepts of clinical pathways. The cardiovascular clinical nurse specialist helps to ensure medical compliance with the path by providing orientation to the house staff as well. They are introduced to the concept and the standard order sets. Emphasis is placed on monitoring patient outcomes as indicators of readiness for discharge.

OUTCOMES

Outcome measurements were initiated with the implementation of the clinical pathway (referred to in our institution as "Clinical Guidelines" to reflect the philosophy that the path is a guideline and that care is individualized based on the patient's response to their illness or hospitalization). Initially, the clinical nurse specialists collected extensive data as illustrated in Exhibit 14–1. A "home-

grown" database configured in Paradox was created. Data were reported quarterly. Every possible variation was carefully recorded. We looked at LOS, patient disposition, reasons for prolonged ICU stay, and variations from expected outcomes on the third and fifth postoperative days. These time frames were selected based on benchmarks with other academic medical centers. Experience revealed that common causes of "falling off the path" occurred at the third and fifth postoperative days. However, it soon became apparent that too much information had been collected. It was difficult to determine if there was a common cause of variation or if the data reflected outliers.

Following the first year the number of variables were reduced by combining categories. In this way we were better able to identify key areas for improvement. We developed criteria for outliers defined as two standard deviations above the mean. Patients were grouped as "fast track" if they were discharged on postoperative day 5 or less. Intermediate LOS was defined as discharge on day 6, 7, or 8. Patients with an LOS greater than 9 days were outliers. In this way we could begin to focus on reasons for variation in the intermediate stay group.

After evaluating variations for the intermediate LOS group, several themes emerged. LOS was increased relative to 30% of patients developing atrial fibrillation on the third postoperative day. In addition, a significant number of patients were awaiting therapeutic warfarin levels. To resolve these issues, several strategies were implemented. The cardiac surgeon recommended atrial fibrillation prophylaxis. Atrial dysrhythmias are a common occurrence postoperatively. Pretreating patients with beta blockers has demonstrated significant reduction in the occurrence of postoperative atrial fibrillation.[4,5] At the same time, the Hospitality House (HH) had expanded its number of beds. This facility is a converted hotel for family members. It provides lodging for families from out of town. Patients can be discharged to the HH with a family member. Community health nurses make "home visits" for lab work and postoperative assessments. In this way, patients can be discharged from the hospital and still have their laboratory tests evaluated prior to returning home. These strategies contributed to the increased number of patients discharged in accordance with the "fast track." Figures 14–3 and 14–4 illustrate the significant reduction in postoperative LOS for DRGs 106 and 107. Currently, the variables that are monitored have been streamlined to reflect major variations in care. These are described in Exhibit 14–2.

The OHS pathway was initially reviewed and revised every 6 months. Now, it is reviewed annually. Outcomes are measured on a shift-to-shift basis. They are recorded on the flowsheet illustrated in Exhibits 14–A–4 and

Exhibit 14–1 Cardiac Surgery Variance Reporting

Sample Report from July 1995
Total patients discharged = 22
Total discharged postoperative day 5 or greater = 12
Patients discharged postoperative day 6, 7, 8 = 4
Outliers postoperative day 9 or greater = 6
Patients deceased = 1
Patients not yet discharged = 4
TOTAL FOR MONTH = 27

Variables measured:
1. Average postoperative time in days
2. Average days in cardiac surgery intensive care
3. Average skin to skin time
4. Average intubation time
5. Average operating room time

Complications tracked:
1. Prolonged intubation; reason (eg, high PCO2**, low PO2♥ saturation, not awake, sleepy)
2. Not transferred to step-down; reason (eg, intubated, not awake; low blood pressure, low cardiac output)
3. Respiratory complications (eg, oxygen saturation < 93%, oxygen required, abnormal breath sounds)
4. Cardiac arrhythmias
5. Renal complications
6. Neurologic complications
7. Not discharged; reason (eg, respiratory, cardiac, neurologic, urologic, wound drainage)
8. Pacer wires still in on postoperative day 3
9. Bleeding
10. Hemodynamic instability (eg, intra-aortic balloon pump, inotropic support)
KEY: **PCO2 = percent carbon dioxide
      ♥ PO2 = percent oxygen

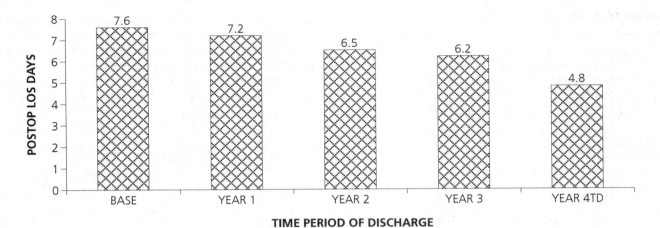

BASE PERIOD: JULY 1993 THROUGH FEBRUARY 1994. GUIDELINES IMPLEMENTED MARCH 1994.
CHARGE AND LOS OUTLIERS AND DEATHS EXCLUDED
INPATIENTS DISCHARGED 07/01/93 THROUGH 06/30/97

Figure 14–3 Postoperative Length of Stay DRG 106 Coronary Bypass with Cardiac Catheterization. *Source:* Copyright © Medical College of Virginia Hospitals.

14–A–5 in the Appendix. Staff nurses record the variations to specific outcomes by selecting the appropriate category. For example, if a patient does not transfer from the cardiac surgery intensive care unit on the first postoperative day, the reason, hemodynamic instability, is checked on the variance record.

The cardiovascular clinical nurse specialist maintains a database on all cardiac surgery patients. The data is summarized monthly and reported quarterly to the depart-

mental quality committee. This is an interdisciplinary committee made up of physicians, nurses, physical therapist, utilization review nurses, advanced practice nurses, social workers, and the cardiovascular product line manager. Data are presented and recommendations are made. Often, a committee will convene to analyze the issues and to design a corrective action plan.

The original variance tracking helped to identify opportunities for improvement. One issue that emerged was

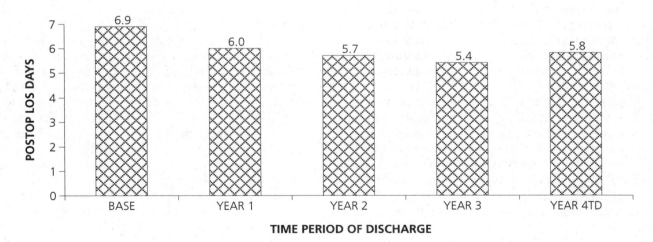

BASE PERIOD: JULY 1993 THROUGH FEBRUARY 1994. GUIDELINES IMPLEMENTED MARCH 1994.
CHARGE AND LOS OUTLIERS AND DEATHS EXCLUDED
INPATIENTS DISCHARGED 07/01/93 THROUGH 06/30/97

Figure 14–4 Postoperative Length of Stay DRG 107 Coronary Bypass without Cardiac Catheterization. *Source:* Copyright © Medical College of Virginia Hospitals.

Exhibit 14–2 Revised Variance Tracking

Hemodynamic instability
- requiring inotropic support
- intra-aortic balloon pump

Prolonged intubation
- not extubated in 8 hours or less

Renal complications
- requires hemodialysis or continuous ultrafiltration

Cardiac arrhythmias
- atrial fibrillation
- ventricular arrhythmias

Respiratory complications
- poor oxygen saturation (< 90%)
- crackles
- requires every 4 hour nebulizers

Neurologic complications
- stroke
- altered mental status

Infection
- fever
- wound drainage

Bleeding
- requiring surgical reexploration
- multiple blood products

prolonged CSICU stay related to intubation time. National standards revealed extubation times of 8 to 12 hours postoperatively. Our extubation time, in many cases, was greater than 20 hours. To address this issue an interdisciplinary team composed of CSICU nurses, the critical care clinical nurse specialist, cardiac surgeon, anesthesiologist, and respiratory therapists convened. The outcome of their work was the development of an extubation protocol. Exhibit 14–3 describes this protocol. Since implementing this protocol in January 1995, intubation time has decreased significantly. As shown in Figure 14–5, early extubation now occurs in approximately 8 hours after the patient arrives in the CSICU.

Another major change in practice occurred as a result of the daily interdisciplinary rounds. Initially, atrioventricular (AV) pacing wires were discontinued on the third postoperative day if the patient was without cardiac arrhythmias. After 1 month of daily rounds, it was discovered that on the fourth postoperative day, many patients did not meet criteria for acute hospital care. Patients were simply remaining in the hospital until the fifth postoperative day, the "designated" day of discharge. The quality improvement/utilization review nurse helped the team to realize that no specific interventions were occurring that necessitated another day of hospitalization. The OHS clinical path was then reduced to 4 days. Moreover, we became aware of measuring outcomes based on patient abilities. House staff were educated to this ap-

proach also. As a result patients are often discharged as early as the third postoperative day.

In addition to reducing hospital LOS, a number of the variables initially identified as opportunities for improvement have been streamlined. As shown in Table 14–1, the number of routine ECGs, chest x-rays, routine labs

Exhibit 14–3 Early Extubation Protocol (Sample)

Patient criteria:
1. Left ventricular function greater than 35%
2. Evaluate preoperative arterial blood gases (ABGs), if available. If no preop ABG, PCO2** less than 45; PO2^ greater than 60% on room air
3. Absence of liver or kidney disease
4. Uncomplicated intubation.
5. Less than 150% ideal body weight. (Obesity contributes to difficulty weaning)

Preweaning criteria:
1. No acute ischemia.
2. Hemodynamically stable: mean arterial pressure (MAP) greater than 65, central venous pressure (CVP) and pulmonary capillary wedge pressure (PCWP) less than 20, cardiac index (CI) greater than 2
3. Absence of new arrhythmias
4. Blood loss less than 2 cc▲/kilogram/hour
5. Urine output greater than 1 cc/kilogram/hour
6. Signs of awakening from anesthesia
7. Core temperature 97.0° Fahrenheit or greater

May wean FIO2▼ from 80% to 60% to 40%; keeping oxygen saturation (SaO2) 96% or greater.

Weaning criteria: (meets above in addition to the following)
1. Awake and cooperative. Follows commands.
2. Able to lift head off pillow.
3. PO2 greater than 100 millimeters mercury with FIO2 less than 60%; positive and expiratory pressure (PEEP) 5 or less; Pressure support ventilation (PSV) 5 or less.
4. Can generate tidal volumes (Vt) greater than 5 cc/kilogram
5. Respiratory rate less than 28

When met, turn intermittent mechanical ventilation (IMV) to 4 for 1 hour. If able to maintain minute ventilation, place on continuous positive airway pressure (CPAP) for 20 to 30 minutes, then draw ABG. Do not leave on CPAP longer than 60 minutes; can extubate from PSV 5 and PEEP of 5.

Extubation criteria:

PCO2 less than 50 and pH greater than 7.35; PO2 greater than 80 with FIO2 no greater than 40%
Respiratory rate less than 28
If unsure of readiness for extubation, weaning parameters may be obtained.

Key: **PCO2 = percent carbon dioxide
^PO2 = percent oxygen
▲ = cubic centimeter
▼FIO2 = inspired oxygen

Source: Copyright © Medical College of Virginia Hospitals.

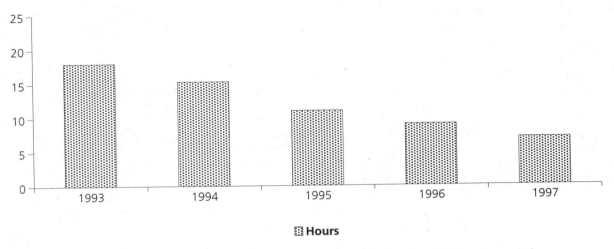

Figure 14–5 Extubation Time after Cardiac Surgery. *Source:* Copyright © Medical College of Virginia Hospitals.

(Chemistry 7 and Heme 8), and arterial blood gases drawn per patient have been reduced. During the initial implementation of the OHS pathway, the clinical nurse specialists carefully monitored each lab that was ordered. Part of the ongoing education and monitoring of the path included encouraging physicians to consider how the lab would be treated when the results were obtained. Helping the house staff to problem solve facilitated the elimination of unnecessary testing.

PATIENT SATISFACTION

The American Hospital Association recently surveyed approximately 37,000 people in regard to their satisfaction with health care. Of those surveyed, nearly 24,000 had been recently discharged from 120 hospitals across the United States. This included large, academic medical centers, small community hospitals, and for-profit health care facilities. In addition, consumer focus groups were held to evaluate consumer satisfaction. Both groups reported similar findings: two thirds of the public were satisfied with their health care services while the other third voiced consistent dissatisfaction with health care services. The survey revealed an overwhelming perception of health care facilities as impersonal. Patients felt little control over decisions made regarding their health care, stating they had been "usurped" by the insurance industry and managed care. Moreover, these decisions did not always seem to be in the best interest of the patient. Nearly 33% reported that they had been discharged before they were ready.[6]

Our patients have not experienced the overwhelming feelings of an impersonal system. However, the realities of managed care are coming. We have maintained a high level of patient satisfaction, due in part to the preoperative workup, which included thorough teaching regarding the expectations and the projected path of care. Patients and families receive a patient guide to open heart surgery, which outlines in lay terms the expected outcomes day by day. In this way, there are no surprises when a patient is discharged 4 days after surgery. Like the national survey results, many of our patients report feeling that "it happened so fast." Others express concern that they are discharged "too soon." However, approximately 70% of patients receive professional nursing services at home. In addition, all patients with phones receive a follow-up call by the cardiovascular clinical nurse specialist 7 to 10 days after discharge. Symptom distress as well as patient satisfaction is evaluated. Each patient receives a discharge information packet with an-

Table 14–1 Routine Postoperative Testing for the Open Heart Surgery Patient. Number of tests per patient.

Test	Prior to the Path (2/93–2/94)	Present Practice (6/96–6/97)
ABGS^	14	4
ECGS+	6	3
CHEM 7▲	7	3
CXRs▼	5	3
HEME 8⊕	8	4

^ABGS = Arterial blood gases
+ECG = Electrocardiogram
▲CHEM 7 profile includes potassium, sodium, chloride, bicarbonate, glucose, blood urea nitrogen and creatinine
▼CXRs = Chest x-ray PA and lateral views
⊕HEME 8 profile includes white blood count, red blood count, hemoglobin, hematocrit, mean corpuscular volume, MCVH, MCHC, and platelet count

swers to commonly asked questions as well as resource numbers of the physician, primary nurse, and cardiovascular clinical nurse specialist. Patients report a high level of satisfaction with the care and experience of hospitalization at our institution. However, one area of continued concern expressed by our patients is the rapid course of hospitalization. Many are discharged home after 2 days on an acute care unit. The elderly in particular often report not feeling ready to go home even though there are no medical conditions to warrant a continued stay in an acute facility. It is often difficult to secure home visits based on the payer's criteria for services. This should be an important observation for insurance companies who are directing the care based on financial reimbursement.

STAFF SATISFACTION

Staff satisfaction was not specifically measured prior to the implementation of the OHS pathway; however, nursing staff embraced the concept in a positive way. Their feedback was incorporated into the revision of the path. Each unit received extensive education from the clinical nurse specialists. Nurses recognized the benefits of a standardized, efficient plan of care and the impact that it could have on patient outcomes. The path describes clear goals for the patient on a daily and shift-to-shift basis. Nurses are able to measure the effectiveness of their interventions. Flow-sheet-style documentation reduced duplicate charting.

House staff find the OHS pathway to be a valuable teaching tool. It provides clear expectations for their role in the management of open heart surgery patients. Their tasks are clearly organized. They are able to function more autonomously since daily outcomes are defined. All open heart surgery patients are placed on the clinical pathway by the physician through the hospital information system. Standard order sets have been developed to guide the physician through the particulars of patient care and to ensure compliance to the pathway. The order sets are grouped into preoperative, postoperative, transfer, and discharge order sets. An example of the preoperative set is shown in Exhibit 14–4.

FINANCIAL OUTCOMES

The approach to realizing financial outcomes for cardiac surgery patients can be enhanced by benchmarking against "best practice" institutions. Who is most successful in reducing cost for CABG procedures while maintaining quality outcomes? What institution is most like yours in terms of mission, characteristics, and location? By speaking to clinicians from like institutions, creative solutions can be developed. Since implementing the OHS guideline, LOS has decreased significantly. This is illus-

Exhibit 14–4 Preoperative Order Sets Open Heart Surgery

Initiate Clinical Guideline for Open Heart Surgery

All orders are to be completed for those preoperative patients who have not had preadmission lab work.

Chemistry 7 profile**
Coagulation screen^
Heme 8▲
Urinalysis
Blood type and cross match. Set up 4 units packed red cells. Notify physician for infusion instructions.

Preop teaching by cardiovascular clinical nurse specialist
View preop video
Instruct patient on use of incentive spirometer

Shower with Hibiclens the evening before and morning of surgery
Routine preoperative electrocardiogram and posterior-anterior and lateral chest x-ray

KEY: **Chemistry 7 profile includes potassium, sodium, chloride, blood urea nitrogen, glucose, creatinine, and bicarbonate. ^Coagulation screen includes prothrombin time, partial thromboplastin time, and international normalized ratio. ▲Heme 8 includes hemoglobin, hematocrit, platelets, red blood count, white blood count, mean corpuscular volume, mean corpuscular hemoglobin concentration, and mean corpuscular hemoglobin.
Source: Copyright © Medical College of Virginia Hospitals.

trated in Figures 14–3 and 14–4. Most recently, the OHS clinical pathway was reduced to a 4-day LOS based on patient outcomes. It would seem that as LOS decreases, cost would also be reduced. However, as demonstrated in Figure 14–6, financial outcomes continue to present us with a challenging opportunity. Initially, we were able to demonstrate a decrease in cost. Most recently, however, cost has increased. This is the result of a number of factors. Physician practice preferences influence supplies, equipment, and even length of ICU stay. In addition, the nature of the academic medical center (AMC) impacts cost. In the AMC, there are multiple levels of physicians. Each has access to order entry. In some cases, additional lab work may be ordered or consultations to other services made. To realize increased financial outcomes, the team must commit to a certain degree of standardization. Our institution has an opportunity for continued improvement in this area. In the near future, an interdisciplinary team will be forming to investigate areas of potential cost savings. Our goal is to realize financial benefits while optimizing quality patient care.

THE CLINICAL PATHWAY: PROS AND CONS

The Open Heart Surgery Pathway has demonstrated several advantages. These are summarized in Exhibit 14–5.

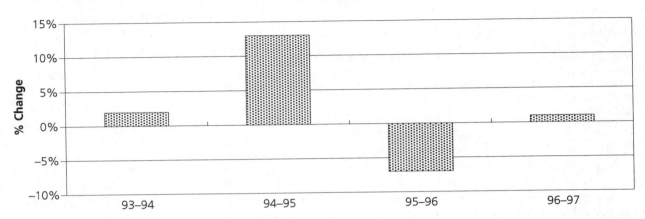

Negative % indicates an increase in cost

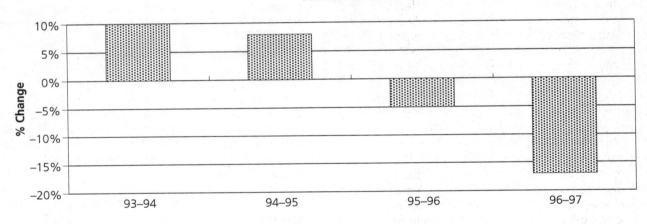

Negative % indicates an increase in cost

Key
    DRG = Diagnosis Related Group
    CABG = Coronary Artery Bypass Graft

Figure 14–6 Cost Reduction Associated with Clinical Guideline for OHS. *Source:* Copyright © Medical College of Virginia Hospitals.

The OHS has become firmly integrated into daily practice. It serves as a communication tool among the disciplines. The documentation flowsheet helps to streamline practice. Goals are clearly articulated and care planning is collaborative.

We have encountered few disadvantages to the use of the OHS clinical pathway. The literature often refers to clinical pathways as "cookbook medicine." This results in a perceived lack of individualized care. We have been for-

tunate to encounter minimal resistance to the implementation of the OHS path. Our physicians utilize the path as a guide. Patients are managed according to individual need.

The key to the successful implementation of a clinical pathway is a physician champion. We were fortunate to have a cardiac surgeon to champion the cause. In addition, our cardiology director was also forward-thinking and supportive. Our cardiac surgeons have similar prac-

Exhibit 14–5 Pros and Cons of the Open Heart Surgery
Pathway

Advantages	Disadvantages
Physician teaching guideline	Perceived "cookbook" medicine
Decreased length of stay	Perceived lack of individualized
Financial cost savings	care
Consistent, quality care	
Streamlined documentation	
Interdisciplinary communication tool	

tices, thus making it easier for agreement on the elements of the path. Common barriers to the successful implementation include lack of physician support and lack of collaborative practice among disciplines. Driving forces for the implementation of clinical pathways include the economic reality of managed care and local competition.

Clinical pathways for the open heart surgery patient facilitate the delivery of efficient, cost-effective patient care. In our region, managed care is slowly evolving and becoming a force in the marketplace. The ability to successfully negotiate with these companies will depend on the ability to demonstrate quality patient outcomes. The OHS clinical pathway puts us in a position to do so. There are, however, several key components that must be present to ensure successful implementation and ongoing compliance with the path.

Prior to the development of an open heart surgery clinical pathway, obtain commitment from the key stakeholders. Most importantly, identify a cardiac surgeon to champion the concept. Other key team members include cardiac anesthesia, intensive care and floor nurses, and cardiac rehabilitation personnel. Working to get "buy in" will help to ensure successful implementation. Next, bring all disciplines to the same table to brainstorm the goals of the pathway. Interdisciplinary planning during the development phase promotes collaborative practice during the execution stage. Identify outcomes for the cardiac surgery patient that must be met prior to discharge. For example, the patient will remain free of cardiac arrhythmias. Patient outcomes then become the focus for each team member. Decide how you plan to achieve these goals. What key indicators must be met to determine compliance?

Assign accountability for monitoring and trending variation from the path. Who will track variations such as

pulmonary complications, wound infection, readmission rates, etc? Set up systems to easily retrieve this data. Develop a reporting mechanism so that the OHS team stays informed. Determine how often to meet and convene subgroups to focus on specific issues.

CONCLUSION

The OHS clinical pathway has demonstrated several positive outcomes. First, postoperative LOS decreased from 7.6 days to 4.8 days for DRG 106 and from 6.9 days to 5.8 days for DRG 107. In addition, the total number of lab tests ordered was significantly reduced. The interdisciplinary pathway promotes collaborative practice. Territorial issues are eliminated. Daily interdisciplinary rounds facilitate discharge planning for cardiac surgery patients. The OHS clinical pathway has established clear goals and patient outcomes. In this way, the path has become a communication tool as well as a teaching tool for the interdisciplinary team. It provides the team with patient progress on a shift-to-shift basis. It serves to educate house staff who rotate onto the cardiac surgery service monthly.

Remember, the patient is the focus of your care. Despite economic pressures to decrease LOS and reduce cost, clinical pathways continue to strengthen our commitment to the delivery of quality patient care. They provide guidance for the collaborative management of the open heart surgery patient, but the health care team must remain focused on the patient as an individual. Continue to search for creative solutions to provide quality patient outcomes.

REFERENCES

1. *1997 Heart and Stroke Statistical Update*. Dallas, TX: American Heart Association; 1996:28.

2. *Quality Action Teams*. Burlington, MA: Organizational Dynamics Inc; 1990:1–3.

3. Delano M, Humphrey J, Collins C, Savage L. *Care of the Adult Patient Undergoing Cardiac Surgery*. Richmond, VA: Medical College of Virginia Hospitals; 1996.

4. Kalman JM, Munawar M, Howes LG, et al. Atrial fibrillation after coronary artery by pass grafting is associated with sympathetic activation. *Ann Thorac Surg*. 1995;60(6):1709–1715.

5. Nur-ozler F, Tokg-ozoglu L, Pasaoglu I. Atrial fibrillation after coronary bypass surgery: predictors and the role of MgSO4 replacement. *J Card Surg*. 1996;11(6):421–427.

6. Grayson M. Get the picture? *Hosp Health Netw*. 1997;71:30, 32.

■ Appendix 14–A ■

Outcomes, Guidelines, Plans of Care, and Pathway Documentation

Exhibit 14–A–1 Patient Outcomes for Open Heart Surgery

Preoperative Day	Postoperative Day 1	Postoperative Day 2	Postoperative Day 3	Postoperative Day 4
1. Patient completes preoperative workup including labs, electrocardiogram, chest x-ray, and physician history/physical.	1. Patient maintains hemodynamic stability.	1. Optimize oxygenation through pulmonary toileting.	1. Patient demonstrates functional bowel status.	1. Patient verbalizes understanding of medication regimen.
2. Patient can state preoperative prep including: when to take shower and what to use.	2. Patient is extubated within 8 hours after surgery.	2. Discontinue oxygen if saturation > 90% and patient is asymptomatic.	2. Patient continues to demonstrate increased incentive spirometer volumes.	2. Patient identifies risk factors to coronary artery disease and strategies to modify them.
3. Patient demonstrates baseline volume with the incentive spirometer.	3. Patient is transferred to the step-down unit.	3. Maintain surgical site integrity.	3. Patient verbalizes understanding of individualized exercise plan.	3. Patient is discharged home.
4. Patient demonstrates effective technique for coughing and deep breathing.	4. Patient demonstrates use of the incentive spirometer every hour while awake.	4. Patient and family verbalize signs and symptoms of infection.	4. Patient demonstrates understanding of dietary restrictions.	
5. Patient states where & when to return for surgery (if same day admission).	5. Patient tolerates being out of bed and ambulates after transfer to the floor.	5. Patient and family demonstrate incision care.	5. Patient identifies supports/resources at home.	
6. Patient verbalizes understanding of nothing by mouth (NPO) status.	6. Patient tolerates a clear liquid diet.	6. Patient can verbalize activity precautions.		
		7. Patient continues to use incentive spirometer with increasing volumes.		
		8. Patient walks in the hall twice a day.		

Source: Copyright © Medical College of Virginia Hospitals.

Exhibit 14–A–2 Medical College of Virginia Hospitals Authority Clinical Guideline for Open Heart Surgery

Revised 11/96	Preoperative	Operative Day	Postoperative Day 1	Postoperative Day 2	Postoperative Day 3	Postoperative Day 4
Care unit	Inpatient or Outpatient testing	Operating room/ CSICU+	CSICU M10C^	M10C	M10C	M10C
Tests	CBC▲, Chem 7● coagulation screen, urinalysis, EKG◆, type & cross, chest x-ray	ABG▼ x2 Chem 7 x2 potassium, chest x-ray, coagulation screen x2 CBC x2, EKG	ABG, Chem 7, CBC, EKG, chest x-ray	Chem 7 CBC	Posterior/anterior and lateral chest x-ray after wires pulled Chem 7, CBC, EKG	
Meds	Continue preoperative medications. Stop aspirin, Coumadin/ heparin 6 hours before surgery Sedation per anesthesia	Operating room: Sedation per anesthesia Prophylactic antibiotics CSICU: Analgesia/sedation, antihypertensives, inotropes, antiarrhythmics, diuretics, antibiotics	Wean drips Convert to oral medications Restart selected preoperative medications Stop prophylactic antibiotics	Medications per standing orders (analgesia, laxatives, etc)	Continue medications	Continue medications
Diet	Nothing by mouth (NPO) after midnight	NPO	Clear liquids Advance to Heart Healthy diet	Heart Healthy	Continue diet	Continue diet
Patient/ family education	Instruct in preoperative preparation and use of incentive spirometer (IS) View preoperative video	Orient family to CSICU Reinforce use of IS and pain management	Transfer teaching Orient to M10C. Give cardiac rehab packet Reinforce use of IS	Instruct on activity level, incision care, and diet	Instruct when to call physician, activity precautions, incision care, risk factors to coronary artery disease	Review instructions Teach about medications
Nursing assessments	Nursing history/physical Hibiclens shower preop	To OR with 2 liters oxygen (O2) CSICU: assess per standard of care Wean vent per protocol	Remove lines prior to transfer M10C Assess per step-down standard	Wean O2 Remove dressings Wire and incision care Weigh daily Pulmonary toilet continue	Evaluate bowel status Assess per standard Discontinue telemetry 12 hours after wire removal	Assess per standard of care
House staff assessment	History/physical Review labs Order preoperative prep Obtain consent	Evaluate hemodynam- ics, bleeding, rhythm, and labs Identify patients for transfer to the floor	Discontinue invasive lines by 8 AM Order EMLA cream prior to chest tube removal	Evaluate BP, heart rate and rhythm, labs, pulmonary status, and activity level Assess O2 needs	Assess as on day 2 Discontinue pacer wires Obtain chest x-ray after wire removal	Make discharge arrangements— follow-up appointment Enter orders for discharge meds
Consults		Physical therapy (PT) Social work	Clinical nurse specialist Dietary Respiratory therapy Community Health as needed			
Patient activities	Up ad lib	Intubated: Bed rest, turn every 2 hours Extubated: out of bed in chair	Out of bed in chair x2 Ambulate with PT	Out of bed in chair for meals Ambulate halls 2–3 times per day	Ambulate 3 times per day Steps with PT	Up ad lib

+CSICU = cardiac surgery intensive care unit
^M10C = intermediate care unit
▲CBC = complete blood count
●Chem 7 profile includes potassium, sodium, chloride, bicarbonate, glucose, blood urea nitrogen, and creatinine
◆EKG = electrocardiogram
▼ABG = arterial blood gases

Source: Copyright © Medical College of Virginia Hospitals.

Exhibit 14–A–3 Open Heart Surgery Plan of Care: A Patient Guide

Therapies	Day before surgery	Day of Surgery	Day 1	Day 2	Day 3	Day 4
Tests	1. Blood tests 2. X-rays 3. Electrocardiogram (ECG)	Blood tests	1. Blood tests 2. X-rays	Blood tests	1. Blood tests 2. X-rays 3. ECG	
Treatments	Hibiclens shower	1. Breathing machine 2. Suctioning and coughing	1. Oxygen 2. Incision dressing change	1. Stop oxygen if tests O.K. 2. Clean incision and leave open to air	1. Clean incision 2. Check oxygen level	Incision care
Lines and tubes		1. Catheter for urine 2. Chest tubes 3. IV** lines 4. Pacer wires 5. Heart monitor	1. Breathing tube out 2. IV lines out 3. Catheter out	Chest tubes out	Pacer wires out	
Activity	Unrestricted	Bed rest, wrists will be loosely tied to protect breathing tube	1. Out of bed to chair 2. Breathing exercises 3. Transfer out of the intensive care 4. Walk in the evening 5. Breathing exercises	1. Out of bed for all meals 2. Walk in hall twice 3. Breathing exercises 20x each hour while awake	1. Continue breathing exercises 2. Continue walking with guidelines from physical therapy	1. Continue breathing exercises 2. Activity guidelines from physical therapy 3. Discharge home
Teaching	Preop video Learn breathing exercises	Orient to ICU^	1. Review breathing exercises 2. Receive cardiac rehab packet	1. Receive instructions on incision care at home 2. Diet information 3. Activity guidelines	1. Instructions on signs and symptoms to report to MD▲ 2. Reinforce activity guidelines 3. Identify risk factors to heart disease	1. Review discharge medications 2. Obtain follow-up appointment

KEY: **IV = intravenous, ^ICU = intensive care unit, ▲MD = physician

Source: Copyright © Medical College of Virginia Hospitals.

Exhibit 14–A–4 Open Heart Surgery Path Documentation

Interdisciplinary Action Plan	Postoperative Day #1 Date:	7–11	11–3	3–7	7–11	11–7
Nursing interventions	Incision care					
	Pacer care per protocol					
	Weigh daily					
	Maintain saline lock					
Consults	Clinical nurse specialist					
	Physical therapy (P.T.)					
	Respiratory therapy					
	Dietician					
	Social work					
Patient activity	Out of bed to chair					
	Ambulate with P.T.					
	Elevate legs when out of bed					
Patient/Family teaching	Reinforce pulmonary toileting—coughing, deep breathing, incentive spirometer (IS)					
	Give rehab packet					
OUTCOMES:						
1. Patient demonstrates use of IS volume_____. (specify)						
2. Patient performs coughing and deep breathing exercises every hour.						
3. Patient tolerates diet without nausea or vomiting.						
4. Patient tolerates activity.						

Source: Copyright © Medical College of Virginia Hospitals.

Exhibit 14–A–5 Intensive Care Documentation Sample

Interventions	Operative Day Date:	7–11	11–3	3–7	7–11	11–7
Nursing assessments	Begin weaning inotropes and vasodilators.					
Patient activity	Bed rest. Out of bed after extubation.					
Respiratory therapy	Wean ventilator per weaning protocol. Extubate per protocol to 40% face mask or nasal cannula.					
Patient/family teaching	Orient family to ICU**. Teach patient about ventilator. Teach patient pulmonary toileting, pain management, and use of IS^.					
Patient outcomes:	Family informed of patient's status.					
	Patient states pain and anxiety adequately controlled.					
	Patient weaned from ventilator and extubated in 8 hours or less.					
	Bleeding is controlled within 5 hours or less.					

Signature:

Variations from the path. Check all that apply.

Intubated greater than 8 hours	
Reason: Not awake/sleepy	
Confused/agitated	
Low PO2▲/low saturations	
Increased PCO2▼	
Hemodynamic instability	
Bleeding not controlled in 5 hours or less	

KEY: **ICU = intensive care unit, ^IS = incentive spirometer, ▲PO2 = percent oxygen, ▼PCO2 = percent carbon dioxide.

Source: Copyright © Medical College of Virginia Hospitals.

■ 15 ■

Achieving Positive Outcomes through Quality Improvement for Cardiac Interventional Procedures

Donna J. Gilski, Jean M. Mau, Delores S. Meyer, and Constance Ryjewski

INTRODUCTION

Cardiac intervention describes a spectrum of nonsurgical invasive procedures to reestablish blood flow in diseased coronary arteries. Case management of the cardiac interventional patient is broadly defined by its goals of enhanced quality of care and improved cost-effectiveness.[1] Care management of this patient population, however, is better described as circular in nature—a continuum that reflects a multidisciplinary and collaborative approach to the efficient movement of patients through the health care system, ensurance of standards for care, evaluation of outcomes (clinical and financial), and implementation of changes that reflect best current practice with the achievement of patient goals.

Lutheran General Hospital-Advocate (LGH-Advocate) is a large community-based medical center that performs over 700 cardiac interventional procedures annually. Recent developments in invasive technology have provided cardiologists with adjunctive therapies to percutaneous transluminal coronary angioplasty (PTCA) for treatment of coronary artery disease. A variety of interventional procedures (coronary stent, rotoblator, atherectomy) have proven effective as modalities for treatment of acute closure and incidence of subsequent restenosis associated with PTCA.[2]

Prior to implementing the care management model for coronary stents, and subsequently all interventional procedures, patient care was determined by the intervening cardiologist with minimal collaboration from other health care professionals. Consequently, case-to-case variability related to patient care management and utilization of resources occurred with regular frequency. In 1994, cost accounting data demonstrated that significant financial losses were associated with the PTCA

population. In addition, more than 50% of the population in this analysis were reimbursed by Medicare. The opportunity to manage resources while providing high-quality, cost-effective care coupled with the advancements in technology was the impetus to care-manage the interventional population. The care management initiative began with the coronary stent population and evolved into a modality that now extends to all cardiac interventional procedures.

CLINICAL PATHWAY DEVELOPMENT

A multidisciplinary team including cardiologists, nurses, and a doctor of pharmacy met to develop a patient management program for the coronary stent population. In addition to members of the team, other disciplines including cardiac rehabilitation, nutrition, and social service were asked for input relative to their pathway components. Together, this team developed a critical pathway, standing orders, nursing protocols and guidelines that would maximize quality of care and achievement of successful outcomes.

The initial pathway, first developed in January 1995 was 5 days in duration and reflected the intensive anticoagulation regimen recommended at the time by stent manufacturers. Pathway duration was influenced by the length of bedrest postprocedure (48 hours) and ability to achieve a therapeutic International Ratio (INR) (2.5–3.0) on warfarin (Coumadin®) before the heparin infusion could be discontinued. Resources utilized in development of the pathway were largely obtained from clinical experience and recommendations from stent manufacturers. Some members of the team attended educational conferences on coronary stents, which also provided a basis for patient management. Final pathway decisions, regarding care management, were obtained by consensus among the team members.

The Quality Management Committee of LGH-Advocate endorses standing orders to be used in conjunction with clinical pathways. The initial set of coronary stent standing orders (preprocedure and postprocedure), developed by the multidisciplinary team, were modeled after the existing PTCA orders for general nursing care (ie, pain management, diet, activity). Additional orders were included that reflected the manufacturer's recommended intensive anticoagulation regimen postprocedure. The anticoagulation regimen was the most dynamic component of the standing orders. A delicate balance needed to be obtained between aggressive management to prevent subacute closure of the vessel and minimizing bleeding complications. The regimen was revised four times based upon medical and manufacturer's recommendations (Exhibit 15–1).

A significant number of bleeding complications observed in phase 1 identified the need to anticoagulate patients based upon a risk for thrombosis. The team revised the protocol (phase 2) with patients being identified by the interventional cardiologist at the end of the procedure as low, moderate, or high risk for thrombosis. In this way, patients who were low risk were less aggressively anticoagulated and had fewer, if any, bleeding complications.

In phase 3 dipyridomole (Persantine®) was replaced with ticlopidine (Ticlid®) as recommended in the literature at that time.[3] Enoxaparin (Lovenox®) was removed as an option largely because of cost and availability in our community pharmacies. Data analysis demonstrated no evidence of major bleeding complications after risk stratification for anticoagulation was instituted. Risk stratification and the anticoagulation regimen revision continued

to demonstrate a decrease in bleeding complications and achievement of desired outcomes. Additional protocol changes in phase 3 combine management of the low-risk or moderate-risk patient into one standard. While there were a few patients identified as moderate risk, the team decided to have a year's worth of experience with the protocol before eliminating this classification altogether.

The current anticoagulation regimen (phase 4) stratifies patients into either standard low-risk or high-risk regimens and eliminates the moderate-risk category. All patients receive the standard protocol with additional anticoagulation for high-risk patients determined at the discretion of the cardiologist. This may include a combination of heparin and/or abciximab (ReoPro®) infusions as well as Coumadin.[4,5]

In addition to changes made in the anticoagulation regimen, data associated with clinical outcomes were also used to make revisions in the pathway. For example, when data demonstrated the anticoagulation regimen significantly reduced bleeding complications and shortened the length of stay, the pathway was revised to 3 days which included a preprocedure day (Appendix 15–A, Exhibit 15–A–1).

Management of coronary stent patients does not differ from patient care associated with other interventional procedures (atherectomy, rotoblator). While the interventional device is different, the clinical management and desired outcomes are the same. During the postprocedure phase, sheaths are removed and progressive ambulation takes place while assessing for symptoms of restenosis and complications. It was noted by both cardiologists and nurses that the management of other interventional procedures could benefit from the coronary stent program. The standing orders and clinical

---

Exhibit 15–1 Postprocedure Anticoagulation Regimen

Phase 1 January 1995	Phase 2 August 1995	Phase 3 January 1996	Phase 4 Current, July 1996

All patients:
  Dextran infusion
  Heparin infusion
  Warfarin
	High Risk:
	  Heparin/ReoPro Infusion
	  Coumadin
	
	Moderate Risk:
	  Aspirin
	  Lovenox
	  Coumadin
	
	Low Risk:
	  Aspirin
	  Persantine
		High Risk:
		  Heparin/ReoPro Infusion
		  Coumadin
		
		Low/Moderate Risk:
		  Aspirin
		  Ticlid
		  Coumadin
			High Risk:
			  Determined by cardiologist
			
			Low Risk:
			  Aspirin
			  Ticlid

pathway were retitled "Cardiac Interventional" by the multidisciplinary team in July 1996. These documents are used to guide patient management in both the telemetry unit and cardiac intensive care unit (CICU). Both low-risk and high-risk patients are managed by the same clinical pathway and standing orders. The major difference in care is the anticoagulation regimen, which is reflected on the pathway under medication. All patient care is guided by one pathway. There have been no changes in the 3-day clinical pathway or standing orders since July 1996.

CARE MANAGEMENT PROCESS

Hospitalwide pathways have emerged as an integral part of patient care management. In order to ensure consistency and a holistic approach, the quality management committee has adopted a set of standard components, which are incorporated into every clinical pathway regardless of medical specialty or diagnosis. These components include assessment/monitoring, medications, nutrition/fluids, treatment, teaching, spiritual/psychosocial, discharge, planning, functional rehabilitation, and outcomes. Progression is reflected over time as the patient moves through the phases of care.

Care management of cardiac interventional procedures coordinates the care of these patients over a continuum. The pathway guides care over a continuum or episode. Coordination of care synchronizes activities of multiple disciplines and facilitates the achievement of desired outcomes.

Admission Day/Preprocedure (Clinical Pathway Phase I)

The clinical pathway is implemented once a patient is identified as a recipient of an interventional cardiac procedure. Patients entering the pathway come from two primary sources. The elective, stable patient is admitted directly to the cardiac catheterization laboratory (CCL) from home for a planned procedure or may be already hospitalized with a related cardiac diagnosis. Patients who present with symptoms of acute myocardial ischemia and are taken to the CCL emergently for an interventional procedure make up the second source of pathway entry.

Regardless of the mechanism of entry, the first phase of the clinical pathway describes those activities that occur before the procedure and are so defined under the heading "Admission Day or Preprocedure." These first-phase activities address all relevant pathway components and incorporate a spectrum of parameters and team accountabilities that include physical assessment, medical history, pharmacological management, patient educa-

tion, and laboratory/test evaluation. When appropriate, preprocedural standing orders are initiated and help ensure standardization of care and pathway compliance. A preprocedural patient checklist, completed by nursing, further augments this process.

Procedure Day/Day I (Clinical Pathway Phase II)

In practice, phases I and II may frequently occur on the same calendar day. Preprocedural versus postprocedural activities, however, necessitate the distinction. Phase II commences upon completion of the cardiac intervention and transfer of the patient to the admitting unit (CICU or telemetry). This phase addresses those acute care activities that occur primarily during the first 12 to 24 hours postintervention. It also introduces discharge evaluation processes and educational needs related to risk-factor modification.

Importantly, this phase of the clinical pathway is further enhanced by several supporting tools and documents. Postprocedure cardiac intervention standing orders are completed by the cardiologist while the patient is still in the CCL. These orders accompany the patient to the admitting unit and mirror the components of the pathway. The process for bedside femoral arterial sheath removal by nursing and the use of a mechanical compression device are guided by written nursing procedures that were developed according to manufacturer guidelines and best clinical practice. The postprocedure anticoagulation regimen, as stated earlier, reflects current literature and clinical experience.

The clinical pathway, standing orders, nursing procedure, and anticoagulation protocols were purposefully designed to complement and support one another, thus serving to reduce case-to-case variability and as a mechanism for variance tracking.

Day 2/Discharge Day (Clinical Pathway Phase III/IV)

"Day 2" and "Discharge Day" may also be the same calendar day. During this phase, opportunities for progressive increases in activity, cardiac rehab evaluation, nutritional assessment, and discharge instruction are included in patient care. Ancillary departments (cardiac rehabilitation and nutritional services) meet independently with each patient and/or provide literature for the patient's review. Prewritten, standardized discharge instructions ensure that the patient receives comprehensive information specific to cardiac intervention prior to going home. Educational booklets are also made available to the patient as well as identification cards for coronary stent procedures. A survey is completed by the patient prior to discharge to assess functional status with a follow-up survey being distributed at 6 months after discharge.

Clinical Practice

Clinical pathways are available on the CICU and telemetry units and placed with the patient's kardex at the time of admission (at present, clinical pathways are not considered part of the permanent medical record). As intended, the pathway serves as a guideline for care and is utilized and referenced throughout the course of the patient's hospital stay. Typically, pathway components are discussed during daily patient work rounds (attended by physicians, nurses, and a pharmacist) to assess patient's progress toward the identified outcomes. During shift-to-shift reports between bedside nurse clinicians, there is discussion of the patient's progress through phases of care.

Patient work rounds provide the opportunity to identify patient progression through the pathway phases as well as the opportunity to investigate patient-specific pathway variation together with the supportive scientific/clinical rationale. The cardiology care manager and care coordinator (unit-based) assist in facilitating pathway adherence and patient progress.

The pathway continues to provide nurses and physicians with a point of reference during the course of the day and throughout the patient's hospitalization. Timeliness and appropriateness of care activities can be tracked and evaluated. The identification of phases of care further assist the caregiver in recognizing that not every patient will progress at exactly the same pace.

During shift report, the pathway serves as a plan for care and checkpoint between nurses. It ensures consistent hand-off of nursing responsibilities, timely follow-up, and continuity of care. Check marks are used to communicate completed activities and chart progress; a signature key denotes ownership. Future plans for automated pathways, computerized documentation of pathway progression, and incorporation into the nursing plan of care are under evaluation.

Accountabilities

During the patient's hospital course and pathway implementation, the cardiology care manager serves as a liaison between involved disciplines as well as a nursing resource and mediator when there is pathway deviation. For example, low-risk patients who take Coumadin for atrial fibrillation, require an adjustment in the anticoagulation therapy postprocedure. The care manager's involvement includes discussion of therapy alternatives with the cardiologists and communication of plan to the staff nurse. Through observation and data collection/analysis, she remains sensitive to developing trends and compliance issues. She seeks feedback and provides input that serves to influence ongoing pathway and supportive tool revision/development.

The clinical pathway falls under nursing's purview when the document is utilized along with the patient kardex. Furthermore, the bedside nurses' unique position as 24-hour caregivers, makes them critically important to the pathway's success. Since the pathway was developed by a multidisciplinary team, content responsibility and compliance accountability is held by each of the respective disciplines.

Reservations and Acceptance

As is common to any new process or significant change in practice, early concerns of "cookbook" medicine were voiced by physicians, nurses, and members of ancillary departments. Over time, however, the standardization of care and the implementation of our clinical pathway demonstrated benefit to both the patient and the caregivers. Satisfaction criteria and measured outcomes support this. Potential disadvantages have not been realized and will surface only if there is a failure to recognize the uniqueness of each individual patient and the intent that a pathway guide, not dictate, patient care. We have been fortunate to experience a positive response with the patient management program. This success is due to ongoing multidisciplinary collaboration as well as to the "can do" attitude of the entire health care team.

OUTCOMES

An outcome, as it relates to health care, is defined as "the goal of a process." The patient management program for cardiac interventional procedures has many goals, which include clinical, financial, and satisfaction. The measurement and analysis of these outcomes influence the maintenance, revision, and refinement of the interventional care management program.

Methods

During the first year, the multidisciplinary team performed a chart review on all coronary stent patients. Specific indicators included extensive length of stay (LOS), complications, vessel reocclusion requiring CCL procedure, and incidence of blood transfusions. Clinical practice, both medical and nursing, were evaluated in conjunction with the chart reviews. A hospital database was established to assist in the collection and measurement of identified outcomes. The care manager performed data entry into this computer program on various components of care, including complications, LOS, and type of procedure. Reports were obtained quarterly and reviewed.

As the program evolved into one that served all interventional procedures, additional methods for measuring outcomes were established. These included follow-up telephone calls made 30 days postprocedure by the cardiology care manager to evaluate outcomes related to procedural complications and to patient education. The cardiac rehabilitation department also followed up with patients whose intervention was performed emergently as related to a myocardial infarction. In 1997, a national database sponsored by the American College of Cardiology (ACC) was approved for use by the hospital. This new database includes all the data points that the current database had and is nationally recognized. It was agreed that the national database would serve as the onsite database. Concurrent data collection has continued throughout program development starting in the CCL and completed at discharge. While participation in the national database has only recently occurred, opportunities to compare patient management of this interventional population on a national scale now exist.

Outcomes assessment and management is important to any patient care program because it serves as a foundation for quality control. The care management program for cardiac interventional patients has benefited from continuous quality improvement. During the developmental phase of the program, data were reviewed monthly by the multidisciplinary team and leaders from cardiology and nursing. Significant issues (ie, variability in clinical practice) were addressed in department or unit meetings. Quarterly data analysis was established when the quality improvement process consistently demonstrated positive outcomes.

Analysis

Achievement of positive outcomes is essential when assessing the merits of any patient management program. The cardiac interventional program demonstrated significant clinical, financial, and satisfaction outcomes.

Clinical

Clinical outcomes of cardiac interventional patients routinely measured are physiologic complications, LOS, and patient education. Complications are further defined into major and minor, shown in Exhibit 15–2.

Clinical outcomes for the interventional program have been measured since January 1995. The median LOS has decreased from 5.75 (phase 1) to 1.76 while the median number of procedures performed per month increased from 8 to 30, shown in Figures 15–1 and 15–2. Data continue to be collected and demonstrate a consistent LOS of approximately 24 hours for most elective cases.

Major and minor complications have demonstrated a decrease when compared to the total number of procedures, shown in Figure 15–3. Revisions in the anticoagulation regimen resulted in a significant decrease in all bleeding complications, shown in Figure 15–4.

Financial

The need for financial improvement is frequently one of the initial outcomes that generates a care management program. LOS, along with the procedure cost per case, constitute a large part of the financial outcomes. The clinical pathway and standing orders were developed to control costs through standardization postprocedure and LOS.

The median hospital costs per case, excluding CCL, have decreased from $2234 to $1400, shown in Figure 15–5. The median diagnostic laboratory costs have decreased from $127 to $64 per hospital admission. The CCL average cost per case has remained essentially un-

---

Exhibit 15–2 Cardiac Interventional Complications

MAJOR
1. Death
*2. Bleeding that requires blood transfusion or surgical intervention
3. Subacute closure requiring return to Cardiac Catheterization Lab or Coronary Artery Bypass Grafting
4. Intervention related myocardial infarction (creatinine kinase >500)

MINOR
*1. Hematoma
*2. Pseudoaneurysm
3. Intervention related myocardial infarction (creatinine kinase <500)

*Denotes bleeding complications

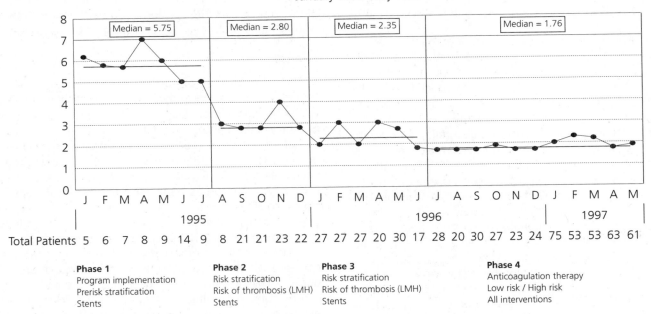

Lutheran General Hospital-Advocate
Coronary Stents/Cardiac Interventions
LOS
January 1995–May 1997

Figure 15–1 Coronary Stents/Cardiac Interventions: LOS. *Source:* Copyright 1996, Lutheran General Hospital-Advocate, Park Ridge, Illinois—used with permission.

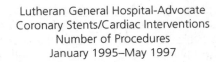

Lutheran General Hospital-Advocate
Coronary Stents/Cardiac Interventions
Number of Procedures
January 1995–May 1997

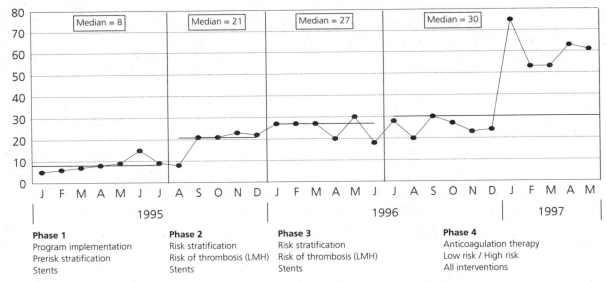

Figure 15–2 Coronary Stents/Cardiac Interventions: Procedures. *Source:* Copyright 1996, Lutheran General Hospital-Advocate, Park Ridge, Illinois—used with permission.

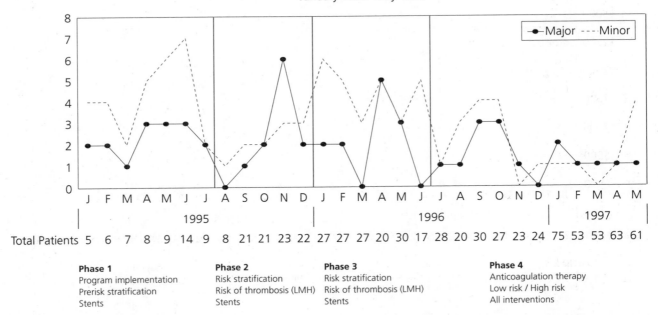

Figure 15–3 Coronary Stents/Cardiac Interventions: Complications. *Source:* Copyright 1996, Lutheran General Hospital-Advocate, Park Ridge, Illinois—used with permission.

Lutheran General Hospital-Advocate
Coronary Stents/Cardiac Interventions
All Bleeding Complications
January 1995–May 1997

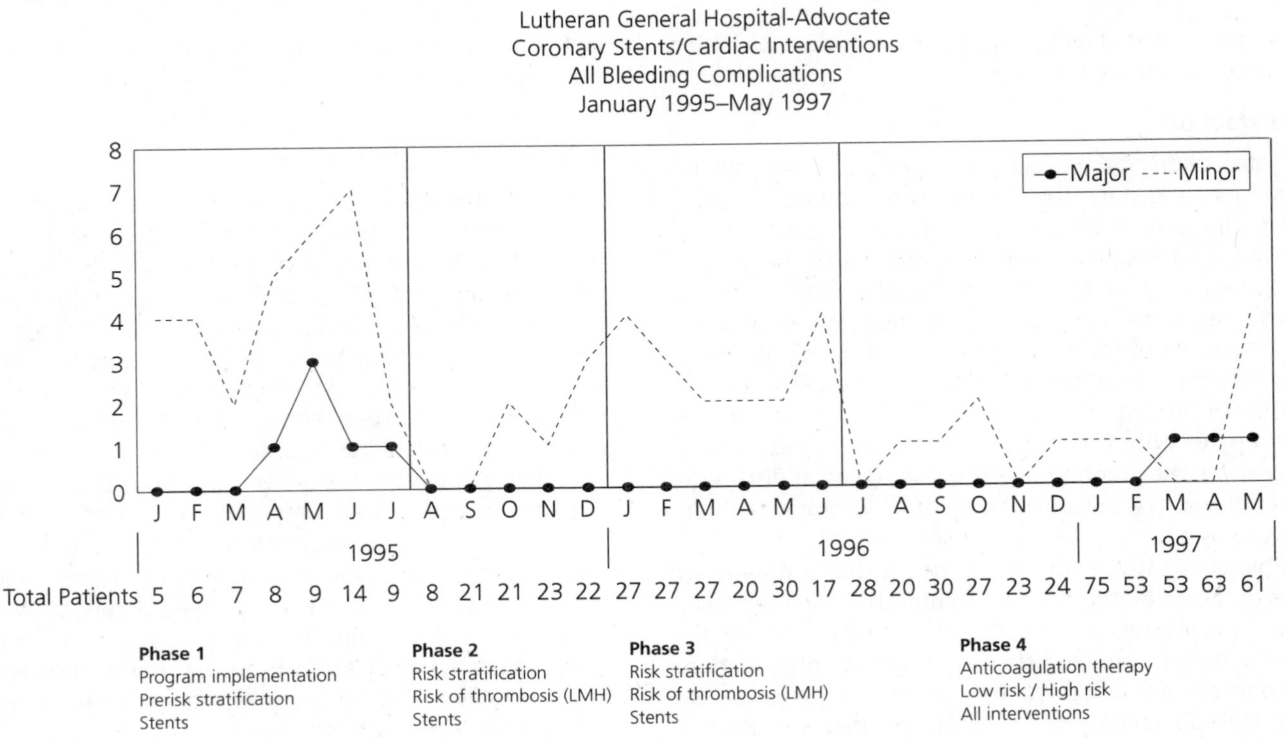

Figure 15–4 Coronary Stents/Cardiac Interventions: Bleeding Complications. *Source:* Copyright 1996, Lutheran General Hospital-Advocate, Park Ridge, Illinois—used with permission.

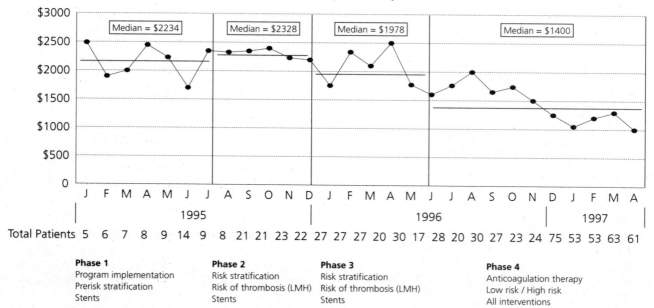

Figure 15–5 Coronary Stents/Cardiac Interventions Average Costs per Case (excluding CCL). *Source:* Copyright 1996, Lutheran General Hospital-Advocate, Park Ridge, Illinois—used with permission.

changed and is directly related to individual physician practice during the procedure.

Satisfaction

Satisfaction of the process is important to the patient, physician, and staff. The staff consists of the CCL, CICU, telemetry, cardiac rehabilitation, and nutrition.

Patient satisfaction is measured by the LGH-Advocate hospital survey, unit survey, and by telephone call by the cardiology care manager. These include very general questions about the hospital experience to specific questions about education provided and the cardiac interventional procedure. It was identified, through the follow-up telephone call, that the patients and families appreciate the preprinted instructions given at discharge. The surveys are tallied quarterly and information is reviewed by each specific area of service.

Physician satisfaction, while not formally measured, was obtained during the first 18 months of the program. The medical director of the CCL discussed issues with the interventional cardiologists, soliciting recommendations for pathway content revisions. These were then shared at the multidisciplinary team meetings. Issues related to

staff satisfaction are discussed at staff meetings and reported to the multidisciplinary team.

VARIANCE TRACKING

The decision to measure specific outcomes is determined by national trends, reviews of the literature, patient mix, market incentives, and a host of other factors. In order to be meaningful and generate change, however, outcome measurements must be analyzed. Outcome analysis requires the ability to examine, investigate, and explain measured outcome results that occur over time or between groups. It also provides the basis for comparison against established standards, criteria, or norms.

Variance is defined as a deviation from the norm. In the case of clinical pathways, variance describes deviation from an expected activity or predicted outcome and may be interpreted as positive or negative. When a more optimal outcome than was anticipated occurs, then the variance is said to be positive; eg, a reduction in average LOS. Negative variance, on the other hand, suggests a less favorable result; eg, noncompliance, an increase in LOS, complications, or cost.[6]

Tracking variances is a critical component of outcome analysis. During the developmental stages of the care management program, pathway compliance together with clinical outcomes were a major focus. In conjunction with the chart review, digression from the prescribed pathway components (clinical practice) was examined on a case-by-case basis. For example, development and management of bleeding complications were evaluated against predetermined criteria. No trends were identified. Today, interventionalists whose clinical practice digresses from the pathway norm document patient specifics and the clinical rationale for deviation in the progress notes. Similarly, nursing staff, because of the clinical situation or response to therapy, may also find it necessary to deviate from prescribed activities. When this occurs, documentation is entered in the nursing notes and communication made with the appropriate cardiologist or clinical discipline. Complications continue to be a measurable outcome.

Once pathway compliance was established, 100% record review was no longer a useful tool. At present, outcome measurement data displayed in run charts is one of the principal mechanisms for variance tracking and outcome analysis. Variations from the norm or mean are investigated so that they may be explained by either an isolated event or pattern (special cause) versus a developing upward or downward trend (common cause).[7] Special cause variation may or may not require independent intervention, but rarely supports changing a process, ie, pathway modification. Conversely, investigation of common cause variation (developing trend) mandates investigation of pathway activities/process. These explanations then provide a source of information that may be used to support future pathway revisions.

The following demonstrate the impact and benefits of variance tracking and analysis that resulted from the care management program:

- Variance tracking and analysis revealed that patients receiving ReoPro (abciximab) as part of their anticoagulation regimen were more likely to experience bleeding complications and/or a longer LOS. This was a pattern, not a trend. Pathway components were not altered. A heightened awareness for complication risk and the judicious use of ReoPro infusion resulted.
- Variance tracking supported removal of the femoral sheath by nursing. Bleeding complications and LOS continued to decline when sheath removal was extended to include telemetry nurses as well as CICU nurses.
- Variance tracking resulted in sequential revisions of the interventional pathway from an original 5-day

hospital course to the present three-phase format. These changes were supported by developing trends (decreased LOS, which paralleled changes in anticoagulation protocols).

The establishment of desirable outcomes or standards is a complex process that requires the examination of current practices as well as the identification of practice norms in like patient populations and clinical environments. Analysis of the Cardiac Interventional Pathway outcomes is dominated by the philosophy of continued quality improvement. Desirable outcomes are not the minimum standards of care but rather the goals for continued improvement and identification of best practice, always acknowledging that clinical judgment takes precedence to pathway compliance.

CONCLUSION

In today's changing health care environment, the challenge to provide quality care that is cost-effective for health care institutions becomes the responsibility of those providing care. In cardiology, rapidly developing high technology coupled with varied payment plans provided the impetus to develop a care management program for the cardiac interventional population.

Utilizing a quality improvement process, Lutheran General Hospital-Advocate developed and implemented a care management program that consisted of a clinical pathway and physician orders. The cardiac interventional population whose LOS and clinical outcomes can be defined is appropriate for the care management delivery model.

Our experience demonstrated maximum utilization of resources, standardization of care for a high volume population, and identification of areas for improvement of quality of care and patient outcomes. As the data demonstrated strong evidence of positive clinical, financial, and satisfaction outcomes, the multidisciplinary team continues to evaluate opportunities to improve the care management program.

Additional outcomes to measure in the future focus on patient education, both preprocedure and postprocedure. Efforts to demonstrate a quantitative reduction in anxiety prior to procedure and assess knowledge base at discharge would contribute to patient satisfaction outcomes. In addition, a patient-focused pathway would also augment education and communication between the health care team and the population.

The care management program for the cardiac intervention population was developed through collaborative efforts of physicians, nurses, and other health care providers to improve the value and quality of patient care.

Our program has demonstrated reduced variation in care, LOS, and cost while achieving positive outcome.

REFERENCES

1. Guiliano K, Poirier C. Nursing care management: critical pathways to desirable outcomes. *Nurs Manage.* 1991;22(3):52–55.
2. Mueller R, Sanborn T. The history of interventional cardiology: cardiac catheterizations, angioplasty and related interventions. *Am Heart J.* 1995;129:146–172.
3. Fajadel J. New coronary stent management. *J Invas Cardiol.* 1995;7:30A–31A.
4. Morice M. Advances in post stenting medication protocol. *J Invas Cardiol.* 1995;7(suppl A):32A–35A.
5. Faulds D, Sorkin E. Abciximab (c7E3 Fab) a review of its pharmacology and therapeutic potential in ischemic heart disease. *Drugs.* 1995;48:583–598.
6. Zander K. Quantifying, managing and improving quality. *New Definition.* 1992;7(4):1–4.
7. Crummer R, Carter V. Critical pathways—the pivotal tool. *J Cardiovasc Nurs.* 1993;7(4):30–37.

Appendix 15–A

Clinical Pathway

Exhibit 15–A–1 Cardiac Interventional Procedure (Stent, PTCA, DCA, Rotoblator, TEC) Clinical Pathway

	Admission Day/Preprocedure	Procedure/Day 1	Day 2	Discharge Day
Date				
Components				
Outcomes	1. **Patient education for procedure** 2. **Patient preparation for procedure**	1. **Absence of complications** 2. **Goals identified for discharge**	1. **Absence of complications with progressive activity** 2. **Participation in teaching for medication, nutrition activity program.**	1. Pt hemodynamically stable and tolerates activity program. 2. Discharge education completed: **Date/Initials** ❑ Activity program _____ ❑ Medication regimen _____ ❑ Dietary restrictions _____ ❑ Procedure site care _____ ❑ Reoccurrence of chest pain _____
Assessment/ Monitoring	❑ VS—unit routine ❑ Pedal pulses ❑ H&P completed ❑ Labs/tests reviewed	❑ Continuous ECG ❑ Unit routine post procedure ❑ Pedal pulses ❑ Groin site ❑ Chest pain	❑ DC continuous tele ❑ VS per unit routine ❑ Pedal pulses/groin site per unit routine ❑ Chest pain	
Tests	❑ ECG ❑ CBC ❑ Chem 7	❑ Cardiac enzymes _____ ❑ ECG (on admission/chest pain) ❑ ACT as ordered (sheath removal) ❑ Additional tests as ordered	❑ ECG in a.m. (if ordered)	3. Follow-up plans completed: ❑ Physician _____ ❑ Cardiac rehab _____ ❑ Community resources _____
Functional Rehab	❑ Preprocedure status	❑ HOB ≤ 30° ❑ Strict bedrest until 6° after sheaths dc'd ❑ Commode 6° after sheaths dc'd ❑ Maintain bedrest for 24 hrs if IV heparin restarted (may be up to commode) ❑ Gradual increases in activity (CR protocol if no groin complication)	❑ Unit ambulation prior to dc	
Nutrition	❑ Preprocedure status ❑ NPO for procedure (May take meds)	❑ Push fluids ❑ Finger food diet until sheaths dc'd ❑ Cardiac diet	❑ Cardiac diet	

continues

Exhibit 15–A–1 continued

	Admission Day/Preprocedure	Procedure/Day 1	Day 2	Discharge Day
Date				
Components				
Medications	Preprocedure (Stents) ❑ ASA 325 mg po ❑ Ticlid 250 mg po q 12 ❑ Premedicate per orders if allergic to contrast medium. (PTCA, stents)	*Anticoagulation Protocol* **Standard** ❑ ASA 325 mg po qd ❑ Ticlid 250 mg q 12 po **Other** ❑ Coumadin ❑ Heparin ❑ ReoPro Cardiac Medications	❑ Continue anticoagulation therapy as ordered ❑ Monitor response to drug therapy.	
Treatment	❑ Foley prn	*Sheath removal* ❑ Remove sheaths 4 hrs after heparin dc'd admission ❑ Follow sheath removal protocol as ordered ❑ Reapply pressure for bleeding/ notify MD ❑ Straight cath x 1 if unable to void ❑ DC Foley when OOB/up to commode ❑ If chest pain occurs, notify MD and obtain stat 12 Lead ECG	❑ DC groin dressing ❑ Cardiac rehab: Phase I Evaluation	
Teaching	❑ Procedural video ❑ Procedural booklet ❑ Procedural pathway ❑ Sign consents	❑ Sheath removal/compression procedure ❑ Bleeding potential ❑ Anticoagulation therapy ❑ Activity progression ❑ Chest pain/pain scale (0–5) ❑ Assess educational needs for risk factors/MI	❑ Nutrition ❑ Cardiac medication ❑ Site assessment ❑ Recovery from MI information (Definite MI)	
Spiritual	❑ Adv directive: Y N ❑ In chart: Y N ❑ Emotional support to Pt/SO ❑ Pastoral care prn			

Psychosocial/ Discharge Planning		❑ Evaluate for discharge needs:
		    Social service
		    Financial needs
		    Home care
			❑ DC plans in place	

Initials	AM _____ PM _____ NOC _____	AM _____ PM _____ NOC _____	AM _____ PM _____ NOC _____	

Source: Copyright 1996, Lutheran General Hospital-Advocate, Park Ridge, Illinois—used with permission.

Improving Outcomes for Coronary Artery Bypass Patients: A Path-Based Initiative

Kelli King Sagehorn, Cynthia L. Russell, and Suzanne M. Burton

The institutional initiative for path-based clinical care at the University of Missouri-Columbia Hospitals and Clinics began in 1991 when the chief nurse executive formed an interdisciplinary team to conceptualize a potential change in the delivery of patient care. The team identified goals based upon Deming's continuous quality improvement philosophy, which involves focusing on consumers, the patients, and families.[1] This initiative provided the philosophy from which path-based clinical care evolved within the institution.

PATIENT SELECTION FOR PATH-BASED CARE

Clinical path development was prioritized by targeting high cost, high volume patient populations by diagnosis-related groups (DRGs). An interdisciplinary approach to patient care was fostered by forming teams through which care was analyzed. The advanced practice nurse (APN) was the designated leader for coordination of care and case management due to expertise with specialty patient populations. Clinical support staff were available to the interdisciplinary team to promote a consistent, quality path product.

The paths initially focused on patients requiring acute care, although work continues to evolve along the health care continuum. Surgical care was targeted because of its predictability. Selecting the coronary artery bypass graft (CABG) population led us to an early success because of a well-established standard of care and enthusiasm from an experienced team. National data for trends and benchmarking were readily available to validate our efforts.

The current system of standardized patient care was complemented by the addition of clinical paths. The paths enabled a more practical approach for organization and utilization of standards at the bedside. Develop-ment of best practices was facilitated through ongoing evaluation of patient outcomes identified on each path.

PATHWAY DEVELOPMENT

Promotion of a consistent method for pathway development was an institutional priority as the paths were to be used by interdisciplinary team members in many different units. After reviewing several examples of pathway formats, a template was selected and modified for use. The template included the patient-need categories of assessment, protocols, tests and labs, consults, nutrition, medications, activity, spiritual, teaching, and discharge planning. Interventions were then organized according to time and sequence. Guidelines were created to specifically delineate the steps of the clinical path process. Exhibit 16–1 depicts the key components of the guidelines.

The interdisciplinary team was established by formalizing the relationships of the care providers for this group of patients. The team included representatives from nursing, cardiac rehabilitation, clinical nutrition, pastoral care, pathology, laboratory, pharmacy, physical therapy, medicine, radiology, anesthesia, respiratory therapy, and social services. Additional members were identified to further promote care along the continuum, such as home health services and ambulatory care staff. Productivity was enhanced by secretarial support and regularly scheduled team meetings.

It was quickly discovered that analysis of data and active participation by a physician champion were essential to the success of the pathway project. With the CABG group, the presence of a physician champion promoted participation by the medical staff and assisted in an early consensus for medical care. In addition, patient care

Exhibit 16–1 Components of Clinical Path Guidelines

> 1. Definition of clinical path
> 2. Institutional goals for clinical path project
> 3. Definition of path changes
> 4. Clinical path format
> 5. Suggested steps for developing a clinical path

charges and their frequencies were analyzed in cooperation with the physician group, which allowed systematic reduction and/or elimination of unnecessary services. Institutional process and outcome data were used for regional and national comparisons. Participation in a national clinical benchmarking project assisted in decision making for best practices. All data sources were selected because of their ongoing accessibility, familiarity, and value to team members.

To capture the current sequence of care, a flowcharting exercise was completed. This enabled the group to validate the patient and family experience of undergoing a coronary artery bypass graft procedure. A draft of the clinical path was created by combining components of the present standard of care with ideal process changes. For example, earlier activity progression and 2 fewer hospital days were proposed.

Interdisciplinary standards of care from ancillary services such as respiratory therapy, social services, and cardiac rehabilitation were integrated into the path. In Appendix 16–A, Exhibit 16–A–1 shows the Uncomplicated CABG clinical path. Standardized physician orders were revised and nursing protocols were incorporated. This provided an opportunity to initiate interdisciplinary protocol development, such as a ventilator management weaning and extubation protocol created by respiratory therapists and nursing staff.

PATHWAY OPERATIONALIZATION

The person responsible for developing and monitoring the clinical paths is the coordinator of care (COC). The coordinator may be any professional with appropriate expertise with the identified clinical population. The COC should be

- empowered to address path changes
- accountable to team members, patient, and family
- knowledgeable about the entire episode of care, including patient goals, from admission to discharge
- able to access resources and support systems across all disciplines

Currently, the majority of COCs are advanced practice nurses (clinical nurse specialists or nurse practitioners) practicing in acute and ambulatory care settings.

A three-tiered educational strategy was utilized in preparing staff for implementation of clinical paths. First, 45-minute educational sessions were provided for staff members across all shifts. Exhibit 16–2 shows the standardized outline used for interdisciplinary staff education. Next, summary posters were prepared for display in patient care areas. Finally, individual mentoring of staff was provided on an ongoing basis by the COC.

The clinical path process is initiated by a physician's order. The COC is responsible for ensuring that all patients meeting criteria for clinical path implementation have a path initiated and placed in the medical record at the time of admission. The COC or interdisciplinary team members initial interventions as they are completed. Signature lines are available at the end of the clinical path. The COC is responsible for ensuring that follow-up, documentation, and clinical path interventions occur as recommended.

The COC or staff nurse designee may revise the clinical path. If a patient deviates from the expected clinical path, a path change occurs. A path change is an unanticipated variance from the interventions outlined on the clinical path. The institution monitors three types of path changes including patient, system, and staff. The COC addresses changes from the path and accesses resources and support systems across all disciplines to resolve the path change. Path changes and rationale are documented in the physician or multidisciplinary progress notes by the COC. For example, if the patient's chest tube is removed early, an entry in the medical record may appear as "the mediastinal chest tubes were removed due to minimal drainage; early progress." This information is also compiled for ongoing tracking of patient outcomes.

Individual patient care areas determined where the clinical path would be located for daily use. Consistency

Exhibit 16–2 Standard Outline for Clinical Path Education

> 1. Introduction
>   National trends in DRG and patient population
>   Changing health care environment
> 2. Impact of managed care
> 3. Definition of clinical path
> 4. Institutional clinical path goals
> 5. Clinical path use
>   Coordination
>   Location and documentation
> 6. Path changes
> 7. Financial data
> 8. Clinical path example
> 9. Interdisciplinary team members
> 10. Implementation plan
> 11. Outcome evaluation plan

in location was promoted in the intensive care units (ICU) and general care areas to facilitate ease of use by interdisciplinary team members. Most clinical paths are kept close to the bedside with the computerized medication administration record.

A component of the institution's case management effort included a role redesign for designated staff nurses, called the patient care coordinator (PCC). This new role involved focused care coordination in an empowered professional environment. The primary responsibilities of the PCC included overseeing the nursing process, delegating nursing interventions, collaborating with the COC and the interdisciplinary team, and serving as the primary communication link for patients and their families. The clinical path is the interdisciplinary documentation record used in this case management initiative. In addition, physician and nurse reporting is often guided by the clinical path.

Implementation of the CABG clinical path began with a 3-month trial period with a goal of at least 12 managed patients. Path changes were closely monitored by the COC during this time. Identified patient path changes assisted us in quantifying physiologic complications of the CABG population. Educational needs were identified through analysis of staff-related path changes. Most importantly, long-standing departmental issues were resolved by documenting system path changes.

Clinical path revisions were instituted by the interdisciplinary team after the 3-month trial period. Changes involved minor adjustments in the sequence of interventions. Additional path revisions are frequently made in an ongoing manner as indicated by the data. The CABG clinical path is reviewed annually by the team to further identify process improvements.

PURPOSE OF OUTCOMES MANAGEMENT

Outcomes management provides support to our institution's path-based initiative by facilitating achievement of expected goals within a fiscally responsible time frame and promoting consistent, high quality patient care.

In the analysis of the process of care for CABG patients prior to the implementation of clinical pathways, the institution discovered a greater degree of practice variability than anticipated. Not only were there differences in physicians' practices, but there were also variations in how staff members interpreted and implemented standards of care. The clinical path facilitates consensus of best practices among all team members, now promoting consistency. The daily monitoring of clinical path data allows the team to critically examine the process of care and assists in making ongoing decisions based on empirical data.

This management process allows for incorporation of future technological advances and practice improvements into the care of CABG patients. These practice changes can be more efficiently evaluated for their impact on selected patient outcomes due to the presence of the program.

OUTCOMES PRIOR TO INITIATION OF THE PROGRAM

The majority of practice changes for the CABG patient group were made concurrently with data analysis and path development. An increased awareness of financial issues and enthusiasm for process improvement led team members to implement practice changes immediately. An initial financial analysis was planned to capture the impact of the clinical path project in the early stages of development. Specifically, length of stay, financial outcomes, and complication rates were monitored prior to initiation of the clinical path and annually thereafter.

MEASUREMENT

Donabedian's model of structure, process, and outcome facilitated our approach to understanding integration of the processes, which were monitored through path changes, and patient outcomes, which were monitored through outcomes tracking.[2] Three types of path changes are followed in the system. The first is a patient-related path change in which there is a physiological, psychological, or cognitive complication. For example, failure to wean from the ventilator. The second is a staff-related path change, which occurs when nurses, physicians, or other health care team members create a situation in which patients vary from the path. For example, failure to carry out orders. The third is a system-related change, which refers to a problem within the health care setting. For example, when there is a delay in diagnostic testing.

Outcomes tracking is a mechanism to pool clinically significant changes from the path for the entire population for careful scrutiny and analysis. The institution defined clinically significant changes as those that interfere with the patient's ability to achieve discharge goals or those that impact length of stay or financial outcomes.

Ideally, the COC documents path changes in the medical record for each individual patient. All data from the patient group are summarized and reported by the COC on a quarterly basis to the interdisciplinary team, service line administration, and quality improvement and clinical outcomes committees. Data are also assimilated across patient groups by combining quarterly reports. The information is used by the team to make needed changes to the path and by administration and quality improvement councils to address broader staff and system issues. Needed corrective actions are taken followed by systematic remonitoring.

The model for outcomes measurement involves annual reporting for DRG 106, Coronary Bypass with Cardiac Catheterization, and DRG 107, Coronary Bypass without Cardiac Catheterization. A financial database is utilized to analyze number of cases, length of stay, ancillary charges, cost, reimbursement, and estimated margin. This systematic review is initiated by the COC with assistance from administration and support staff.

Since the beginning of the program, average length of stay has decreased a total of 2.27 days while number of cases remained stable. Figure 16–1 demonstrates the decrease in length of stay over a 3-year period for DRG 107 by medical-surgical and ICU days. Initial investigation by the interdisciplinary team revealed that hospital costs and patient charges were distinct financial issues. Efforts to influence hospital costs involved system efficiencies while patient charges were aggressively managed through decreased utilization.

Table 16–1 presents financial trends for DRG 107. Average ancillary charges per case are depicted over a 3-year time period. In addition, although room rates tended to increase on a yearly basis, a decline in total room charges was noted related to shorter lengths of stay. Average cost and estimated margin are also illustrated. Reimbursement levels were followed during the same time period with little fluctuation.

SATISFACTION

Physician satisfaction with the project has been strongly positive. Even though a physician champion was identified, the other three physicians in the department remained involved and interested over time. This positive response is attributed to continued use of data-based reporting and availability of standardized physicians' orders.

Ability to anticipate day of discharge, decreased need to contact physicians, and less duplication of services are the issues that have contributed to overall acceptance and satisfaction of the staff. Compliance with documentation of interventions has presented a challenge; however, our staff will continue to struggle with completion of documentation until clinical paths are more widely used throughout the institution.

In addition, availability of clinical outcomes data has been beneficial to a number of interdisciplinary groups. For example, respiratory care practitioners (RCP) examined information regarding early extubation to support a redesign effort within their department. Physician assistants have also improved daily documentation as a result of receiving outcomes data.

ACCEPTABILITY OF OUTCOMES

Our process for outcomes monitoring is based upon a continuous quality improvement (CQI) philosophy. Al-

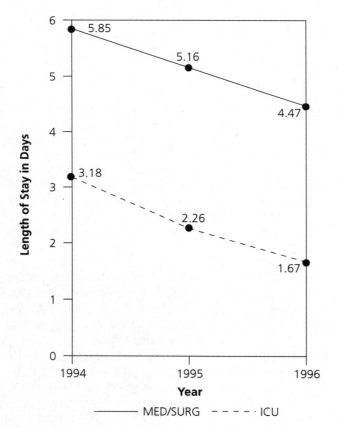

Figure 16–1 DRG 107: Coronary Bypass Graft without Cardiac Catheterization

Table 16–1 Financial Outcomes for DRG 107: Coronary Bypass without Cardiac Catheterization in Dollars

	Fiscal Year		
	1994	1995	1996
Average charges	39,211	34,781	31,720
Ancillary	32,371	28,707	26,718
Room	6840	6074	5002
Average hospital cost	20,664	18,300	16,375
Average reimbursement	24,432	25,156	24,902
Average estimated margin	3769	6856	8527

though thresholds of acceptability have not been formally established, the team monitors continued progress toward identified outcomes while preserving desired patient care quality. Any identified negative trends are carefully scrutinized by appropriate team members for opportunities to improve care. Participation in a national clinical benchmarking project has allowed valuable comparison of database elements with similar institutions.

RESULTS OF OUTCOMES MONITORING

As a result of clinical path initiation, outcomes measurement formally began in April of 1995. Prior to that time, outcomes monitoring was less systematic, with a focus on structure and process through quality assurance activities. Additional benefits that have been realized since beginning the project include continued decreased utilization; increased financial awareness for physicians, team members, and staff; and greater appreciation for interdisciplinary activities. Clinical path development for this patient population has also served as a model for broad practice changes across all settings.

EVALUATION OF A PATIENT-FAMILY PATHWAY

In an effort to assist patients and families to anticipate care and shorter lengths of stay, the interdisciplinary team considered utilizing a patient-family version of the clinical path. However, there was an absence of empirical evidence in the literature to support this strategy. A research project was undertaken by Sagehorn et al (unpublished data, 1997) to discover if implementation of a patient-family pathway with patients and families undergoing CABG surgery impacted their anxiety, information with care planning, and length of stay. Using an experimental design, a sample of 60 patients and 60 family members were included who met the following criteria: elective, isolated, first-time CABG surgery, significant other present, and CABG clinical path implemented. The patient and a designated family member received either the patient-family pathway, Exhibit 16–A–2 in the Appendix, or our standard method of care planning. The Patient-Family Information Questionnaire was developed by the authors to measure degree of information perceived by the patient and family about care planning. Anxiety was measured by the State-Trait Anxiety Inventory test.[3] Both instruments were administered within 48 hours of patient discharge from the hospital by trained staff nurse researchers. Findings indicated no statistically significant difference in state ($t = .6280$) or trait ($t = .3370$) anxiety between those patients receiving the patient-family pathway and those receiving our standard

method of care planning. There was also no difference in state ($t = .3107$) or trait ($t = .4886$) anxiety for both family groups. Results also showed no difference on the Patient-Family Information Questionnaire between patients receiving the patient-family pathway and those receiving our standard care planning ($t = .8399$). Family members' results from the Patient-Family Information Questionnaire indicated no statistically significant difference between the two groups ($t = .6888$). Results indicated all groups generally reported being "very informed." Finally, there was no statistically significant difference between length of stay between the two patient groups ($t = .2710$).

The results of this 18-month study demonstrated to the interdisciplinary team that the CABG patient-family pathway had limited value to patients and families. The team felt that resources should be reallocated to other uses that could have a more positive impact on the patient and family experience. Therefore, the patient-family pathway is not currently used with the CABG population, although the format is being tested for use in other patient groups. In addition, the CABG clinical pathway is also not reviewed by the patient or family members because the belief is that a well-defined standard of care can be easily described by nurses providing care at the bedside.

FUTURE WORK

After standards are defined and agreed upon, the natural progression is to begin automation to facilitate standards access, documentation, and data analysis. This work complements the 3- to 5-year institutional plan for development and implementation of an electronic medical record and hospital database system. To date, the clinical paths and associated standards are not automated.

Future plans include piloting a CQI model organized by service line. Coronary artery bypass surgery data would be reviewed by the appropriate surgery service line with clinical problems referred back to the interdisciplinary team for resolution. The surgery service line would forward this data to the Clinical Outcomes Committee for integration into a hospitalwide clinical outcomes summary. This model would eliminate multiple, fragmented CQI committee efforts while promoting greater sophistication in the area of clinical outcomes measurement.

Clinical directions for the future will involve a focus on quality issues from the patients' perspectives. Specifically, the plan includes addressing patient and family satisfaction and utilizing valid and reliable measurement tools to determine functional status over time.

CONCLUSION

This chapter has described the structure, process, and outcomes used in the institution's path-based collaborative practice. Due to realized clinical improvements, hospital administration, physicians, and staff continue to support the clinical outcomes initiative. Finally, the institution has advanced to a program of decision making based on clinical data analysis that allows for resolution of long-standing system issues that directly influence patient results.

REFERENCES

1. Gillem TR. Deming's 14 points and hospital quality: responding to the consumer's demand for the best value health care. *J Nurs Qual Assur.* 1988;2(3):70–78.

2. Donabedian A. Measuring the effectiveness of medical interventions: new expectations of health services research. *Health Serv Res.* 1966;25:697–708.

3. Speilberger CD, McDonald RJ. Measuring anxiety in hospitalized geriatric patients. *Series in Clinical and Community Psychological Stress and Anxiety.* 1993;2:135–143.

■ Appendix 16–A ■
Clinical and Patient/Family Pathways

Exhibit 16–A–1 Coronary Artery Bypass Graft (Uncomplicated) Clinical Path

Anticipated Length of Stay:　7 days
Coordinator of Care:　Cardiothoracic Clinical Nurse Specialist—Beeper #1337
Patient Discharge Outcomes:
- manageable pain/anxiety
- tolerating diet
- daily progressive activity
- no physiologic complications
- self-care activities performed at manageable level
- able to verbalize discharge teaching instructions and medications

Patient Identification
This pathway is intended to be used as a guideline.
Various circumstances may necessitate a deviation from the pathway.

PATIENT NEEDS	DAY 1 Preop (Inpatient/Outpatient) Date	DAY 2 D.O.S. Date	DAY 3 P.O.D. #1 Date	DAY 4 P.O.D. #2 Date
ASSESSMENT	• Nursing admission • History and physical • Anesthesia	• Nursing assessment in OR		
PROTOCOLS	• Initiate CABG Standard of Care	• Continue CABG Standard of Care • Wean ventilator as tolerated	• Continue CABG Standard of Care • Assess for transfer to stepdown	• Continue CABG Standard of Care • Transfer to 5W monitored bed

TESTS/LABS	• Type and cross 2 units
	• HPD
	• Chem 7
	• EKG
	• CXR
	• PT/PTT
		• Routine postop labs including:
		  –Chem 7
		  –HPD
		• ABG
		• CXR (portable)
		• K+ per protocol
			• Chem 7
			• HPD
			• CXR (portable)
			• K+ per protocol	

CONSULTS	• Social Services • Clinical Nutrition • Diabetes CNS if diabetic patient on insulin or oral agent			
DIET	• Regular or ADA	• NPO after midnight	• Advance as tolerated to regular (or ADA) 4 gm sodium	• Advance as tolerated to regular (or ADA) 4 gm sodium
MEDICATIONS	• Home medications • PRN hs sedative	• Kefzol 1 gm/2 gm IVPB on chart to OR	• ASA 325 mg daily	• ASA 325 mg daily
ACTIVITY	• As tolerated	• Bedrest	• UOOB-chair after extubation	• Activity progression
TREATMENTS	• Hibiclens shower or bath		• Extubate • Oxygen therapy—wean as tolerated	• Remove chest tubes (wait till patient has been out of bed x1)
SPIRITUAL	• Assessment of patient/family including: –Pastoral Care	• Pastoral Care follow-up as needed	• Pastoral Care follow-up as needed	• Pastoral Care follow-up as needed
TEACHING	• Preop instruction		• Begin discharge teaching if transferred to step-down	• Continue discharge teaching
DISCHARGE PLANNING	• Assess for special discharge needs and make consults according to needs			

continues

Exhibit 16–A–1 continued

Anticipated Length of Stay: 7 days
Coordinator of Care: Cardiothoracic Clinical Nurse Specialist—Beeper #1337
Patient Discharge Outcomes:
- manageable pain/anxiety
- tolerating diet
- daily progressive activity
- no physiologic complications
- self-care activities performed at manageable level
- able to verbalize discharge teaching instructions and medications

Patient Identification

PATIENT NEEDS	DAY 5 P.O.D. #3 Date ____	DAY 6 P.O.D. #4 Date ____	DAY 7 P.O.D. #5 Date ____
ASSESSMENT			
PROTOCOLS	• Continue CABG Standard of Care	• Continue CABG Standard of Care	• Continue CABG Standard of Care • Remove chest and leg staples AM
TESTS/LABS	• Chem 7 • HPD • Chest x-ray (PA and LAT) (after pacer wires removed) • EKG		
CONSULTS			
DIET	• Advance as tolerated to regular (or ADA) 4 gm sodium	• Advance as tolerated to regular (or ADA) 4 gm sodium	• Advance as tolerated to regular (or ADA) 4 gm sodium
MEDICATIONS	• ASA 325 mg daily	• ASA 325 mg daily	• ASA 325 mg daily
ACTIVITY	• Activity progression • Bedrest 20 minutes post pacer wire removal	• Activity progression	• Activity progression
TREATMENTS	• Remove pacer wires		• Remove chest tube sutures
SPIRITUAL	• Pastoral Care follow-up as needed	• Pastoral Care follow-up as needed	• Pastoral Care follow-up as needed
TEACHING	• Continue discharge teaching	• Continue discharge teaching • Medication teaching • Dietary instruction by dietitian • Discharge instruction related to activity progression	• Continue discharge teaching • Medication teaching with schedule • Review of summary sheet by nursing staff
DISCHARGE PLANNING	• Assess for need for visiting nurses	• Arrange for visiting nurses as needed	
SIGNATURE BLOCK:			
	/	/	
	/	/	
	/	/	
	/	/	
	/	/	

Source: Copyright © The Children's Hospital.

Exhibit 16–A–2 Coronary Artery Bypass Graft Patient/Family Path

Your Name: _____

Physician's Name: _____

This plan is a guide for your care. The care that you receive may vary from the path because of your unique needs. If you have any questions about this path or your progress, please talk to your doctor or nurse.

FIRST DAY (Day before Surgery)

This is the day before your surgery. It will be a busy day. Please ask any questions that you have along the way.

ASSESSMENT:

- You will see doctors, nurses, and others from your heart surgery team. They will be asking you questions about your health history. They may examine you and take your vital signs.

TESTS/LABS:

- You will have the following tests: EKG, Chest X-ray, Blood Work, and Urine Sample.

NUTRITION:

- You will be able to eat the foods that you normally eat until midnight. *After midnight* you will be asked not to eat or drink anything. The nurse may give you some of your medicines *or* the nurse may ask you to take your medicines at home before coming to the hospital.

MEDICATIONS:

- You will be taking your regular medication up to the time for your surgery. You will be asked to stop taking all blood thinning medicines (Coumadin and aspirin) before surgery. During your hospital stay all your medications will be evaluated. A sleeping pill will be prescribed for you if needed the evening before surgery.

ACTIVITY:

- You may walk and move about as you normally do.

TREATMENTS:

- You will be asked to take a shower with special antiseptic soap the evening before surgery.

SPIRITUAL:

- A hospital chaplain will be available to you for a preoperative visit and continued spiritual support if you would like.

TEACHING:

- Your surgical procedure and its risks will be explained to you. You will be asked to sign some consent forms.
- Your nurse will teach you about your surgery and what you and your family can expect during and after. You will be taught how to use a deep breathing device (incentive spirometer) for use after your surgery.

DISCHARGE PLANNING:

- Your family is welcome to visit during visiting hours and to assist in your care as you desire.

Please let your nurse know if there is anything about you, your family, or your home situation that we should know in order to plan for your return home.

SECOND DAY (Day of Surgery)

Your operation will occur today at its scheduled time. Your family will carry a beeper during surgery. They will receive messages about you and will be waiting in the main lobby area.

MEDICATIONS:

- Your nurse will give you medicine to help you relax before surgery.

NUTRITION:

- You will not be able to eat or drink anything today. You will be sleeping most of the time.

TREATMENTS:

- During your surgery and following, you will be assisted with your breathing by a tube in your throat. You will not be able to talk until the tube is removed.
- Your blood pressure, temperature, breathing, and pulse will be checked frequently. Your surgery site dressing will also be checked. You will be receiving pain medication through your IV.

ASSESSMENT:

- When you wake up from surgery you will be in the Thoracic Intensive Care Unit (TICU). Your doctor and heart surgery team members will continue to visit you following surgery.

ACTIVITY:

- You will be sleeping most of the day. The nurses will help you turn from side to side.

THIRD DAY (Move to 5 West)

TREATMENTS:

- The breathing tube will be removed. The nurses will encourage you to use the incentive spirometer (IS) at least 10 times every hour.
- You will be moved out of the TICU to a room on 5 West.

TESTS/LABS:

- Someone will draw your blood and you will have a chest x-ray.

ASSESSMENT:

- Your doctor and nurses will be in to check on you.

NUTRITION:

- You will have clear liquids such as broth or juice. Your doctor will advance your diet to low cholesterol and low saturated fat over time.

MEDICATIONS:

- You will be asked frequently about your pain and given medications as you need them. The medications will be given as ordered by your doctor.

continues

Exhibit 16–A–2 continued

ACTIVITY:

- You will be assisted to sit up in the chair.

TEACHING:

- Your nurses will begin to teach you how to care for yourself at home.

FOURTH DAY (Removal of Dressings/Chest Tubes)

TREATMENTS:

- Your dressings over your incisions will be removed.
- The chest tubes will be removed.
- Continue to use your incentive spirometer (IS) 10 times every 2 hours.

MEDICATIONS:

- Your medications will continue as ordered by your doctor.

ACTIVITY:

- A cardiac rehab nurse will visit you and begin to talk about your walking and exercise plan.
- Your nurses will assist you to walk daily and increase your activity as you tolerate it.

TEACHING:

- Your nurses will continue to teach you about your home care.

FIFTH DAY (Removal of Pacer Wires)

MEDICATIONS:

- If you have not had a bowel movement since your hospitalization, you may be given a laxative.

ACTIVITY:

- You will continue walking in the hallway with assistance.

TREATMENTS:

- Continue to use your incentive spirometer (IS) 10 times every 2 hours.
- Pacer wires will be removed today.

TEACHING:

- Your nurse will continue to teach you about caring for yourself at home and will give you information about your medication.

Source: Copyright © The Children's Hospital.

SIXTH DAY (Day Before Going Home)

ACTIVITY:

- Your activity will increase to walking in the hall 3 times today or as directed by your nurse.
- A cardiac rehab nurse will talk with you about your home walking and exercise plan.

TREATMENTS:

- Continue to use your incentive spirometer (IS) 10 times every 2 hours.

TEACHING:

- A dietitian will give you information about your diet.
- A nurse will continue teaching you about caring for yourself at home.

DISCHARGE PLANNING:

- You will need to arrange to have someone pick you up from the hospital tomorrow.

SEVENTH DAY (Going Home Day!)

TREATMENTS:

- Your IV will be removed. The staples will be removed from your incision.

ACTIVITY:

- You will continue to be able to get out of bed, go for short walks, and maybe even take a shower, as long as you are able to without making yourself too tired.

MEDICINES:

- You will have medicine prescribed to take at home. Your nurse will help you plan a schedule for your medications and answer any questions you may have about them.

DISCHARGE PLANNING:

- Your nurse will discuss your plans for returning home and any assistance you may need at home. This may include a home nurse or aide if needed.

■ 17 ■

Abdominal Aortic Aneurysm: Management and Outcomes

Sherri L. Stevens

The concept of case management had been in existence at Saint Thomas Hospital since the early 1990s. Different methods for utilization of case management were being practiced in various areas. Some areas were developing critical pathways for specific patient populations. Those areas implementing case management were doing so independently of one another, and there were no standards for development or implementation of the critical pathways. Various areas had implemented critical pathways to prepare for managed care and the development of a case management model. The use of critical pathways and the concept of case management at Saint Thomas was influenced by the work of Karen Zander and the Center for Case Management.[1]

In the fall of 1995 Saint Thomas Hospital was utilizing critical pathways for many diagnosis-related groups (DRGs) throughout the system. Vascular surgery had been involved with case management since 1993. The nursing staff, along with the vascular surgeon's clinical nurse specialists, had designed three vascular critical pathways for the vascular patient population. Other areas of the system such as cardiac, pulmonary, and orthopaedic/neurological units began initiating the case management concept, and in some areas the concept was more successful than in others.

ROLE OF THE CLINICAL NURSE SPECIALIST/CASE MANAGER

Clinical nurse specialist/case managers (CNS/CM) were assigned the task of coordinating the case management process in the organization. As utilization of critical pathways increased, so did nursing staff involvement with the case management process. Certain clinical areas had designated case managers and others did not. Throughout the system, areas experimenting with case management were functioning independently of one another. As time passed, it was deemed necessary to engineer a structure and a model for the management of patient outcomes. The vascular service did not have a case manager or unit-specific CNS. This area did have administrators ready to commit to a case management approach. A vascular CNS/CM was hired for the vascular unit in 1995. The case management program at Saint Thomas Hospital began to organize and identify common needs and opportunities for improvement.

The vascular CNS/CM investigated the hospitalization processes for the abdominal revascularization population. Initially, data were extracted by the finance department and analyzed into various components by fiscal year population. Components of care were categorized into pharmacy, operative day expenses, unit charges, medical surgical supplies, laboratory, etc, to gain insight into how various departments of the organization functioned independently to provide services for the abdominal revascularization patient population. The economics of the components also were utilized to determine areas of high cost. During this period of information seeking, improvements for efficiency and quality often surfaced. For example, changes in the methods of ordering diagnostic laboratory tests have been improved. Habitual patterns of ordering excessive laboratory tests are not necessarily the wave of the future nor best for the patient. Health care must become focused on the quality and efficiency of the care provided. Saint Thomas does not discriminate on financial outcomes nor does the organization make decisions solely on the basis of financial data. The organization does focus on the patient as a holistic being and the benefits of treating the patient and family.

To prepare for CareMap® development, the vascular CNS/CM analyzed the traditional plan of care for the abdominal revascularization patient. The care process of the abdominal revascularization patient included collecting and reviewing data at every level of the institution. The CNS/CM followed the caseload of the abdominal revascularization patients observing interactions and transactions of all areas. Rehlman[2] discussed health care as entering an era of assessment and accountability with a focus on quality. As the CNS/CM assessed the current use of the vascular critical pathways, it was determined that a more interactive plan of care would benefit the case management model. Interviews with patients and families were done to examine detailed components of care. Thus the role of the CNS/CM contributes significantly to the plan of care for the abdominal revascularization variant patient.

THE PROCESS

In 1995 a group known as the documentation task force was formed to change the use of critical pathways and documentation. The goal was to create a case management model for the institution that would encompass and benefit all patients. This task force consisting of representatives from most areas of the organization included staff nurses, managers, clinical nurse specialists, case managers, discharge planners, rehabilitation therapists, representatives from medical records, and others. Members from all departments were invited to share in developing the future model. The goal was to develop a model that would decrease actual documentation and focus on patient outcomes. Months of preparation and consultation occurred among the task force members. Each discipline took information back to the units for more analysis and input. Ultimately a decision was made to convert from the first generation critical pathways to standardized CareMaps®. CareMaps® would be an updated version or second generation component for case management. The critical pathways were organized plans of care for the vascular patients and required little documentation from the staff. The abdominal revascularization CareMaps® were designed to guide a plan of care, require daily interaction and documentation, and be placed in the patient's medical record. As the months passed, many members of the original group lost interest and attendance decreased due to individual work assignments. The clinical nurse specialists and the case managers remained throughout the entire project. In hindsight, our group should have worked harder at maintaining membership and representation from all areas. As the final project was unveiled some disciplines felt that they did not have equal representation and participation with

the project. After months of preparations, it was decided that all patients at Saint Thomas Hospital would be placed on a CareMap®. All CareMaps® were designed to follow a standard format, utilizing the same grammatical content and structure when applicable.

Global educational sessions were offered to educate the staff from all departments in the use of CareMaps® and the new documentation. Sessions were offered on all shifts as well as weekends to reach and include everyone. When the CareMaps® were executed (May 6, 1996), it was a tremendous task. As new patients were admitted the goal was to place them on the appropriate CareMap®. After one month, most patients were being placed onto the appropriate CareMap®. Realization of a learning curve did have an impact on this enormous project. There were small glitches in the system. Not all DRGs had a specific CareMap®. This lack of specificity in CareMaps® created problems among the nursing staff and clerical staff as to the uncertainty of which CareMaps® to use for some patients. The CNS/CM assisted with the decision-making process for the complex patients.

The new CareMap® was used for approximately 6 weeks on several pilot units. A telephone hotline was established for the staff to call and voice concerns, problems, and suggestions regarding the new format of the CareMap®. All suggestions were discussed during task force meetings and most were accepted and incorporated into the final version of the abdominal revascularization CareMap®.

The CNS/CM has functioned as the coordinator during the introduction of the abdominal revascularization CareMaps® and the other changes related to the case management model. Clinical experience as well as an understanding of the organizational culture have been contributing factors to the success and use of abdominal revascularization CareMaps® and the case management model.

Systems theory can be utilized to describe the implementation of the CareMap® process throughout the organization. As the concept of case management and use of CareMaps® permeated into the system, the daily interactions among the multiple teams displayed characteristic phases of change, including resistance as well as cooperation. Each department contributed to the whole and each profession contributed to the outcomes and goals. Both positive and negative experiences resulted from the change. Managers were encouraged to participate in the change. Administration was in favor of the new abdominal revascularization CareMaps® and the potential for outcomes tracking. Individual unit education was planned and implemented. The CNS/CM has been involved in educating staff about the use of abdominal revascularization CareMaps®. While the organizational changes were unsettling in many respects, they had the positive effect of bringing many teams together, opening the doors of collaboration, and creating an avenue to improve system variances.

VASCULAR CRITICAL PATHWAYS

There are many differences between the old vascular critical pathways and the new abdominal revascularization CareMaps®. The critical pathways identified global variance codes for system, patient/family, and provider and community variances. The abdominal revascularization CareMap® would create an avenue, by virtue of design, to guide nurses to document by checking reasons for variances on the outcome tracking tool. Explanations of variances were to be detailed in the progress notes of the patient's medical record. This created a charting by exception method for documentation. The vascular critical pathways did not contain a critical care component. The critical care staff were actively involved with this component of the abdominal revascularization CareMap®. Most abdominal revascularization patients spent 24 hours in the critical care area and transferred to the vascular care unit afterwards. It was determined that inclusion of a critical care phase would decrease fragmentation of the plan of care and increase documentation on the CareMap®. According to Willoughby and colleagues,[3] system variances can offer the greatest opportunity for quality improvement in an institution. The true impact of system variances such as delay times for procedures, staffing issues, delayed bed placement, and transfer orders have not been identified in the vascular population due to the lack of data. Such information would be vital to direct our most efficient efforts in the quality and patient satisfaction issues.[4]

CAREMAP® CONTENT

The abdominal revascularization CareMap® is designed to include the following surgical procedures: abdominal revascularization, renal artery bypass, aortofemoral bypass, iliac aneurysm repair, aortoiliac bypass, and mesenteric artery repair (Exhibit 17–A–1 in Appendix 17–A). The first page or cover page of the abdominal revascularization CareMap® lists these procedures so staff will know what vascular surgical procedures can be included in the CareMap®. The second and third pages of the abdominal revascularization CareMap® are designated as the outcome tracking tool. This is where the variances can be identified when the specified outcomes have not been met (Exhibit 17–A–2). This format was designed to facilitate data collection. All the numbered outcomes are listed in order on the outcome tracking tool. There are 32 items listed on the outcome tracking tool pages of the abdominal revascularization CareMap®.

The abdominal revascularization CareMap® is designed to incorporate the preadmission, operative, postoperative and discharge processes experienced by the patient. Each page of the abdominal revascularization CareMap® guides or maps the chain of events for the typical abdominal revascularization patient. The abdominal revascularization CareMap® has specific categories of needs to be addressed on a daily basis. The categories included are nutrition, activity, medications, assessment, treatments, diagnostics, teaching, discharge planning, psychosocial, individual patient needs, individual patient outcomes, and the established outcomes for the DRG. Outcomes are addressed in preset time frames as being met or not met. Incorporated into the postoperative section of the abdominal revascularization CareMap® are components designed to be used while the patient is in critical care as well as after transfer to the vascular unit.

The abdominal revascularization CareMap® is designed to accommodate the patient with a length of stay (LOS) up to 10 days. The last two pages are actually designed for the patient experiencing variances. This would allow individualized care to be added to the document if needed. The average length of stay for DRG 110 at Saint Thomas Hospital is 8 to 9 days. Phase I for the admission preoperative section can accommodate the patient on admission to the hospital (Exhibit 17–A–3). The following page of the CareMap® contains the operative day (Exhibit 17–A–4). This phase of care has been designed to incorporate the interventions that occur during a 24-hour operative day period. It contains areas of transfer for the patient postsurgery. Page 6 of the abdominal revascularization CareMap® is the critical care component of the patient's stay (Exhibit 17–A–5). Throughout the CareMap®, outcomes such as daily weights, telemetry, and patient education have been distributed to facilitate data collection as well as quality improvement. The remaining pages of the CareMap® pertain to the identified postoperative days. Some of the pages contain a 48-hour time interval for the patient to meet an outcome. The final page of the abdominal revascularization CareMap® is the discharge phase (Exhibit 17–A–6).

The goal of the abdominal revascularization CareMap® has been to improve documentation and focus on the outcomes of the patient. Consistency was a major concern of the documentation task force. Therefore, the same admission assessment form, the same format of CareMaps®, and the same flowsheets would allow an easier transition for the patient and the staff throughout the system. Consistency was needed to help the staff during this major change. The group focused as much as possible on tailoring the CareMaps® to the system instead of creating individual documents that would cause frustration and confusion.

Staff nurses from the vascular unit had input on the critical pathways that were initially developed. They were also asked to participate in the new design of CareMaps®. Some of the staff were more actively involved with the project than others. Many of the staff had mixed emotions pertaining to changing the critical pathways and the documentation system. Many staff nurses offered helpful suggestions for the CareMaps®. In time, involvement with CareMaps® and case management would reach everyone in the building.

Systems theory can be applied to the introduction and implementation of the case management model. Each area adapted and contributed to the project. All areas interacted with each other via patient transfer through the system. According to Bertrand[5] systems receive and process information. The manner in which each area responded affected the pieces of the project. The CNS/CMs were assessing the project internally as members of the documentation task force as well as externally in their specialty areas. As problems surfaced, the manner in which details were solved or improved affected integral pieces of the whole. One would have to step outside the area and evaluate the process to determine solutions. For example, if one area felt exemption of the CareMaps® should apply to them, explanations of how each component fit together in the system were offered to solidify the whole. Education was a critical piece of the transition. Utilization of the CareMaps® encouraged collaboration among various team members in a manner that would enhance communication among the team. Engineering and designing the CareMaps® created an avenue for discussion that would bring professions together.

PHYSICIAN PARTICIPATION

An integral component of the abdominal revascularization CareMap® process has been the physician participation. The vascular surgeons have been involved with the content of the abdominal revascularization CareMap®, and they approved the use of the CareMaps® on all of the vascular population. Once abdominal revascularization CareMaps® were appropriately designed to meet the criteria selected by the documentation task force, each had to go through an approval process. This process included discussions with the Chief of Surgery and the Chief of the Vascular Surgery Division. The Chief of the Vascular Division has been very helpful in participating with this project. Also, the involvement of the advanced practice nurses from the Edwards-Eve Clinic contributed to the teamwork of this project. They attended some of the meetings and assisted with the revisions. Total team involvement has contributed to the success of the implementation of the abdominal revascularization CareMap®. Naturally, some physicians were more involved than others.

Some physicians felt the use of CareMaps® was similar to cookbook medicine. The goal of the vascular CNS/CM was to work closely with the vascular surgeons and encourage participation with the process. The physicians do not utilize the CareMaps® for information. They view them as a nursing document and communicate their information in the progress notes of the patient's medical record. Attempts have been made to develop standardized order forms. The goal was to create standard orders that would decrease the amount of unnecessary procedures such as laboratory work, and thus decrease costs by ensuring that standards were met. Input was obtained from the vascular surgeons. Standard order sets were devised after much discussion with the vascular surgeons. For example, laboratory work, diagnostic procedures, and the use of antibiotics were incorporated into the standard order sets as well as into the abdominal revascularization CareMaps®. The standard orders were approved, printed, and distributed to the vascular surgical unit and the EMA unit. It was later identified that the orders would need to begin in the physician's office and in preadmission testing (PAT). Implementation of standard orders was not successful. Due to the variety of diagnostic procedures and ports of entry into the system, orders were being placed in the patients' medical records. Systems variances were identified as lack of education with clerical associates and the nursing staff as to when to place the orders on the medical record. Some patients were being admitted for surgery not necessarily requiring all steps of the standard orders. This omission in steps caused clerical staff difficulty with decision-making processes.

The clerical staff in the postanesthesia care unit (PACU) were unsure of what order sheets to place in the patient's medical record postoperatively. Eventually, due to the system problem of storing and using the orders, the physicians became frustrated with the lack of availability of the standard orders and began to write them as in the past. This created two systems, a written set of orders and some preprinted standard orders. This increased confusion for the clerical staff as well as the nursing staff as to the usage of the standard orders. The patient with comorbidities often required additional orders and eventually the preprinted standard orders were not placed on the patient's medical record.

As a result of the efforts of using the preprinted standard orders and the problems experienced, it was decided to delay this project. However, standard admission orders are still used and are started when the patient arrives in the physician's office. The orders are sent to the PAT center. Once the patient goes to surgery, no other standard orders are used. The initial preoperative and admission portion of the standard orders were successful and less confusing to use due to the ease of implementation. It has also been recognized that perhaps introducing the new CareMaps® and standard orders simultaneously was not the appropriate strategic plan.

DATA

The variance tracking tool contains all of the numbered outcomes that are listed throughout the abdominal revascularization CareMap®. All bold, numbered items are extricated and added to the cardiovascular database. Outcomes are measured not only from the abdominal revascularization CareMaps® but by quality indicators selected by the database staff. Indicators such as LOS, charges, and infections are monitored by the database staff and presented at the quarterly Vascular Division meetings. According to Cole and Houston,[6] organizations must be able to monitor outcomes and benchmark data to improve quality issues and maintain information technology. However, patient care outcomes or clinical issues must be included in data collection to ascertain the true quality as experienced by the patient.

Nurses are responsible for documenting the outcomes of the patients. The CNS/CM makes daily rounds on the units to determine the variant patients and to coordinate care. The staff are key players in determining when the patient has met the outcomes and/or what the variances may be. The CNS/CM has been active in developing the outcomes for disease-specific populations as well as educating and supporting the health care team. The dual role of the CNS/CM has been a springboard for the case management model at Saint Thomas and is mostly CNS-driven.

Over the last 5 years the average length of stay (ALOS) of the abdominal revascularization patient has been reduced from a 9- to 10-day LOS to an 8- to 9-day LOS (Figure 17–1). The ALOS has been reduced by 10%. The reduced LOS can be attributed not to the use of

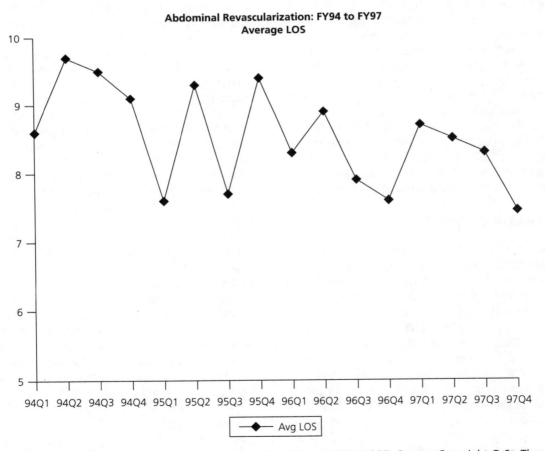

Figure 17–1 Quarterly Variation of the Average Length of Stay, Fiscal Years: 1994–1997. *Source:* Copyright © St. Thomas Hospital, graphs created by Ginny Tacker, RN, MBA.

CareMaps® but to the changes in physician practice. Decreased LOS has contributed to decreased charges for DRG 110. These decreased charges may be attributed to the CareMaps® or to streamlining the postoperative care of the patient. For example, the ALOS in critical care for abdominal revascularization patients has been reduced from a 2- to 3-day LOS to 24 hours. This has resulted in decreased costs. The average charge per case has decreased by 10% (Figure 17–2). Abdominal revascularization patients typically spend 24 to 48 hours in critical care and transfer to the vascular surgical monitoring unit for short-term telemetry. Occasionally patients require further monitoring before transferring to the vascular unit. Should this need arise, transfer to the special care unit for 24 to 48 hours may be needed. The average use of telemetry has been about 2 days (Figure 17–3).

When case management was first implemented in 1993, the vascular unit was designed to receive the patients 24 to 48 hours postoperatively from the critical care units. Installation of telemetry monitors was provided for the vascular unit to accommodate patients requiring short-term telemetry postoperatively. The staff were educated and prepared for the change in the care management of the postoperative vascular patients. All RNs had to become telemetry-trained to work on the vascular surgical unit.

Abdominal revascularization patients are extubated in the operating room prior to transferring to the PACU. After recovery the patient transfers to critical care for approximately 24 hours. Most patients have a Swan Ganz in place, oxygen per nasal cannula, epidural anesthesia for pain management, and intravenous fluids. After the 24-hour stay, the majority of patients are transferred to the vascular monitoring unit, barring any complications. Patient satisfaction outcomes have not been specifically measured for DRG 110. The organization has used a global survey to detect patient satisfaction in all aspects of care delivery, environment, and services. This survey captures some of the vascular surgical patient population, but not all. Currently each profession collects specific

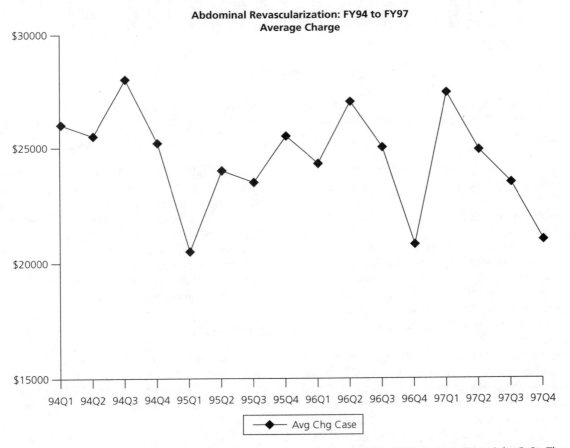

Figure 17–2 Quarterly Variation of the Average Charge per Case, Fiscal Years: 1994–1997. *Source:* Copyright © St. Thomas Hospital, graphs created by Ginny Tacker, RN, MBA.

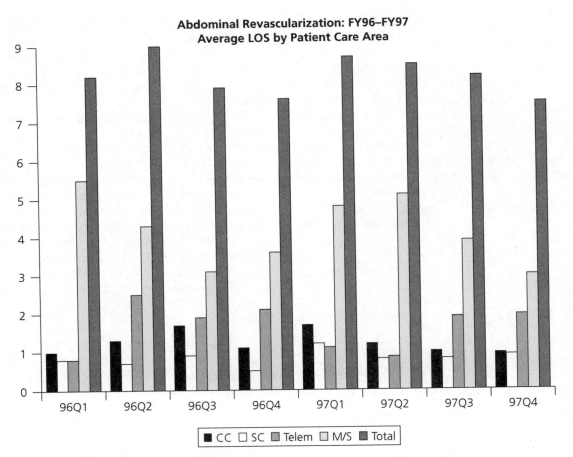

Figure 17–3 Abdominal Revascularization, Comparison of Average Length of Stay by Patient Care Area, Fiscal Years: 1996–1997. *Source:* Copyright © St. Thomas Hospital, graphs created by Ginny Tacker, RN, MBA.

data and determines quality indicators. Often these indicators are shared at quality forums or other collaborative meetings.

Improvements are needed in the area of capturing and collecting patient outcomes. Some clinical outcomes such as pain management and patient education are collected by the staff as a quality focus for improving the care of the patients. These outcomes are reported to the unit practice councils by the nurses participating in the quality council. Unit specific indicators and outcomes are posted on the vascular unit for continuous quality improvement results.

Outcomes are measured in 24- to 48-hour time frames on the abdominal revascularization CareMaps®. Abdominal revascularization CareMap® outcomes are not reported to the Vascular Surgery Division meetings. Established indicators are reported at division meetings by the database staff. Data collected by the database staff are shared with the vascular surgeons. There has not been a mechanism in place for reporting clinical outcomes from the abdominal

revascularization CareMaps® to the staff. Organizational conflict has surfaced intermittently relating to the lack of informational support provided from the database personnel. It has been recognized that in the future data must be presented to the staff to incorporate quality improvement. Staff involvement and knowledge of tracking outcomes and the leadership of the CNS/CM will be needed in the future for clinical outcomes management of the abdominal revascularization patients.

IMPROVEMENTS

During the development of the abdominal revascularization CareMaps® several variances were identified for areas of improvement. Some examples of these variances were utilization of telemetry and daily weights. It was identified that use of telemetry should be monitored for frequency and duration. A pilot study was conducted by placing reminder sheets in the patient medical records for the physicians to assess the utilization of telemetry. After

48 hours if the patient did not develop new cardiac arrhythmias or was stable, the telemetry would be discontinued. These reminder sheets continue to be used to prevent prolonged monitoring. This strategy has decreased the amount of unnecessary monitoring, thus decreasing costs. A telemetry committee has been established systemwide to define admission and discharge criteria for telemetry monitoring in all areas. The committee was established to review the levels of telemetry and focus on the aspects of quality improvement.

Twenty-four hours postoperatively, abdominal revascularization patients are transferred to the floor. In the past, patients were in the critical care unit for 2 to 3 days. Assessment of the patient for cardiopulmonary status, as well as fluid volume, resulted in adherence to strict daily weights. Dramatic fluid shifts can occur on approximately the third postoperative day resulting in pulmonary congestion. Clinically, the patients must be assessed closely for myocardial ischemia, fluid status, and renal function. Daily weights were incorporated into the CareMap® as part of the daily plan of care for this patient population. Fluid overload resulted in variances that contributed to increased LOS and poor outcomes.

MULTIDISCIPLINARY FOCUS

The CNS/CM reviews patient variances on fact sheets collected from the patient medical records. All information is entered into the cardiovascular database. The CNS/CM coordinates multidisciplinary discharge planning rounds biweekly. The team collaborates during discharge planning rounds to determine the needs of the patient and family posthospitalization. There are specific discharge criteria incorporated into the abdominal revascularization CareMaps® to ensure that patients leave the system in optimal condition. Social workers contribute to the plan of care by arranging needed services. Often times socioeconomic barriers exist as well as noncompliance to medical care with many of the chronic vascular patients, and discharge planning addresses these issues.

Educational barriers are common among many of the patients as well as failure to change lifestyle. For example, many of the patients smoke, are diabetic, and have a history of cardiac disease. Patients are encouraged to stop smoking and smoking cessation classes are available at Saint Thomas Hospital. Dietary personnel are consulted for diet modification and teaching as needed. Diabetic educators are consulted for many of the patients to discuss the importance of insulin administration, foot care, and reinforcement of proper nutrition and diet. Many factors are discussed by the multidisciplinary team during discharge rounds to prepare the patient for optimal self care outside the institution.

Utilization of the abdominal revascularization CareMap® has been an advantage in patient education and preparation for discharge. The abdominal revascularization CareMap® prompts the staff to consult the CNS to educate the patient and family. Social workers are consulted early on to facilitate discharge needs beyond the hospital walls and into the home. Surgical services hold multidisciplinary meetings monthly to identify problems in the surgical areas as well as issues for quality improvement. Collaboration through use of multidisciplinary team members has resulted from the use of the case management model. The abdominal revascularization CareMap® has been the tool to provide the guide or plan of care for the patient. It has been the vehicle to put necessary components in place, but it has not directly caused the changes.

DISADVANTAGES

One of the disadvantages of using the abdominal revascularization CareMaps® has been management of the complex patient and the reengineering required of the abdominal revascularization CareMap® for the staff. Clinical nurse specialist/case managers play an active role in education of the staff with CareMaps®, as well as coordinate the care of complex patients. They are consulted frequently for complex patients, which is the goal of case management. Those patients that develop complications from underlying comorbidities can be assisted by individualizing the plan of care on the abdominal revascularization CareMap®. When an outcome has not been met in the specified time frame, the reason must be reflected on the variance tracking tool as well as in the progress notes. Some of the most common comorbidities of the vascular population are coronary artery disease, diabetes mellitus, and hypertension. Some of the most frequently occurring variances postoperatively are arrhythmias, acute renal failure, and fluid overload. The designated areas titled "individual patient needs" and "outcomes" were added to the abdominal revascularization CareMap® format to create an individual plan of care for patients with variances. This allows the staff to quickly identify the major problems of the patients during report without scanning through the medical record.

Other disadvantages noted for the vascular disease population beyond the hospital walls was a lack of ongoing education. More follow-up posthospitalization with many of the vascular patients may result in decreased admissions. Many of the physicians do not feel that home health is the best alternative for the patient. Some of the physicians feel that home health is overutilized for the patients and feel that family should assist more often. Patients could benefit from having other programs; however, there are not many resources available.

There have been focus studies conducted by the vascular CNS reflecting CareMap® use. Many times the focus studies have been done manually to capture data. The vascular CNS/CM has been following the use of abdominal revascularization CareMaps® with the vascular patients for 2 years. There are changes that need to be made as result of trial and error. The case management process has resulted in the progression of team members collaborating for the patient. The abdominal revascularization CareMap® has not caused this change, but the process and the focus of the team has been on the patient and family. Nurses can utilize the abdominal revascularization CareMaps® to vision the expected norm for the patient and to obtain resources when needed. The focus on the continuum of care has resulted from the use of the case management process. Besides decreased LOS, the vascular CNS/CM has been consulted more often to assist complex patients.

Currently patients do not review their abdominal revascularization CareMaps®. When vascular critical pathways were utilized, there was a patient pathway designed for the patient to keep during the hospitalization. With the evolution of the abdominal revascularization CareMaps®, the terminology changed and the patient CareMap® has not been revised. There is a patient education committee that is reviewing all materials distributed to patient and families. The future goal is to have a patient abdominal revascularization CareMap® for the patient and family.

The abdominal revascularization CareMaps® are designed with a worksheet cover that provides a means of facilitating report from nurse to nurse (Exhibit 17–A–7 in the Appendix). This was a new change with the implementation of abdominal revascularization CareMaps®. Prior to CareMaps®, the organization had patient profiles that were used for patient report and for clerks to transcribe orders.

The multidisciplinary education record was designed to address the lack of teaching documentation on the abdominal revascularization CareMap®. Designated areas were designed to document teaching on the CareMap®. After auditing the patient education documentation, it was decided that patient education must be consistently documented by all disciplines. Thus, the multidisciplinary education record was developed. All disciplines document teaching on this tool, and if great detail is required, the progress notes reflect further explanations. Various booklets are given to the abdominal revascularization patients for educational support. Documentation of these materials are recorded on the multidisciplinary education record.

In the future, greater emphasis will need to be placed on the posthospitalization component for patients. Redesign of the CareMaps® will allow us to focus on the needs of the complex patients. Initially, when the abdominal revascularization CareMaps® were designed, the goal was to capture all data related to the patient hospitalization. Realistically, this has been impossible due to the numerous outcomes incorporated into the abdominal revascularization CareMap®. The organization has not been able to measure CareMap® outcomes for all DRGs. Another problem identified over the past year has been the lack of completion of the CareMaps® by the staff. Numerous chart reviews have been done to audit the abdominal revascularization CareMaps®. Audit results are reported to the unit managers. Outcomes management is important because the future calls for specific data related to patient care. Continuous quality improvement can only occur through good outcomes data collection.

CONCLUSION

The abdominal revascularization CareMaps® have contributed to creating change in the organization. New ways of thinking and delivering patient care have been introduced. There are no data to explain a cause-and-effect relationship between abdominal revascularization CareMaps® and outcomes. An increase in collaboration of the various professions has resulted in better care for the patient by providing earlier services for the patient and family. In the future, when abdominal revascularization CareMaps® are revised, all disciplines will need to identify specific outcomes for quality care. Data collection and management of the outcomes will need to occur in a more timely manner to remain focused on identification. Information systems are needed to monitor clinical outcomes. Since the abdominal revascularization CareMaps® were started in May 1996, many things have been learned. The outcomes currently collected are helpful, but revisions need to be made to focus on a small number of outcomes during a specified time. Once data have been reported and outcomes are identified through a quality process, then continuous revision of current outcomes can be managed.

The abdominal revascularization CareMaps® themselves have not caused decreases in LOS or charges. The total process of preparation for managed care and the evolving model of case management has been evolutionary in facilitating these changes. The case management model has moved the focus to the patients and has created a new emphasis on quality. In the future, the case management model of care needs to include the full continuum of care. Patient satisfaction data as well as patient progression should be monitored for quality. The nursing staff should be surveyed to assess their satisfaction of patient care delivery. Physicians should also be surveyed to assess their satisfaction with case management. Outcomes from all sources can contribute to research desperately needed to quantify the changing health care environment.

REFERENCES

1. Zander K. Nursing case management: strategic management of cost and quality outcomes. *J Nurs Admin.* 1997;18(5):23–30.

2. Rehlman AS. Assessment and accountability: the third revolution in medical care. *N Engl J Med.* 1997;319:1220–1222.

3. Willoughby C, Budreau G, Livingston D. A framework for integrated quality improvement. *J Nurs Care Qual.* 1997;11(3):44–53.

4. Micek WT, Berry L, Gilski D, Kallenbach A, Link D, Scharer K. Patient outcomes: the link between nursing diagnoses and interventions. *J Nurs Admin.* 1996;26(11):29–35.

5. Bertrand A. *Social Organization: A General Systems and Role Therapy Perspective.* Philadelphia: FA Davis; 1972.

6. Cole L, Houston S. Integrating information technology with an outcomes management program. *Crit Care Nurs Q.* 1997;19(4):71–79.

■ Appendix 17–A ■
CareMaps® and Patient Worksheet

Exhibit 17–A–1 Abdominal Revascularization CareMap®

CareMap®: Abdominal Revascularization	Addressograph

Inclusion: Includes all patients undergoing any of the following procedures.

- **Abdominal Revascularization**
- **AAA**
- **Renal Artery Bypass**
- **Aortofemoral Bypass**
- **Iliac Aneurysm Repair**
- **Aortoiliac Bypass**
- **Mesenteric Artery Repair**

Printed Name	Initial	Printed Name	Initial	Printed Name	Initial

All signatures are those individuals recording information in the CareMap®

Source: Copyright © Saint Thomas Health Services.

Exhibit 17–A–2 Abdominal Revascularization CareMap® Outcomes Tracking Tool

Outcome Tracking Tool CareMap®: Abdominal Revascularization	Addressograph

If outcome unmet, check reason(s); if reason is not listed, check "Other" and document in Progress Notes. Actions taken as a result of unmet outcomes also are charted in the Progress Notes.

#1 Patient Understanding

❏ Patient too sick for discussion
❏ Communication barrier
❏ Information not provided at this time
❏ Other _____

#2 Family Verbalizes Understanding of Plan of Care

❏ Family unavailable
❏ Has no family support
❏ Communication barrier
❏ Information not provided at this time
❏ Other _____

#3   Hemodynamic Stability

❏ SBP > 180
❏ SBP < 100
❏ Rhythm unstable
❏ Low CVP or wedge pressure
❏ Return to OR
❏ Other _____

#4 Urine Output > 240 ml per 8 hours

❏ Undiagnosed condition
❏ Chronic renal failure
❏ Dehydration
❏ Low cardiac output state
❏ Renal function impaired
❏ Other _____

#5 Extubated Prior to Transfer to CC

❏ ABGs inadequate
❏ Not sufficiently awake
❏ Bleeding
❏ Hemodynamically unstable
❏ Rhythm unstable
❏ Patient remains on mechanical ventilation
❏ Unable to maintain adequate saturation
❏ Other _____

#6 Pedal Pulses Present

❏ Unable to palpate left Dorsalis Pedis pulse
❏ Unable to palpate left Posterior Tibial pulse
❏ Unable to obtain left Dorsalis Pedis pulse per Doppler
❏ Unable to obtain left Posterior Tibial pulse per Doppler
❏ Unable to palpate Right Dorsalis Pedis pulse
❏ Unable to palpate Right Posterior Tibial pulse
❏ Unable to obtain Right Dorsalis Pedis pulse per Doppler
❏ Unable to obtain Right Posterior Tibial pulse per Doppler
❏ Other _____

#7 Daily Weight

❏ Patient not weighed
❏ No order to weigh patient
❏ Other _____

#8 Hemodynamic stability with SBP <180 or >100 x24 hours

❏ SBP > 180
❏ SBP < 100
❏ Rhythm unstable
❏ Low CVP or wedge pressure
❏ Return to OR
❏ Other _____

#9 SPO2 > 92%

❏ SPO2 < 92% on nasal cannula < 6 l/min
❏ Face tent required for adequate oxygenation
❏ Postop SpO2 less than 92
❏ Other _____

#10 Extremity Pulses Palpable

❏ Unable to palpate left Dorsalis Pedis pulse
❏ Unable to palpate left Posterior Tibial pulse
❏ Unable to obtain left Dorsalis Pedis pulse per Doppler
❏ Unable to obtain left Posterior Tibial pulse per Doppler
❏ Unable to palpate Right Dorsalis Pedis pulse
❏ Unable to palpate Right Posterior Tibial pulse
❏ Unable to obtain Right Dorsalis Pedis pulse per Doppler
❏ Unable to obtain Right Posterior Tibial pulse per Doppler
❏ Other _____

#11 Urine Output > 240 ml per 8 hours

❏ Undiagnosed condition
❏ Chronic renal failure
❏ Dehydration
❏ Low cardiac output state
❏ Renal function impaired
❏ Other _____

#12 Transfer to 6C Vascular Unit

❏ Patient to remain in CC per order
❏ Patient to transfer to Special Care per order
❏ Not hemodynamically stable
❏ Dobutamine/Dopamine > 6 mic/kg/min
❏ Pulmonary Artery Catheter
❏ Other _____

#13 Daily Weight

❏ Patient not weighed
❏ No order to weigh patient
❏ Other _____

#14 Urine Output 240 cc per 8 hours

❏ Undiagnosed condition
❏ Chronic renal failure
❏ Dehydration
❏ Low cardiac output state
❏ Renal function impaired
❏ Other _____

#15 Pedal Pulses Present

❏ Unable to palpate left Dorsalis Pedis pulse
❏ Unable to palpate left Posterior Tibial pulse
❏ Unable to obtain left Dorsalis Pedis pulse per Doppler
❏ Unable to obtain left Posterior Tibial pulse per Doppler
❏ Unable to palpate Right Dorsalis Pedis pulse
❏ Unable to palpate Right Posterior Tibial pulse
❏ Unable to obtain Right Dorsalis Pedis pulse per Doppler
❏ Unable to obtain Right Posterior Tibial pulse per Doppler
❏ Other _____

continues

Exhibit 17–A–2 continued

> **If outcome unmet, check reason(s); if reason is not listed, check "Other" and document in Progress Notes.**
> **Actions taken as a result of unmet outcomes also are charted in the Progress Notes.**

#16 Patient Verbalizes Understanding of Plan of Care
- ❏ Family unavailable
- ❏ Has no support system
- ❏ Communication barrier
- ❏ Information not provided at this time
- ❏ Other _____

#17 Daily Weight
- ❏ Patient not weighed
- ❏ No order to weigh patient
- ❏ Other _____

#18a Telemetry
- ❏ Normal sinus rhythm
- ❏ Atrial fibrillation, controlled
- ❏ Atrial flutter
- ❏ Rapid atrial arrhythmias
- ❏ Paced rhythm
- ❏ Frequent pvc's
- ❏ Other _____

#18b Telemetry
- ❏ Normal sinus rhythm
- ❏ Atrial fibrillation, controlled
- ❏ Atrial flutter
- ❏ Rapid atrial arrhythmias
- ❏ Paced rhythm
- ❏ Frequent pvc's
- ❏ Other _____

#19 Ambulate Length of Hall
- ❏ Patient unable to tolerate ambulation
- ❏ Patient ambulate 1/2 length of hall
- ❏ Patient refused
- ❏ Other _____

#20 Pain < 5 of 0–10 Scale
- ❏ Patient states intervention ineffective
- ❏ Medication held
- ❏ Patient refuses medication
- ❏ Knowledge deficit
- ❏ Communication barrier
- ❏ By patient report pain level is acceptable
- ❏ Med not ordered
- ❏ Other _____

#21 Bowel Sounds Present
- ❏ Unable to auscultate bowel sounds
- ❏ Abdomen distended
- ❏ Other _____

#22 Tolerating po Intake without N/V
- ❏ Unable to take po fluids
- ❏ Patient nauseated
- ❏ Episode of emesis
- ❏ Refuses po
- ❏ Patient still has NG tube
- ❏ Other _____

#23 Lungs Clear to Auscultation
- ❏ Lungs are congested
- ❏ Lungs have diminished breath sounds
- ❏ Crackles
- ❏ Wheezes
- ❏ Other _____

#24a Daily Weight
- ❏ Patient not weighed
- ❏ No order to weigh patient
- ❏ Other _____

#24b Daily Weight
- ❏ Patient not weighed
- ❏ No order to weigh patient
- ❏ Other _____

#25a Telemetry
- ❏ Normal sinus rhythm
- ❏ Atrial fibrillation, controlled
- ❏ Atrial flutter
- ❏ Rapid atrial arrhythmias
- ❏ Paced rhythm
- ❏ Frequent pvc's
- ❏ Other _____

#25b Telemetry
- ❏ Normal sinus rhythm
- ❏ Atrial fibrillation, controlled
- ❏ Atrial flutter
- ❏ Rapid atrial arrhythmias
- ❏ Paced rhythm
- ❏ Frequent pvc's
- ❏ Other _____

#26 Abdomen Nondistended with BS Present
- ❏ Abdomen distended
- ❏ Unable to auscultate bowel sounds
- ❏ Other _____

#27 Temperature < 100.0° x 24 Hours
- ❏ Temperature 100.8° to 102.0° po
- ❏ Temperature ≥ 102.0° po
- ❏ Other _____

#28 Weight within 5 lbs. of Admission Weight
- ❏ Weight > 10 lbs. over admission weight
- ❏ Weight < admission weight
- ❏ Weight > 5 lbs. over admission weight
- ❏ Other _____

#29 Temperature < 100.0° x 24 Hours
- ❏ Temperature 100.8° to 102.8° po
- ❏ Temperature ≥ 102.0° po
- ❏ Other _____

#30 Incision without Signs of Infection
- ❏ Incision has areas of erythema
- ❏ Incision has erythema and drainage
- ❏ Incision has open area
- ❏ Other _____

#31 Patient Verbalizes Understanding of Discharge Instructions
- ❏ Patient unable to understand
- ❏ Communication barrier
- ❏ Patient will be transferred to other facility
- ❏ Other _____

#32 Discharge Disposition
- ❏ Home with self care/family care
- ❏ Home with home health
- ❏ Home with hospice
- ❏ Rehab facility
- ❏ Skilled nursing facility
- ❏ Patient expired
- ❏ Other _____

Source: Copyright © Saint Thomas Health Services.

Exhibit 17–A–3 Phase I, Admission, Preoperative Section

CareMap®: Abdominal Revascularization	Addressograph

Time Interval	PHASE I Preoperative Date _____ Admission_____
Nutrition	Clear liquids
Activity	Ad lib
IVs **Medication**	IV per order Mg Citrate as ordered Evaluate for food/drug interactions
Assessment	Vital signs per order Admission weight recorded _____ I & O if ordered Pain assessment every 8 hours and PRN Skin assessment daily and PRN
Treatments	Scrubs as ordered
Diagnostics	Lab Type, screen
Teaching	Evaluate learning needs; document on Education Record STEPS Preop packet Orient patient and family to environment
Discharge Planning	Evaluate Discharge Planning needs Social Service Consult Indicated: **YES** ❑ **NO** ❑ Requested: **Date**_____ **Initial** _____ Consult Initiated: **Date**_____ **Consultant's Initial**_____ Encourage need for CNS/CM participation in education of patient family
Psychosocial	Pastoral Care Consult Indicated: **YES** ❑ **NO** ❑ Requested: **Date**_____ **Initial** _____ Consult Initiated: **Date**_____ **Consultant's Initial**_____ Encourage verbalizing of feelings and questions about surgery and hospital routine
Individual Patient Needs	
Individual Patient Outcomes	

Outcomes		Date	Time	Initial
1. Patient verbalizes understanding of plan of care **YES** ❑ **NO** ❑	1			
2. Family verbalizes understanding of plan of care **YES** ❑ **NO** ❑	2			

Treatments documented on the CareMap® do not need to be duplicated on the Flowsheet.

Source: Copyright © Saint Thomas Health Services.

Exhibit 17–A–4 Operative Day

CareMap®: Abdominal Revascularization	Addressograph

Time Interval	**PHASE I OPERATIVE DAY** Date _____ PACU _____ Critical Care _____
Nutrition	NPO
Activity	Bed rest
IVs **Medication**	IV per order Epidural/PCA as ordered for pain Ancef 1 gm IV every 8 hour x 3 doses or per order Vasoactive drips per order Evaluate for food/drug interactions
Assessment	Vital signs every 5 minutes PACU or as ordered Vital signs every 15 minutes after transfer to CC or as ordered I & O every 1 hour Daily weight Pain assessment every 8 hours and PRN to assess comfort Skin assessment daily and PRN to assess circulation and surgical incision Vascular checks per order to assess peripheral tissue perfusion Neuro checks per order to assess neuro/motor function Intubation in OR, wean to extubate as ordered
Treatments	Oxygen as ordered postextubation Foley catheter Intubation in OR, wean to extubate as ordered Swan Ganz for assessment of filling pressures Arterial line for hemodynamic monitoring
Diagnostics	CXR for Swan placement as ordered Pulse oximeter to maintain $SpO_2 > 92\%$ CV monitoring Monitor lab values and report abnormals
Teaching	Evaluate learning needs; document on Education Record Evaluate need for CNS/CM participation in D/C plan, education of patient and family
Discharge Planning	Evaluate Discharge Planning needs
Psychosocial	
Individual Patient Needs	

Individual Patient Outcomes		Date	Time	Initial

Outcomes	**Outcomes should be addressed as they are achieved and up to 24 hours postop**			
		Date	Time	Initial
	3. Hemodynamic stability YES ❑ NO ❑ 3			
	4. Urine Ouput > 240 ml per 8 hours YES ❑ NO ❑ 4			
	5. Extubated prior to transfer to CC YES ❑ NO ❑ 5			
	6. Pedal pulses palpable YES ❑ NO ❑ 6			

Treatments documented on the CareMap® do not need to be duplicated on the Flowsheet.

Exhibit 17–A–5 Critical Care Phase

CareMap®: Abdominal Revascularization	Addressograph		

PHASE I   POSTOP DAY 1
Critical Care—Special Care _____

Time Interval	Date _____			
Nutrition				
Activity	Bed rest			
IVs Medication	IV per order IV antibiotics per order Vasoactive drips, wean per parameters Epidural/PCA per order Evaluate for food/drug interactions			
Assessment	Vital signs every 1 hour or per order I & O every 1 hour **7. Daily weight** YES ☐ NO ☐ Pain assessment every 8 hours and PRN Skin assessment daily and PRN Vascular checks every 2 hours or as ordered to assess tissue perfusion Neuro checks every 4 hours or as ordered to assess neuro/motor function Oxygen saturation > 92% Continuous cardiovascular monitoring Swan for assessment of filling pressures, critical care per order Arterial line per order	Date / Time / Initial 7		
Treatments	Wean to extubate, as ordered Oxygen as ordered Incentive spirometer every 2 hours while awake NG Tube			
Diagnostics	Lab, as ordered Pulse oximeter to maintain SpO_2 > 92% CV monitoring Monitor lab values and report abnormals			
Teaching	Evaluate learning needs; document on Education Record Evaluate need for CNS/CM participation in D/C plan, education of patient and family			
Discharge Planning	Evaluate Discharge Planning needs			
Psychosocial				
Individual Patient Needs				
Individual Patient Outcomes		Date / Time / Initial		
Outcomes	**Outcomes should be addressed as they are achieved and up to 24 hours postop** **8. Hemodynamic stability with SBP < 180 > 100** YES ☐ NO ☐ **9. SpO_2 > 92% on nasal cannula** YES ☐ NO ☐ **10. Extremity pulses palpable** YES ☐ NO ☐ **11. Urine output 240 ml/8 hours** YES ☐ NO ☐ **12. Transfer to 6C Vascular Unit** YES ☐ NO ☐ ***If patient remains in CC/SC contact CNS/CM for vascular surgery**	Date / Time / Initial 8 9 10 11 12		

Treatments documented on the CareMap® do not need to be duplicated on the Flowsheet.

Exhibit 17–A–6 Discharge Phase

CareMap®: Abdominal Revascularization	Addressograph

Time Interval	DISCHARGE PHASE POSTOP DAY 9 and 10 Date _____
Nutrition	Diet per order
Activity	Ambulate TID
IVs **Medication**	IV to saline lock Evaluate for food/drug interactions
Assessment	Vital signs every per routine I & O Daily weight Pain assessment every 8 hours and PRN Skin assessment as ordered
Treatments	
Diagnostics	
Teaching	Evaluate learning needs; document on Education Record Ensure all education is documented on the Education Record Evaluate need for CNS/CM for education of patient and family
Discharge Planning	Evaluate Discharge Planning needs Social Service Consult Indicated: **YES ❑ NO ❑** Requested: **Date_____ Initial_____** Consult Initiated: **Date_____ Consultant's Initial_____**
Psychosocial	
Individual Patient Needs	
Individual Patient Outcomes	
Outcomes	**Outcomes should be addressed as they are achieved**

			Date	Time	Initial
31. Patient verbalizes understanding of discharge instructions YES ❑ NO ❑		31			
32. Discharge disposition (document on Outcome Tracking Tool) YES ❑ NO ❑		32			

*If patient length of stay > 10 days—contact CNS/CM and begin Phase II of CareMap®

Treatments documented on the CareMap® do not need to be duplicated on the Flowsheet.

Source: Copyright © Saint Thomas Health Services.

Exhibit 17–A–7 Front of Worksheet Cover

Patient Care Worksheet	**Addressograph**
Allergies:	**Admission Date/Diagnosis:**

Advance Directives:

Durable Power of Attorney	Yes	No
Living Will	Yes	No
Organ Donor	Yes	No
DNR _____	Yes	No

Date   Procedure

_____

_____

_____

_____

Concurrent Medical Conditions:

ICD on	Yes	No	NA
Permanent pacemaker	Yes	No	
Pregnant	Yes	No	NA
Diabetic	Yes	No	
On Metformin/Glucophage?	Yes	No	NA

_____

_____

Room Numbers/Transfers

#_____   #_____   #_____   #_____
Date   Date   Date   Date

#_____   #_____   #_____   #_____
Date   Date   Date   Date

Pertinent Medical History:

Primary Physician:

Consulting Physicians:

Case Manager:

Miscellaneous: _____

_____

Wears glasses	❑ _____
Hard of hearing	❑ _____
Uses cane	❑ _____
Uses walker	❑ _____

SPECIAL ORDERS:
Level 1 Extubation: YES ❑   NO ❑
PCA/Epidural Information: _____

_____

_____

_____

Started on _____ CareMap; Date _____
Transferred to _____ CareMap; Date _____
Transferred to _____ CareMap; Date _____
Transferred to _____ CareMap; Date _____

Other: _____

_____

_____

_____

Isolation: YES ❑   NO ❑
Type _____

_____

Restraints on   YES ❑   NO ❑
If restraints ordered, check patient every hour and prn
Obtain reorder for restraints every 24 hours

Date Started _____ Date Ended _____

Contact Person: _____
Name: _____

CAPS: YES ❑   NO ❑

_____

Date Started _____ Date Ended _____

Phone: _____

_____

This worksheet is not part of the permanent medical record and will be discarded upon discharge.

Source: Copyright © Saint Thomas Health Services.

■ 18 ■

Outcomes of a Critical Pathway for Uncomplicated Cardiac Surgical Patients

Donna M. Rosborough, Dorothy Goulart Fisher, and Lawrence H. Cohn

Managed care and changes in health care payment have led to efforts to increase efficiency and decrease costs while maintaining high quality care. Cardiac surgery at Brigham and Women's Hospital in Boston, Massachusetts, was not immune to these pressures. In April of 1993, a cardiac surgical care improvement team made up of representatives from cardiac surgery, anesthesia, perfusion, nursing, administration, and finance were charged with identifying improvement opportunities and lowering the cost of treating coronary artery bypass graft (CABG) patients while maintaining or improving the quality of care.

INTRODUCTION

This chapter describes how that process led to the development of our pathway for uncomplicated cardiac surgical patients. A consultant reviewed the medical records of 100 patients undergoing CABG with cardiac catheterization (DRG 106) and CABG without cardiac catheterization (DRG 107) and provided the team with the cost profile including the cost of room and care, surgery, supplies, drugs, intravenous therapy; laboratory tests, x-rays, and other tests; blood utilization; and therapies such as physical, occupational, and respiratory therapy. The consultant presented benchmark data from four other similar institutions and constructed a "best demonstrated practice" by combining best practice from all five institutions. Critical pathways, with measurable targets and patient outcomes, were suggested as a method of lowering costs while maintaining high quality care. Dr. Thomas Lee, director of the hospital's clinical initiatives development program, described how critical pathways were being used successfully in other cardiac surgical programs in the country. He also presented the

hospital's intention to initiate critical pathways in several patient populations. The team decided to construct a critical pathway for CABG patients.

CRITICAL PATHWAY DEVELOPMENT

The critical pathway was developed by the interdisciplinary care improvement team in four 2-hour meetings. During the meetings, the task force compared the hospital's cost profile to the cost profiles of four similar institutions and the combined best demonstrated practice. From this the team identified opportunities for improvement in the areas of length of stay, laboratory tests (electrolytes, blood gases, hematocrit), and other tests (chest x-ray, electrocardiogram). From the sample charts reviewed, patients had an average preoperative length of stay (LOS) of 1.5 days, an intensive care unit (ICU) length of stay of 2 days, and a postoperative length of stay of 4.5 days. The length of stay targets for the critical pathway were set as an ICU stay of 24 hours or less, and a postoperative LOS of 5 days (ie, discharge on postoperative day 5). One key factor in ICU LOS was the amount of time that the patient was intubated. In 1993 very few same-day extubations were being done; the majority of patients were extubated on their first postoperative day. It was determined that same day extubation should be a goal. An additional LOS goal was to increase the number of same-day admit patients.

Telemetry, laboratory tests, and chest x-ray utilization were improvement opportunities. From the chart review, patients spent an average of 3.5 days on telemetry and 20% of the patients remained on telemetry until the day of discharge. The team determined that the goal should be 3 days of post-ICU telemetry. The main areas of expenditure within labs and tests were hematology, chemistry, EKG, and chest x-rays. The team reviewed the current standing orders, the March 1993 utilization, and a sample critical pathway from a California institution. The standing orders were then revised to reflect the minimum required for the uncomplicated CABG patient within the new LOS targets. These were incorporated into the critical pathway.

Once all of the care processes were listed, the team developed daily expected patient outcomes. Although there were several outcomes identified for each day, the "key" outcomes to be achieved were

1. extubation on the day of surgery
2. transfer to intermediate unit within 24 hours postoperatively
3. tolerating solid foods on postoperative day 2
4. removal of pacing wires on postoperative day 3
5. tolerating one flight of stairs on postoperative day 4
6. discharge on postoperative day 5

After reviewing a draft of the critical pathway, the team discussed that the care of patients recovering from an uncomplicated valve repair or replacement, atrial septal defect repair, and myxoma resection was virtually the same as the care of the CABG patients. The decision was made to utilize the pathway for all uncomplicated cardiac surgical patients. The initial eligibility criteria for pathway use included age less than 80, elective or urgent cardiac surgery, except combined CABG/mitral valve replacement or aortic dissection repair. Exclusion criteria included a left ventricular ejection fraction of less than 30%, uncontrolled hypertension, severe chronic obstructive pulmonary disease, or chronic renal insufficiency.

CRITICAL PATHWAY FORMAT

The critical pathway was initially constructed as a task-time matrix with the task categories of assessment, consult, tests, activity, medications, treatments, diet, patient/family education, and discharge planning (Exhibit 18–A–1 in Appendix 18–A). The pathway unit of time was in days, starting with a preoperative day of presurgery testing through discharge on the fifth postoperative day. The care tasks for each day were listed on the matrix. Although this format served as a good overview display of the pathway, the team decided to construct a format that could be used for documentation of the care as well. Exhibit 18-A–2 demonstrates the documentation format that was designed. Each of the patient care tasks and expected outcomes are listed in the left column. The next three columns are for charting on the day, evening, and night shifts (D, E, N). The caregivers enter their initials in the appropriate columns to indicate task completion or outcome achievement. The column labeled "VAR" is for indicating a variance, either that

the task was not completed or the expected outcome was not achieved. The caregivers enter their initials in the variance column and then use the "progress notes" column to describe the variance and the corrective action. Additional progress note space was placed on the back of each sheet. A signature box for entering the caregivers' initials, signature, and licensure is located at the bottom of the day's pathway. In addition there is a section to indicate the reason for pathway termination if it occurred.

Although this format served nurses well in their documentation, variance tracking was difficult and time consuming. The other major drawbacks to the paper format were the number of pages (2 to 3 per day) and the difficulty and timeliness of revisions. This led to the decision to automate the pathway.

AUTOMATED CRITICAL PATHWAY

A programmer, the program manager for patient care information systems, and the cardiac surgical case managers collaborated on development and testing of the pathway software. The software application has two portions, one for authoring and revising the pathway and one for documentation. The documentation portion was designed to closely resemble the paper format. Each patient care category is designed as a file folder (Figure 18–1) that can be selected by the documenter. The care tasks and outcomes are listed on the left of the screen display with the shift and variance columns displayed on the right. At the bottom of the screen are "action" buttons for

- signing
- selecting a patient and/or pathway
- writing additional progress notes
- printing the pathway
- removing a patient from the pathway
- review of variance information
- accessing any of the rest of the patient's online clinical record
- invoking the outcome achievement graph
- exiting the pathway application

Documenters are first prompted to sign in with their computer keys. The patient is selected, placed on the appropriate pathway, and then placed on the appropriate day of the pathway. Documentation of task completion and outcome achievement is done simply by clicking in the appropriate column and cell. This automatically enters the documenter's initials. Clicking on another file folder brings up that section's content. A "check mark" appears on the tab of the file folders that have been documented on (Figure 18–2).

To document variance, one clicks on the variance column that corresponds with the task or outcome. This

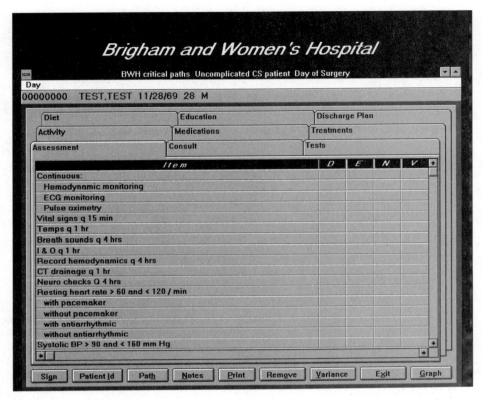

Figure 18–1 Automated Day of Surgery Critical Path. *Source:* Copyright © Brigham and Women's Hospital.

brings up the variance code choices for selection as well as space to write additional notes (Figure 18–3). A "V" appears on the tab of the file folder indicating that a variance was documented. To review the variances the caregiver clicks on the "Variance" button and is able to read the documentation. The "Graph" button displays the outcome achievement graph, which shows the patient's actual progress on achieving the key expected outcomes of the pathway in comparison to the expected progression. This new feature will provide caregivers with a quick and simple method to review progress and individual patient variance.

In the original version of the software, admission, transfer, discharge, and progress notes were written by selecting the "notes" button and typing in free text. In the most recent upgrade, templates for the admission, transfer, and discharge notes have been added. Progress notes still require free text typing. The notes screen has both write and read-only sections so that previously written notes cannot be altered. All of the documentation is saved by using the "Sign" button and entering the computer key. When a patient is to be taken off the pathway, the "Remove" button is selected and the documenter is prompted to provide an explanation for termination.

The paper version of the pathway was initiated in 1993 and the automated pathway in 1995. The automated version offers several advantages, such as:

- ease of authoring and revising the pathway
- online documentation rather than multiple pieces of paper
- less time- and personnel-intensive variance data collection and reporting
- instantaneous view of individual patient variance and progress

CARE/CASE MANAGEMENT

Patients are assessed postoperatively in the intensive care unit by the nursing staff for critical pathway eligibility using the previously mentioned criteria. Nursing documentation occurs on the automated pathway with a printed copy placed in the patient's medical record daily. The care coordination team manager for cardiac services documents a weekly list of all patients placed on the pathway, which is faxed to the research assistant who is responsible for tabulating outcome data and variances. Quarterly reports are generated and provided to the care

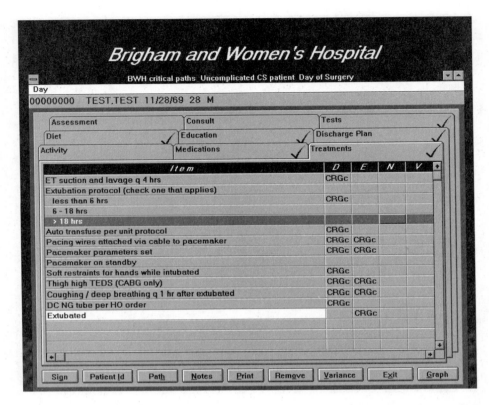

Figure 18–2 Automated Documentation on Day of Surgery Critical Path. *Source:* Copyright © Brigham and Women's Hospital.

coordination team manager, the division chief, and the nurse manager of cardiac surgery. The care coordination team manager also serves as an educational resource to nursing and physician staff in regard to critical pathway eligibility criteria, documentation, and variance analysis.

The care coordination team for cardiac services assesses all patients for discharge planning needs. Most patients on the critical pathway are able to be discharged home with home health services although a small percentage have required an extended care facility stay. Two home care agencies and two extended care facilities have collaborated with the care coordination team manager in developing a critical pathway for these patients in their settings.

CRITICAL PATHWAY OUTCOME MEASUREMENT

Key outcomes routinely measured on the critical pathway identified by the cardiac surgery care improvement team include extubation time, transfer to the intermediate care unit within 24 hours of admission to the intensive care unit following cardiac surgery, tolerating a diet, removal of temporary epicardial pacing wires, and tolerating stair climbing. Expected time frames are designated

for each outcome. The outcomes are measured concurrently by the nursing staff as patients progress along the critical pathway. A graphic display was developed for the automated critical pathway that measures key outcomes achieved from day of surgery to discharge. This graph will be utilized to determine readiness for discharge regardless of actual days on the critical pathway (Figure 18–4). Variances are documented for outcomes not achieved. Retrospective review of the medical record by a research assistant occurs after discharge for key outcome and variance data collection. Quarterly reports are distributed to members of the cardiac surgery care improvement team for analysis and review (Exhibit 18–1).

Managed care networks are interested in critical path development for specific patient diagnoses secondary to financial savings and quality of care issues. One managed care plan on site at Brigham and Women's Hospital monitors their patient population for variances in care and LOS. Additionally, managed care plans reap a cost savings by not having to utilize extended care facilities as frequently.

Patient outcomes also measured and analyzed by the cardiac surgery team include LOS in the intensive care unit, total hospital LOS, arrhythmias, wound infections, and readmissions to the hospital within 30 days of dis-

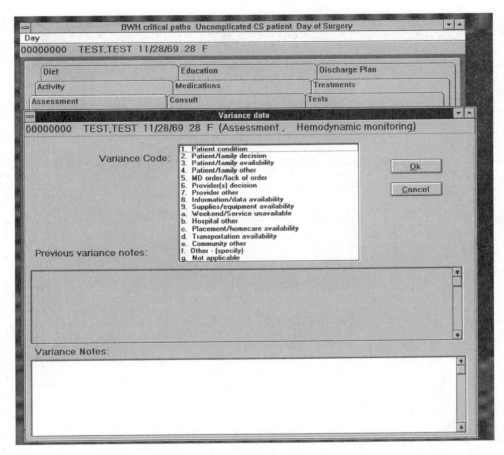

Figure 18–3 Automated Documentation Critical Path Variance. *Source:* Copyright © Brigham and Women's Hospital.

charge. A nurse in the cardiac surgery office contacts all patients who are either discharged or transferred to an extended care facility within 48 hours to monitor any variances from expected outcomes. Weekly morbidity/mortality statistics are incorporated into weekly rounds each Wednesday with all members of the cardiac surgery team present. Changes in practice and standards of patient care are developed as a result of trends in variances from expected patient outcomes. An example is the recent change to minimally invasive saphenous vein harvesting in coronary artery bypass patients to minimize leg wound cellulitis and infections.

CRITICAL PATHWAY VARIANCE MEASUREMENT

Caregivers are able to document on all variances from the pathway both for care processes and the expected outcomes. As previously described, variance documentation is done on the automated pathway by selecting the variance code and adding additional free text describing the variance and/or the corrective action. Figure 18–3 displays the variance documentation screen and the vari-

ance codes. Key expected outcomes were identified by the team to be tracked for variance. These are reported quarterly to the team for review. Individual patient variance is discussed on daily resident rounds.

CRITICAL PATHWAY UTILIZATION

The design and utilization of the critical pathway has led to several positive outcomes. First, it required that an interdisciplinary team do a systematic analysis of care processes and examine the rationale for the care provided. Frequently tasks are done because of history or tradition ("it's always been done that way") rather than solid scientific rationale. Second, making decisions on the care protocols within the pathway led to a higher degree of interdisciplinary consensus and less individual provider variability. Third, identifying expected outcomes forced the team to think about how to assist patients in their recovery progression rather than thinking only in terms of care processes. Fourth, having the expected outcomes in the pathway led to improved documentation of patient outcome achievement. Fifth, utilizing the pathway pro-

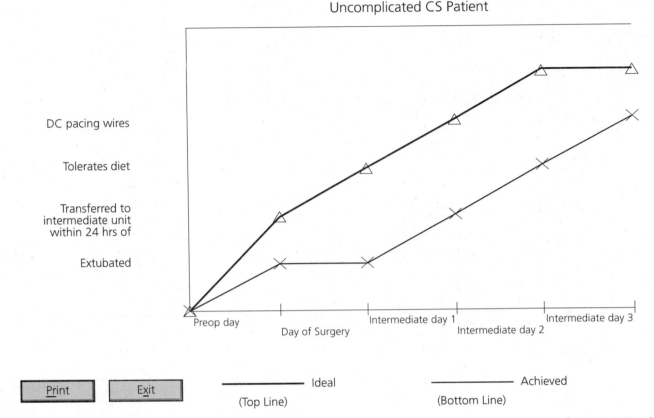

Figure 18–4 Outcome Achievement Graph from Day of Surgery to Discharge. *Source:* Copyright © Brigham and Women's Hospital.

vided the team with the opportunity to examine patient outcomes other than morbidity and mortality. Finally, changes in care protocols such as early extubation were based on review of data rather than anecdotal impressions of care delivery.

Although there have been positive outcomes from pathway utilization, it has not been used to its full potential. Initially the team envisioned that the pathway would be utilized for reporting from nurse to nurse or nurse to physician; this has occurred in only a limited fashion. It is primarily used by nursing to document their care while physicians and other care providers continue to document in the progress note section of the medical record.

While residents do use the order template designed for the critical pathway, additional testing is ordered without consideration of the pathway. The outcome achievement graph is a relatively new addition to the software and is only beginning to be used by the physician staff. Despite the fact that the pathway was designed by an interdisciplinary team, the only professions that utilize it are nursing and, to a lesser degree, the physicians.

Resistance to the pathway was initially high and was dealt with through educational sessions for the nursing and

Exhibit 18–1 Clinical Pathway Outcomes for 1996

- Financial outcomes: 35% decrease in charges for DRGs 104–108 for fiscal year 1996 compared to base fiscal year 1994
- 819 patients placed on the critical pathway during fiscal year 1996
- Mean postoperative length of stay was 8.1 days for all patients placed on the pathway with a postoperative length of stay of 5.3 days (50% of the patients met the pathway length-of-stay goal)
- Extubation analysis revealed that 95% of the patients placed on the pathway were extubated within 18 hours of cardiac surgery
- 88% of the patients placed on the pathway were discharged home with services versus transfer to an extended care facility
- Patient satisfaction outcomes measured quarterly for fiscal year 1996 demonstrated 95% overall satisfaction with the care received at Brigham and Women's Hospital
- Nursing care was rated excellent/good by 95% of the patients
- Physician care was rated excellent/good by 93% of the patients
- Overall satisfaction for recommendation of Brigham and Women's Hospital to others was 96%

Source: Copyright © Brigham and Women's Hospital.

physician staff and clear communication by the cardiac surgery division chief and nurse manager in support of the pathway and on expectation of usage. Providing staff with data on positive patient outcomes lowered the resistance further. Our hospital continues to strategize on ways to increase physician involvement with the pathway.

A patient/family version of the pathway was developed to outline the key activities and/or outcomes for each day. This was incorporated into the preoperative instruction provided to patients and families to give them a more realistic expectation of recovery and anticipated stay in the hospital. What was not built in was a mechanism or expectation for nursing or physician staff to formally review progression on the pathway with the patient or family on a daily basis.

CRITICAL PATHWAY OUTCOME MANAGEMENT

Outcome assessment and management is important for several reasons.

- Outcome assessment can identify changes that are needed in care protocols.
- Outcome assessment can identify improvement opportunities in hospital systems and processes.
- Positive clinical, LOS, and cost outcomes make the program more attractive to payers.

The cardiac surgery team holds a monthly conference to discuss patient morbidity and mortality outcomes. The early extubation outcome of the critical pathway is one that is reviewed and discussed at the weekly team meetings. The charges per case data are reviewed by the division chief and used in contracting with payers. The team, however, has not established a regular forum for review of variance and outcome data or pathway revision. Revisions were made early on in the implementation but a formal comprehensive review and revision has not occurred. There are plans being discussed by hospital administration to revitalize the care improvement initiative including a refocus on outcomes and pathway revision.

CONCLUSION

Critical pathways have been instituted nationally in cardiac surgery programs to decrease costs while maintaining high quality patient care. An interdisciplinary cardiac surgical care improvement team at Brigham and Women's Hospital developed a critical pathway for uncomplicated cardiac surgical patients in 1993 with identification of key outcomes to be achieved daily. Initially, a paper format was implemented for documentation of patient care and expected outcomes. Difficulties in variance tracking and documentation led to the development of an automated version of the critical pathway in 1995. The automated version offered several advantages: (1) ease of revising the pathway, (2) online documentation versus the paper format, (3) more efficient variance data collection, and (4) an instantaneous view of

Exhibit 18–2 Gantt Chart

Multidisciplinary staff actions	Day 1 (Day of Surgery)	Day 2 (POD1)
Consults	Anesthesia for early extubation protocol	
Tests	EKG and CXR on admission to ICU	CXR in the AM or after CT removal
Medications	Analgesia/sedation per MD orders	1. PO analgesic 2. Adequate pain control achieved
Monitoring	Telemetry	Telemetry
Treatments	ET suction and lavage every 4 hours	Heated nebulizer to keep oxygen saturation ≥ 94%
Diet	Ice chips after extubation	1. Clear liquids 2. Patient tolerates clear liquids well
Patient/Family Education	Explain ICU routine to patient and family	Provide "Moving Right Along After Heart Surgery" booklet
Discharge Planning		Contact case management nurse

Note: CT = chest tube; CXR = chest radiograph; EKG = electrocardiogram; ET = endotracheal tube; ICU = intensive care unit; MD = physician; PO = by mouth; POD1 = first postoperative day.
Source: Copyright © Brigham and Women's Hospital.

an individual patient's progress on the pathway. Patients are assessed postoperatively in the intensive care unit by the nursing staff for critical pathway eligibility.

A research assistant is responsible for tabulating outcome data and variances on all critical pathway patients. Quarterly reports provide feedback to members of the cardiac surgery care improvement team for analysis and review. Changes in practice and standards of care are developed as a result of trends in variances. Since implementation of the critical pathway, the following key outcomes have been achieved: (1) 35% decrease in charges for DRGs 104–108, (2) reduction in postoperative length of stay to 5.3 days for patients on the pathway, (3) extubation within 18 hours of cardiac surgery for 95% of the patients on the pathway, (4) 88% of the patients on the pathway are discharged home with services versus an extended care facility, and (5) continued high patient satisfaction with the care received at Brigham and Women's Hospital.

Areas for future critical pathway development include documentation on the pathway by physicians and other members of the health care team, implementation of the outcome achievement graph to monitor patient progress and achievement of expected outcomes in preparation for discharge, inclusion of the patient and family on a daily basis in their progression on the pathway, and development of a critical pathway for minimally invasive cardiac surgery patients due to the need for less tests and a postoperative LOS under 5 days. Additionally, critical pathways are being implemented in home care agencies and extended care facilities as an extension of the acute care hospitalization. The care coordination team at Brigham and Women's Hospital continues to meet with area home care agencies and extended care facilities to develop critical pathways in an effort to enhance the quality of patient care provided to cardiac surgery patients.

Economic pressures in health care have forced hospitals to closely monitor their costs in relationship to the quality of patient care provided. Critical pathways are an excellent example of decreasing costs while maintaining quality patient care. Pathway development forces health care team members to work in a collaborative manner in order to achieve key patient outcomes. Successful critical pathways have a great impact administratively as well in bargaining for insurance contracts and in building a reputation for clinical excellence in the ever increasing competition in the health care arena.

■ Appendix 18–A ■
Clinical Pathways

Exhibit 18–A–1 BWH Clinical Systems—Critical Paths (Path: Uncomplicated CS patient)

Day 2—Day of Surgery	N	D	E	V
Assessment				
Continuous:				
Hemodynamic monitoring				
ECG monitoring				
Pulse oximetry				
Vital signs q 15 min				
Temps q 1 hr				
Breath sounds q 4 hrs				
I & O q 1 hr				
Co/CI/SVR/PCWP q 4 hrs				
CT drainage q 1 hr				
NGT ph, irrigate, drainage q 8 hrs				
Resting heart rate > 60 and < 120/min				
with pacemaker				
without pacemaker				
with antiarrhythmic				
without antiarrhythmic				
Systolic BP > 90 and < 160 mm Hg				
with vasopressor				
without vasopressor				
CT-output < 150 cc for previous 8 hrs				
Urine output > 30 cc/hr				

continues

Exhibit 18–A–1 continued

	N	D	E	V
Extubated for 1 hr prior to transfer				
Resting O_2 sat > 94% on oxygen				
Effective cough and airway clearance				
Temp > 98°F and < 101°F				
Bibasilar rales no more than 1/3 of way up				
Moves all extremities				
Follows commands				
Consult				
Anesthesia for early extubation protocol				
Tests				
ABGs: on admission to ICU				
after each vent change				
if O_2 sat < 95%				
ionized calcium from ABG				
Magnesium on admit to ICU				
EKG on admit to ICU				
CXR on admit to ICU				
CK-MB: on admit to ICU, at 8 hrs and 16 hrs				
CBC: on admit to ICU, at 8 hrs and 16 hrs				
PT/PTT on admit to ICU				
Profile 7 on admit to ICU				
K on admit to ICU and q 4 hrs				
Activity				
Bedrest				
HOB elevated 30 degrees when hemodynamically stable				
Turn side to side, reposition q 2 hrs				
Elevate harvest leg on pillow (CABG only)				
Dangle when extubated				
Dorsiflexion q 1 hr while awake				
Medications				
Per MD order:				
vasoactive meds				
antiarrhythmics				
analgesics/sedation meds				
antibiotics				
K, Ca, Mg replacement per protocol				
Adequate pain control with analgesics				

continues

Exhibit 18–A–1 continued

	N	D	E	V
Treatments				
ET suction and lavage q 4 hrs				
Extubation protocol (check one that applies)				
less than 6 hrs				
6–18 hrs				
> 18 hrs				
Auto transfuse per unit protocol				
Pacing wires attached via cable to pacemaker				
Pacemaker parameters et				
Pacemaker on standby				
Soft restraints while intubated				
Thigh-high TEDS				
or compression boots (CABG or DVT study only)				
Coughing/deep breathing q 1 hr after extubated				
Diet				
NPO				
Ice chips after extubation				
Education				
Instruct on:				
coughing/deep breathing/supporting incision				
dorsiflexion				
Explain ICU routine to family/significant other				
Provide ICu number to family/significant other				
Verify contact person's phone number				
Family expresses receiving adequate information about patient's condition				

Source: Copyright © Brigham and Women's Hospital.

Exhibit 18–A–2 Initial Paper Format of Critical Path

| KEY: | Dotted lines = Interventions |
| | Bolded lines = Outcomes |

CRITICAL PATHWAYS—UNCOMPLICATED CARDIAC SURGICAL PATIENT

INTERMEDIATE DAY 3   Postop Day _____

DATE:

		N	D	E	VAR	PROGRESS NOTES
ASSESSMENT	DC Telemetry (if no arrhythmia)					
	Vital signs Q4h					
	Vital signs 1 hr after pacing wires removed					
	I & O q shirt					
	O₂ sat BID ☐ ☐					
	Breath sounds QID ☐ ☐ ☐ ☐					
	Resting heart rate > 60 and < 100/min					
	Systolic BP > 90 and < 160 mm Hg					
	Resting O₂ sat ≥ 90% off O₂					
	Effective cough and airway clearance					
	Temp < 101°F					
	Lungs clear					
	Decreased breath sounds at left base no more than ¼ of way up (CABG with LIMA only)					
	Alert and oriented x3					
TESTS	Profile 7					
	CBC					
	Mg					
	PT (if on Coumadin)					
	BS: fasting, 3pm, HS (if diabetic)					
ACTIVITY	Ambulate in pod independently at least QID ☐ ☐ ☐ ☐					
	Assist with ADLs					
	Limit sitting with legs down to meals and BR only (CABG only)					
	Remain on Unit 2 hrs after pacing wire removal					
	Tolerates ambulating in pod independently at least QID					
	Assists with ADLs					

Source: Copyright © Brigham and Women's Hospital.

Index